THE EXTENDED PHENOTYPE

Richard Dawkins, FRS, was Charles Simonyi Professor for the Public Understanding of Science at Oxford University from 1995 to 2008. Born in Nairobi of British parents, he was educated at Oxford and did his doctorate under the Nobel Prize-winning ethologist Niko Tinbergen. From 1967 to 1969 he was an Assistant Professor at the University of California at Berkeley, returning as University Lecturer and later Reader in Zoology at New College, Oxford, before becoming the first holder of the Simonyi Chair. He is an Emeritus Fellow of New College.

The Selfish Gene (1976; second edition 1989) catapulted Richard Dawkins to fame, and remains his most famous and widely read work. It was followed by a string of bestselling books: *The Extended Phenotype* (1982), *The Blind Watchmaker* (1986), *River Out of Eden* (1995), *Climbing Mount Improbable* (1996), *Unweaving the Rainbow* (1998), *A Devil's Chaplain* (2003), *The Ancestor's Tale* (2004), *The God Delusion* (2006), and *The Greatest Show on Earth* (2009). He has also published a science book for children, *The Magic of Reality* (2011), and two volumes of memoirs, *An Appetite for Wonder* (2013) and *Brief Candle in the Dark* (2015). Dawkins is a Fellow of both the Royal Society and the Royal Society of Literature. He is the recipient of numerous honours and awards, including the 1987 Royal Society of Literature Award, the *Los Angeles Times* Literary Prize of the same year, the 1990 Michael Faraday Award of the Royal

Society, the 1994 Nakayama Prize, the 1997 International Cosmos Prize for Achievement in Human Science, the Kistler Prize in 2001, and the Shakespeare Prize in 2005, the 2006 Lewis Thomas Prize for Writing About Science, and the Nierenberg Prize for Science in the Public Interest in 2009.

Daniel Dennett is Distinguished Arts and Sciences Professor of Philosophy, and Director of the Center for Cognitive Studies at Tufts University. His first book, *Content and Consciousness*, appeared in 1969, followed by *Brainstorms* (1978), *Elbow Room* (1984), *The Intentional Stance* (1987), *Consciousness Explained* (1991), *Darwin's Dangerous Idea* (1995), and *Kinds of Minds* (1996). He co-edited *The Mind's I* with Douglas Hofstadter in 1981. He is the author of over a hundred scholarly articles on various aspects of the mind, published in journals ranging from *Artificial Intelligence* and *Behavioural and Brain Sciences* to *Poetics Today* and the *Journal of Aesthetics and Art Criticism*. His most recent book is *Brainchildren: Essays on Designing Minds* (MIT Press and Penguin, 1998).

The Extended Phenotype

The Long Reach of the Gene

RICHARD DAWKINS

With an afterword by Daniel Dennett

OXFORD UNIVERSITY PRESS

OXFORD
UNIVERSITY PRESS

Great Clarendon Street, Oxford, OX2 6DP,
United Kingdom

Oxford University Press is a department of the University of Oxford.
It furthers the University's objective of excellence in research, scholarship,
and education by publishing worldwide. Oxford is a registered trade mark of
Oxford University Press in the UK and in certain other countries

First published 1982
Revised edition with new Afterword and Further Reading 1999
Reprinted with corrections 2008
Revised impression, as Oxford Landmark Science 2016

Published in the United States of America by Oxford University Press
198 Madison Avenue, New York, NY 10016, United States of America

British Library Cataloguing in Publication Data
Data available

Library of Congress Control Number: 2016935437

ISBN 978-0-19-878891-1

Printed in Great Britain by
Clays Ltd, Elcograf S.p.A.

PREFACE

The first chapter does some of the work of a Preface, in explaining what the book does and does not set out to accomplish, so I can be brief here. It is not a textbook, nor an introduction to an established field. It is a personal look at the evolution of life, and in particular at the logic of natural selection and the level in the hierarchy of life at which natural selection can be said to act. I happen to be an ethologist, but I hope preoccupations with animal behaviour will not be too noticeable. The intended scope of the book is wider.

The readers for whom I am mainly writing are my professional colleagues, evolutionary biologists, ethologists and sociobiologists, ecologists, and philosophers and humanists interested in evolutionary science, including, of course, graduate and undergraduate students in all these disciplines. Therefore, although this book is in some ways the sequel to my previous book, *The Selfish Gene*, it assumes that the reader has professional knowledge of evolutionary biology and its technical terms. On the other hand it is possible to enjoy a professional book as a spectator, even if not a participant in the profession. Some laypeople who read this book in draft have been kind enough, or polite enough, to claim to have liked it. It would give me great satisfaction to believe them, and I have added a glossary of technical terms which I hope may help. I have also tried to make the book as near as possible to being enjoyable to read. The resulting tone may possibly irritate some serious professionals. I very much hope not, because serious professionals are the primary audience to whom I wish to speak. It is as impossible to please everybody in literary style as it is in any other matter of taste, and styles that give the most positive pleasure to some people are often the most annoying to others.

Certainly the tone of the book is not conciliatory or apologetic—such is not the way of an advocate that sincerely believes in his case—and I must pack all apology into the Preface. Some of the earlier chapters reply to criticisms of my previous book, which might recur in response to the present one. I am sorry that this is necessary, and I am sorry if a note of exasperation creeps in from time to time. I trust, at least, that my exasperation remains good humoured. It is necessary to point to past misunderstandings and try to forestall their repetition, but I would not wish to give an aggrieved impression that misunderstanding has been widespread. It has been confined to numerically very limited quarters, but in some cases rather vocal ones. I am grateful to my critics for forcing me to think again about how to express difficult matters more clearly.

I apologize to readers who may find a favourite and relevant work missing from the bibliography. There are those capable of comprehensively and exhaustively surveying the literature of a large field, but I have never been able to understand how they manage it. I know that the examples I have cited are a small subset of those that could have been cited, and are sometimes the writings or recommendations of my friends. If the result appears biased, well, of course it is biased, and I am sorry. I think nearly everybody must be somewhat biased in this kind of way.

A book inevitably reflects the current preoccupations of the author, and these preoccupations are likely to have been among the topics of his most recent papers. When those papers are so recent that it would be an artificial contrivance to change the words, I have not hesitated to reproduce a paragraph almost verbatim here and there. These paragraphs, which will be found in Chapters 4, 5, 6, and 14, are an integral part of the message of this book, and to omit them would be just as artificial as to make gratuitous changes in their wording.

The opening sentence of Chapter 1 describes the book as a work of unabashed advocacy but, well, perhaps I am just a little bit abashed! Wilson (1975, pp. 28–9) has rightly castigated the 'advocacy method' in any search for scientific truth, and I have therefore devoted some of my first chapter to a plea of mitigation. I certainly would not want science to adopt the legal system in which professional advocates make the best case they can for a position, even if they believe it to be false. I believe deeply in the view of life that this book advocates, and have done so, at least in part, for a long time, certainly since the time of my first published paper, in which I characterized adaptations as favouring 'the survival of the animal's genes...' (Dawkins 1968). This belief—that if adaptations are to be treated as 'for the good of' something, that something is the gene—was the fundamental assumption of my previous book. The present book goes further. To dramatize it a bit, it attempts to free the selfish gene from the individual organism which has been its conceptual prison. The phenotypic effects of a gene are the tools by which it levers itself into the next generation, and these tools may 'extend' far outside the body in which the gene sits, even reaching deep into the nervous systems of other organisms. Since it is not a factual position I am advocating, but a way of seeing facts, I wanted to warn the reader not to expect 'evidence' in the normal sense of the word. I announced that the book was a work of advocacy, because I was anxious not to disappoint the reader, not to lead her on under false pretences and waste her time.

The linguistic experiment of the last sentence reminds me that I wish I had had the courage to instruct the computer to feminize personal pronouns at random throughout the text. This is not only because I admire the current awareness of the masculine bias in our language. Whenever I write I have a particular imaginary reader in mind (different imaginary readers oversee and 'filter' the same passage in numerous successive revisions) and at least half my imaginary readers are, like at least half my friends, female. Unfortunately it is still true in English that the unexpectedness of a feminine pronoun, where a neutral meaning is intended, seriously distracts the attention of most readers, of either sex. I believe the experiment of the previous paragraph will substantiate this. With regret, therefore, I have followed the standard convention in this book.

For me, writing is almost a social activity, and I am grateful to the many friends who have, sometimes unwittingly, participated through discussion, argument, and moral support. I cannot thank them all by name. Marian Stamp Dawkins has not only provided sensitive and knowledgeable criticism of the whole book in several drafts. She has also kept me going by believing in the project even through the times when I lost my own confidence. Alan Grafen and Mark Ridley, officially my graduate students, really, in their different ways, my mentors and guides through difficult theoretical territory, have influenced the book immeasurably. In the first draft their names seemed to creep in on almost every page, and it was only the pardonable grumblings of a referee that compelled me to banish to the Preface my acknowledgment of debt to them. Cathy Kennedy manages to combine close friendship for me with deep sympathy for my bitterest critics. This has put her in a unique position to advise me, especially over the earlier chapters which attempt to reply to criticism. I fear that she will still not like the tone of these chapters, but such improvement as there may be is largely due to her influence and I am very grateful to her.

I was privileged to have the first draft criticized in its entirety by John Maynard Smith, David C. Smith, John Krebs, Paul Harvey, and Ric Charnov, and the final draft owes much to all of them. In all cases I acted on their advice, even if I did not always take it. Others kindly criticized chapters in their own special fields: Michael Hansell the chapter on artefacts, Pauline Lawrence that on parasites, Egbert Leigh that on fitness, Anthony Hallam the section on punctuated equilibria, W. Ford Doolittle that on selfish DNA, and Diane De Steven botanical sections. The book was finished at Oxford, but begun during a visit to the University of Florida at Gainesville on a sabbatical leave kindly granted by the University of Oxford and the Warden and Fellows of New College. I am grateful to my many Floridan friends for giving me such a pleasant atmosphere in which to work, especially Jane Brockmann, who also provided helpful criticism of preliminary drafts, and Donna Gillis, who also did much of the typing. I benefited, too, from a month's exposure to tropical biology as the grateful guest of the Smithsonian Institution in Panama during the writing of the book. Finally, it is a pleasure once again to thank Michael Rodgers, formerly of Oxford University Press and now of W. H. Freeman and Company, a 'K-selected' editor who really believes in his books and is their tireless advocate.

RICHARD DAWKINS
Oxford, June 1981

NOTE TO OXFORD PAPERBACK EDITION

I suppose most scientists—most authors—have one piece of work of which they would say: It doesn't matter if you never read anything else of mine, please at least read *this*. For me, it is *The Extended Phenotype*. In particular, the last four chapters constitute the best candidate for the title 'innovative' that I have to offer. The rest of the book does some necessary sorting out on the way. Chapters 2 and 3 are replies to criticisms of the now widely accepted 'selfish gene' view of evolution. The middle chapters deal with the 'units of selection' controversy currently fashionable among philosophers of biology, taking the gene's eye view; perhaps the most useful contribution here is the 'Replicators and Vehicles' distinction. My intention was that this piece of sorting out should put paid to the whole controversy once and for all!

As for the extended phenotype proper, I have never seen any alternative to placing it at the end of the book. Nevertheless, this policy has a disadvantage. The earlier chapters inevitably draw attention to the general 'units of selection' issue, and away from the more novel idea of the extended phenotype itself. It is for this reason that I have dropped the original subtitle, 'The Gene as the Unit of Selection', from this edition. The replacement, 'The Long Reach of the Gene', captures the idea of the gene as the centre of a web of radiating power. The book is otherwise unchanged apart from minor corrections.

RICHARD DAWKINS
Oxford, May 1989

CONTENTS

ACKNOWLEDGEMENTS

pp. 215–16 'The Fifth Philosopher's Song' from *The Collected Poetry of Aldous Huxley* edited by Donald Watt. Copyright © 1971 by Laura Huxley. Reproduced by permission of Mrs. Laura Huxley and Chatto & Windus Ltd, also by permission of Harper & Row, Publishers, Inc. (USA)

p. 24 'McAndrew's Hymn' by Richard Kipling. Extract reproduced by permission of Doubleday & Company Inc.

1

NECKER CUBES AND BUFFALOES

This is a work of unabashed advocacy. I want to argue in favour of a particular way of looking at animals and plants, and a particular way of wondering why they do the things that they do. What I am advocating is not a new theory, not a hypothesis which can be verified or falsified, not a model which can be judged by its predictions. If it were any of those things, I agree with Wilson (1975, p. 28) that the 'advocacy method' would be inappropriate and reprehensible. But it is not any of those things. What I am advocating is a point of view, a way of looking at familiar facts and ideas, and a way of asking new questions about them. Any reader who expects a convincing new theory in the conventional sense of the word is bound to be left, therefore, with a disappointed 'so what?' feeling. But I am not trying to convince anyone of the truth of any factual proposition. Rather, I am trying to show the reader a way of seeing biological facts.

There is a well-known visual illusion called the Necker Cube. It consists of a line drawing which the brain interprets as a three-dimensional cube. But there are two possible orientations of the perceived cube, and both are equally compatible with the two-dimensional image on the paper. We usually begin by seeing one of the two orientations, but if we look for several seconds the cube 'flips over' in the mind, and we see the other apparent orientation. After a few more seconds the mental image flips back and it continues to alternate as long as we look at the picture. The point is that neither of the two perceptions of the cube is the correct or 'true' one. They are equally correct. Similarly the vision of life that I advocate, and label with the name of the extended phenotype, is not provably more correct than the orthodox view. It is a different view and I suspect that, at least in some respects, it provides a deeper understanding. But I doubt that there is any experiment that could be done to prove my claim.

The phenomena that I shall consider—coevolution, arms races, manipulation of hosts by parasites, manipulation of the inanimate world by living things, economic 'strategies' for minimizing costs and maximizing benefits—are all familiar enough, and are already the subject of intensive study. Why, then, should the busy reader bother to go on? It is tempting to borrow Stephen Gould's winningly ingenuous appeal at the beginning of a more substantial volume (1977a) and simply say, 'Please read the book' and you will find out why it was worth bothering to do so. Unfortunately I do not have the same grounds for confidence. I can only say that, as one ordinary biologist studying animal behaviour, I have found that the viewpoint represented by the label 'extended phenotype' has made me see animals and their behaviour differently, and I think I understand them better for it. The extended phenotype may not constitute a testable hypothesis in itself, but it so far changes the way we see animals and plants that it may cause us to think of testable hypotheses that we would otherwise never have dreamed of.

Lorenz's (1937) discovery that a behaviour pattern can be treated like an anatomical organ was not a discovery in the ordinary sense. No experimental results were adduced in its support. It was simply a new way of seeing facts that were already commonplace, yet it dominates modern ethology (Tinbergen 1963), and it seems to us today so obvious that it is hard to understand that it ever needed 'discovering'. Similarly, D'Arcy Thompson's (1917) celebrated chapter 'On the theory of transformations...' is widely regarded as a work of importance although it does not advance or test a hypothesis. In a sense it is obviously necessarily true that any animal form can be turned into a related form by a mathematical transformation, although it is not obvious that the transformation will be a simple one. In actually doing it for a number of specific examples, D'Arcy Thompson invited a 'so what?' reaction from anyone fastidious enough to insist that science proceeds only by the falsifying of specific hypotheses. If we read D'Arcy Thompson's chapter and then ask ourselves what we now know that we did not know before, the answer may well be not much. But our imagination is fired. We go back and look at animals in a new way; and we think about theoretical problems, in this case those of embryology and phylogeny and their interrelations, in a new way. I am, of course, not so presumptuous as to compare the present modest work with the masterpiece of a great biologist. I use the example simply to demonstrate that it is *possible* for a theoretical book to be worth reading even if it does not advance testable hypotheses but seeks, instead, to change the way we see.

Another great biologist once recommended that to understand the actual we must contemplate the possible: 'No practical biologist interested in sexual reproduction would be led to work out the detailed consequences experienced by organisms having three or more sexes; yet what else should he do if he wishes to understand why the sexes are, in fact, always two?' (Fisher 1930a, p. ix). Williams (1975), Maynard Smith (1978a), and others have taught us that one of the commonest, most universal features of life on

Earth, sexuality itself, should not be accepted without question. Indeed, its existence turns out to be positively surprising when set against the imagined possibility of asexual reproduction. To imagine asexual reproduction as a hypothetical possibility is not difficult, since we know it is a reality in some animals and plants. But are there other cases where our imagination receives no such prompting? Are there important facts about life that we hardly notice simply because we lack the imagination to visualise alternatives which, like Fisher's three sexes, might have existed in some possible world? I shall try to show that the answer is yes.

Playing with an imaginary world, in order to increase our understanding of the actual world, is the technique of 'thought experiment'. It is much used by philosophers. For instance in a collection of essays on *The Philosophy of Mind* (ed. Glover 1976), various authors imagine surgical operations in which one person's brain is transplanted into another person's body, and they use the thought experiment to clarify the meaning of 'personal identity'. At times philosophers' thought experiments are purely imaginary and wildly improbable, but this doesn't matter given the purpose for which they are made. At other times they are informed, to a greater or lesser extent, by facts from the real world, for instance the facts of split-brain experiments.

Consider another thought experiment, this time from evolutionary biology. When I was an undergraduate obliged to write speculative essays on 'the origin of the Chordates' and other topics of remote phylogeny, one of my tutors rightly tried to shake my faith in the value of such speculations by suggesting that anything could, in principle, evolve into anything else. Even insects could evolve into mammals, if only the right sequence of selection pressures were provided in the right order. At the time, as most zoologists would, I dismissed the idea as obvious nonsense, and I still, of course, don't believe that the right sequence of selection pressures ever would be provided. Nor did my tutor. But as far as the principle is concerned, a simple thought experiment shows it to be nearly incontrovertible. We need only prove that there exists a continuous series of small steps leading from an insect, say a stag beetle, to a mammal, say a stag. By this I mean that, starting with the beetle, we could lay out a sequence of hypothetical animals, each one as similar to the previous member of the series as a pair of brothers might be, and the sequence would culminate in a red deer stag.

The proof is easy, provided only that we accept, as everyone does, that beetle and deer have a common ancestor, however far back. Even if there is no other sequence of steps from beetle to deer, we know that at least one sequence must be obtained by simply tracing the beetle's ancestors back to the common ancestor, then working forwards down the other line to the deer.

We have proved that there exists a trajectory of stepwise change connecting beetle to deer and, by implication, a similar trajectory from any modern animal to any other modern animal. In principle, therefore, we may presume that a series of selection pressures could be artificially contrived to propel a lineage along one of these trajectories. It was a quick thought experiment along these lines that enabled me to say, when discussing D'Arcy Thompson's transformations, that 'In a sense it is obviously necessarily true that any animal form can be turned into a related form by a mathematical transformation, although it is not obvious that the transformation will be a simple one.' In this book I shall make frequent use of the thought-experiment technique. I warn the reader of this in advance, since scientists are sometimes annoyed by the lack of realism in such forms of reasoning. Thought experiments are not supposed to be realistic. They are supposed to clarify our thinking about reality.

One feature of life in this world which, like sex, we have taken for granted and maybe should not, is that living matter comes in discrete packages called organisms. In particular, biologists interested in functional explanation usually assume that the appropriate unit for discussion is the individual organism. To us, 'conflict' usually means conflict between organisms, each one striving to maximize its own individual 'fitness'. We recognize smaller units such as cells and genes, and larger units such as populations, societies, and ecosystems, but there is no doubt that the individual body, as a discrete unit of action, exerts a powerful hold over the minds of zoologists, especially those interested in the adaptive significance of animal behaviour. One of my aims in this book is to break that hold. I want to switch emphasis from the individual body as focal unit of functional discussion. At the very least I want to make us aware of how much we take for granted when we look at life as a collection of discrete individual organisms.

The thesis that I shall support is this. It is legitimate to speak of adaptations as being 'for the benefit of' something, but that something is best not seen as the individual organism. It is a smaller unit which I call the active, germ-line replicator. The most important kind of replicator is the 'gene' or small genetic fragment. Replicators are not, of course, selected directly, but by proxy; they are judged by their phenotypic effects. Although for some purposes it is convenient to think of these phenotypic effects as being packaged together in discrete 'vehicles' such as individual organisms, this is not fundamentally necessary. Rather, the replicator should be thought of as having *extended* phenotypic effects, consisting of all its effects on the world at large, not just its effects on the individual body in which it happens to be sitting.

To return to the analogy of the Necker Cube, the mental flip that I want to encourage can be characterized as follows. We look at life and begin by seeing a collection of interacting individual organisms. We know that they contain smaller units, and we know that they are, in turn, parts of larger composite units, but we fix our gaze on the whole organisms. Then suddenly the image flips. The individual bodies are still there; they have not moved, but they seem to have gone transparent. We see through them to the replicating fragments of DNA within, and we see the wider world as an arena in which these genetic fragments play out their tournaments of manipulative skill. Genes manipulate the world and shape it to assist their replication. It happens that they have 'chosen' to do so largely by moulding matter into large multicellular chunks which we call organisms, but this might not have been so. Fundamentally, what is going on is that replicating molecules ensure their survival by means of phenotypic effects on the world. It is only incidentally true that those phenotypic effects happen to be packaged up into units called individual organisms.

We do not at present appreciate the organism for the remarkable phenomenon it is. We are accustomed to asking, of any widespread biological phenomenon, 'What is its survival value?' But we do not say, 'What is the survival value of packaging life up into discrete units called organisms?' We accept it as a given feature of the way life is. As I have already noted, the organism becomes the automatic subject of our questions about the survival value of other things: 'In what way does this behaviour pattern benefit the individual doing it? In what way does this morphological structure benefit the individual it is attached to?'

It has become a kind of 'central theorem' (Barash 1977) of modern ethology that organisms are expected to behave in such a way as to benefit their own inclusive fitness (Hamilton 1964a,b), rather than to benefit anyone, or anything, else. We do not ask in what way the behaviour of the left hind leg benefits the left hind leg. Nor, nowadays, do most of us ask how the behaviour of a group of organisms, or the structure of an ecosystem, benefits that group or ecosystem. We treat groups and ecosystems as collections of warring, or uneasily cohabiting, organisms; and we treat legs, kidneys, and cells as cooperating components of a single organism. I am not necessarily objecting to this focus of attention on individual organisms, merely calling attention to it as something that we take for granted. Perhaps we should stop taking it for granted and start wondering about the individual organism, as something that needs explaining in its own right, just as we found sexual reproduction to be something that needs explaining in its own right.

At this point an accident of the history of biology necessitates a tiresome digression. The prevailing orthodoxy of my previous paragraph, the central dogma of individual organisms working to maximize their own reproductive success, the paradigm of 'the selfish organism', was Darwin's paradigm, and it is dominant today. One might imagine, therefore, that it had had a good run for its money and should, by now, be ripe for revolution, or at least constitute a solid-enough bastion to withstand iconoclastic pinpricks such as any that this book might deliver. Unfortunately, and this is the historical accident I mentioned, although it is true that there has seldom been any temptation to treat units smaller than the organism as agents working for their own benefit, the same has not always been true of larger units. The intervening years since Darwin have seen an astonishing retreat from his individual-centred stand, a lapse into sloppily unconscious group selectionism, ably documented by Williams (1966), Ghiselin (1974a), and others. As Hamilton (1975a) put it, 'almost the whole field of biology stampeded in the direction where Darwin had gone circumspectly or not at all'. It is only in recent years, roughly coinciding with the belated rise to fashion of Hamilton's own ideas (Dawkins 1979b), that the stampede has been halted and turned. We painfully struggled back, harassed by sniping from a Jesuitically sophisticated and dedicated neo-group-selectionist rearguard, until we finally regained Darwin's ground, the position that I am characterizing by the label 'the selfish organism', the position which, in its modern form, is dominated by the concept of inclusive fitness. Yet it is this hard-won fastness that I may seem to be abandoning here, abandoning almost before it is properly secured; and for what? For a flickering Necker Cube, a metaphysical chimera called the extended phenotype?

No, to renounce those gains is far from my intention. The paradigm of the selfish organism is vastly preferable to what Hamilton (1977) has called 'the old, departing paradigm of adaptation for the benefit of the species'. 'Extended phenotype' is misunderstood if it is taken to have any connection with adaptation at the level of the group. The selfish organism, and the selfish gene with its extended phenotype, are two views of the same Necker Cube. The reader will not experience the conceptual flip-over that I seek to assist unless he begins by looking at the right cube. This book is addressed to those that already accept the currently fashionable selfish organism view of life, rather than any form of 'group benefit' view.

I am not saying that the selfish organism view is necessarily wrong, but my argument, in its strong form, is that it is looking at the matter the wrong way up. I once overheard an eminent Cambridge ethologist say to an eminent Austrian ethologist (they were arguing about behaviour development): 'You know, we really agree. It is just that you *say* it wrong.' Gentle 'individual selectionist', we really do almost agree, at least in comparison to the group selectionists. It is just that you *see* it wrong!

Bonner (1958), discussing single-celled organisms, said, 'what special use to these organisms are nuclear genes? How did they arise by selection?' This is a good example of the kind of imaginative, radical question that I think we ought to ask about life. But if the thesis of this book is accepted, the particular question should be turned upside down. Instead of asking of what use nuclear genes are to *organisms,* we should ask why *genes* chose to group themselves together in nuclei, and in organisms. In the opening lines of the same work, Bonner says: 'I do not propose to say anything new or original in these lectures. But I am a great believer in saying familiar, well-known things backwards and inside out, hoping that from some new vantage point the old facts will take on a deeper significance. It is like holding an abstract painting upside down; I do not say that the meaning of the picture will suddenly be clear, but some of the structure of the composition that was hidden may show itself' (p. 1). I came across this after writing my own Necker Cube passage, and was delighted to find the same views expressed by so respected an author.

The trouble with my Necker Cubes, and with Bonner's abstract painting, is that, as analogies, they may be too timid and unambitious. The analogy of the Necker Cube expresses my *minimum* hope for this book. I am pretty confident that to look at life in terms of genetic replicators preserving themselves by means of their extended phenotypes is at least as satisfactory as to look at it in terms of selfish organisms maximizing their inclusive fitness. In many cases the two ways of looking at life will, indeed, be equivalent. As I shall show, 'inclusive fitness' was defined in such a way as to tend to make 'the individual maximizes its inclusive fitness' equivalent to 'the genetic replicators maximize their survival'. At least, therefore, the biologist should try both ways of thinking, and choose the one he or she prefers. But I said this was a minimum hope. I shall discuss phenomena, 'meiotic drive' for instance, whose explanation is lucidly written on the second face of the cube, but which make no sense at all if we keep our mental gaze firmly fixed on the other face, that of the selfish organism. Moving from my minimum hope to my wildest daydream, it is that whole areas of biology, the study of animal communication, animal artefacts, parasitism and symbiosis, community ecology, indeed all interactions between and within organisms, will eventually be illuminated in new ways by the doctrine of the extended phenotype. As is the way with advocates, I shall try to make the strongest case I can, and this means the case for the wilder hopes rather than the more cautious minimum expectations.

If these grandiose hopes are eventually realized, perhaps a less modest analogy than the Necker Cube will be pardoned. Colin Turnbull (1961) took a pygmy friend, Kenge, out of the forest for the first time in his life, and they climbed a mountain together and looked out over the plains. Kenge saw some buffalo 'grazing lazily several miles away, far down below. He turned to me and said, "What insects are those?"... At first I hardly understood, then I realized that in the forest vision is so limited that there is no great need to make an automatic allowance for distance when judging size. Out here in the plains, Kenge was looking for the first time over apparently unending miles of unfamiliar grasslands, with not a tree worth the name to give him any basis for comparison... When I told Kenge that the insects were buffalo, he roared with laughter and told me not to tell such stupid lies' (pp. 227–8).

This book as a whole, then, is a work of advocacy, but it is a poor advocate that leaps precipitately to his conclusion when the jury are sceptical. The second face of my Necker Cube is unlikely to click into clear focus until near the end of the book. Earlier chapters prepare the ground, attempt to forestall certain risks of misunderstanding, dissect the first face of the Necker Cube in various ways, point up reasons why the paradigm of the selfish individual, if not actually incorrect, can lead to difficulties.

Parts of some early chapters are frankly retrospective and even defensive. Reaction to a previous work (Dawkins 1976a) suggests that this book is likely to raise needless fears that it promulgates two unpopular '-isms'—'genetic determinism' and 'adaptationism'. I myself admit to being irritated by a book that provokes me into muttering 'Yes but...' on every page, when the author could easily have forestalled my worry by a little considerate explanation early on. Chapters 2 and 3 try to remove at least two major sources of 'yes-buttery' at the outset.

Chapter 4 opens the case for the prosecution against the selfish organism, and begins to hint at the second aspect of the Necker Cube. Chapter 5 opens the case for the 'replicator' as the fundamental unit of natural selection. Chapter 6 returns to the individual organism and shows how neither it nor any other major candidate except the small genetic fragment qualifies as a true replicator. Rather, the individual organism should be thought of as a 'vehicle' for replicators. Chapter 7 is a digression on research methodology. Chapter 8 raises some awkward anomalies for the selfish organism, and Chapter 9 continues the theme. Chapter 10 discusses various notions of 'individual fitness', and concludes that they are confusing, and probably dispensable.

Chapters 11, 12, and 13 are the heart of the book. They develop, by gradual degrees, the idea of the extended phenotype itself, the second face of the Necker Cube. Finally, in Chapter 14, we turn back with refreshed curiosity to the individual organism and ask why, after all, it is such a salient level in the hierarchy of life.

2

GENETIC DETERMINISM AND GENE SELECTIONISM

Long after his death, tenacious rumours persisted that Adolf Hitler had been seen alive and well in South America, or in Denmark, and for years a surprising number of people with no love for the man only reluctantly accepted that he was dead (Trevor-Roper 1972). In the First World War a story that a hundred thousand Russian troops had been seen landing in Scotland 'with snow on their boots' became widely current, apparently because of the memorable vividness of that snow (Taylor 1963). In our own time myths such as that of computers persistently sending householders electricity bills for a million pounds (Evans 1979), or of well-heeled welfare-scroungers with two expensive cars parked outside their government-subsidized council houses, are familiar to the point of cliché. There are some falsehoods, or half-truths, that seem to engender in us an active desire to believe them and pass them on even if we find them unpleasant, maybe in part, perversely, *because* we find them unpleasant.

Computers and electronic 'chips' provoke more than their fair share of such myth-making, perhaps because computer technology advances at a speed which is literally frightening. I know an old person who has it on good authority that 'chips' are usurping human functions to the extent not only of 'driving tractors' but even of 'fertilizing women'. Genes, as I shall show, are the source of what may be an even larger mythology than computers. Imagine the result of combining these two powerful myths, the gene myth and the computer myth! I believe that I may have inadvertently achieved some such unfortunate synthesis in the minds of a few readers of my previous book, and the result was comic misunderstanding. Happily, such misunderstanding was not widespread, but it is worth trying to avoid a repeat of it here, and that is one purpose of the present chapter. I shall expose the myth of genetic determinism, and explain why it is necessary to use language that can be unfortunately misunderstood as genetic determinism.

A reviewer of Wilson's (1978) *On Human Nature*, wrote: 'although he does not go as far as Richard Dawkins (*The Selfish Gene*...) in proposing sex-linked genes for "philandering", for Wilson human males have a genetic tendency towards polygyny, females towards constancy (don't blame your mates for sleeping around, ladies, it's not their fault they are genetically programmed). Genetic determinism constantly creeps in at the back door' (Rose 1978). The reviewer's clear implication is that the authors he is criticizing believe in the existence of genes that force human males to be irremediable philanderers who cannot therefore be blamed for marital infidelity. The reader is left with the impression that those authors are protagonists in the 'nature or nurture' debate, and, moreover, died-in-the-wool hereditarians with male chauvinist leanings.

In fact my original passage about 'philanderer males' was not about humans. It was a simple mathematical model of some unspecified animal (not that it matters, I had a bird in mind). It was not explicitly (see below) a model of genes, and if it had been about genes they would have been sex-limited, not sex-linked! It was a model of 'strategies' in the sense of Maynard Smith (1974). The 'philanderer' strategy was postulated, not as *the* way males behave, but as one of two hypothetical alternatives, the other being the 'faithful' strategy. The purpose of this very simple model was to illustrate the kinds of conditions under which philandering might be favoured by natural selection, and the kinds of conditions under which faithfulness might be favoured. There was no presumption that philandering was more likely in males than faithfulness. Indeed, the particular run of the simulation that I published culminated in a mixed male population in which faithfulness slightly predominated (Dawkins 1976a, p. 165, although see Schuster & Sigmund 1981). There is not just one misunderstanding in Rose's remarks, but multiple compounded misunderstanding. There is a wanton eagerness to misunderstand. It bears the stamp of snow-covered Russian jackboots, of little black microchips marching to usurp the male role and steal our tractor-drivers' jobs. It is a manifestation of a powerful myth, in this case the great gene myth.

The gene myth is epitomized in Rose's parenthetic little joke about ladies not blaming their mates for sleeping around. It is the myth of 'genetic determinism'. Evidently, for Rose, genetic determinism is determinism in the full philosophical sense of irreversible inevitability. He assumes that the existence of a gene 'for' X implies that X cannot be escaped. In the words of another critic of 'genetic determinism', Gould (1978, p. 238), 'If we are programmed to be what we are, then these traits are ineluctable. We may, at best, channel them, but we cannot change them either by will, education, or culture.'

The validity of the determinist point of view and, separately, its bearing on an individual's moral responsibility for his actions, has been debated by philosophers and theologians for centuries past, and no doubt will be for centuries to come. I suspect that both Rose and Gould are determinists in that they believe in a physical, materialistic basis for all our actions. So am I. We would also probably all three agree that human nervous systems are so complex that in practice we can forget about determinism and behave as if we had free will. Neurones may be amplifiers of fundamentally indeterminate physical events. The only point I wish to make is that, whatever view one takes on the question of determinism, the insertion of the word 'genetic' is not going to make any difference. If you are a full-blooded determinist you will believe that all your actions are predetermined by physical causes in the past, and you may or may not also believe that you therefore cannot be held responsible for your sexual infidelities. But, be that as it may, what difference can it possibly make whether some of those physical causes are *genetic*? Why are genetic determinants thought to be any more ineluctable, or blame-absolving, than 'environmental' ones?

The belief that genes are somehow super-deterministic, in comparison with environmental causes, is a myth of extraordinary tenacity, and it can give rise to real emotional distress. I was only dimly aware of this until it was movingly brought home to me in a question session at a meeting of the American Association for the Advancement of Science in 1978. A young woman asked the lecturer, a prominent 'sociobiologist', whether there was any evidence for genetic sex differences in human psychology. I hardly heard the lecturer's answer, so astonished was I by the emotion with which the question was put. The woman seemed to set great store by the answer and was almost in tears. After a moment of genuine and innocent bafflement the explanation hit me. Something or somebody, certainly not the eminent sociobiologist himself, had misled her into thinking that genetic determination is for keeps; she seriously believed that a 'yes' answer to her question would, if correct, condemn her as a female individual to a life of feminine pursuits, chained to the nursery and the kitchen sink. But if, unlike most of us, she is a determinist in that strong Calvinistic sense, she should be equally upset whether the causal factors concerned are genetic or 'environmental'.

What does it ever mean to say that something determines something? Philosophers, possibly with justification, make heavy weather of the concept of causation, but to a working biologist causation is a rather simple statistical concept. Operationally we can never demonstrate that a particular observed event C caused a particular result R, although it will often be judged highly likely. What biologists in practice usually do is to establish *statistically* that events of class R reliably follow events of class C. They need a number of paired instances of the two classes of events in order to do so: one anecdote is not enough.

Even the observation that R events reliably tend to follow C events after a relatively fixed time interval provides only a working hypothesis that C events cause R events. The hypothesis is confirmed, within the limits of the statistical method, only if the C events are delivered by an *experimenter* rather than simply noted by an observer, and are still reliably followed by R events. It is not necessary that every C should be followed by an R, nor that every R should be preceded by a C (who has not had to contend with arguments such as 'smoking cannot cause lung cancer, because I knew a non-smoker who died of it, and a heavy smoker who is still going strong at ninety'?). Statistical methods are designed to help us assess, to any specified level of probabilistic confidence, whether the results we obtain really indicate a causal relationship.

If, then, it were true that the possession of a Y chromosome had a causal influence on, say, musical ability or fondness for knitting, what would this mean? It would mean that, in some specified population and in some specified environment, an observer in possession of information about an individual's sex would be able to make a statistically more accurate prediction as to the person's musical ability than an observer ignorant of the person's sex. The emphasis is on the word 'statistically', and let us throw in an 'other things being equal' for good measure. The observer might be provided with some additional information, say on the person's education or upbringing, which would lead him to revise, or even reverse, his prediction based on sex. If females are statistically more likely than males to enjoy knitting, this does not mean that all females enjoy knitting, nor even that a majority do.

It is also fully compatible with the view that the reason females enjoy knitting is that society brings them up to enjoy knitting. If society systematically trains children without penises to knit and play with dolls, and trains children with penises to play with guns and toy soldiers, any resulting differences in male and female preferences are strictly speaking genetically determined differences! They are determined, through the medium of societal custom, by the fact of possession or non-possession of a penis, and that is determined (in a normal environment and in the absence of ingenious plastic surgery or hormone therapy) by sex chromosomes.

Obviously, on this view, if we experimentally brought up a sample of boys to play with dolls and a sample of girls to play with guns, we would expect easily to reverse the normal preferences. This might be an interesting experiment to do, for the result just might turn out to be that girls *still* prefer dolls and boys still prefer guns. If so, this might tell us something about the tenacity, in the face of a *particular* environmental manipulation, of a genetic difference. But all genetic causes have to work in the context of an environment of some kind. If a genetic sex difference makes itself felt through the medium of a sex-biased education system, it is still a genetic difference. If it makes itself felt through some other medium, such that manipulations of the education system do not perturb it, it is, in principle, no more and no less a genetic difference than in the former, education-sensitive case: no doubt some other environmental manipulation could be found which *did* perturb it.

Human psychological attributes vary along almost as many dimensions as psychologists can measure. It is difficult in practice (Kempthorne 1978), but in principle we could partition this variation among such putative causal factors as age, height, years of education, type of education classified in many different ways, number of siblings, birth order, colour of mother's eyes, father's skill in shoeing horses, and, of course, sex chromosomes. We could also examine two-way and multi-way interactions between such factors. For present purposes the important point is that the variance we seek to explain will have many causes, which interact in complex ways. Undoubtedly genetic variance is a significant cause of much phenotypic variance in observed populations, but its effects may be overridden, modified, enhanced, or reversed by other causes. Genes may modify the effects of other genes, and may modify the effects of the environment. Environmental events, both internal and external, may modify the effects of genes, and may modify the effects of other environmental events.

People seem to have little difficulty in accepting the modifiability of 'environmental' effects on human development. If a child has had bad teaching in mathematics, it is accepted that the resulting deficiency can be remedied by extra good teaching the following year. But any suggestion that the child's mathematical deficiency might have a genetic origin is likely to be greeted with something approaching despair: if it is in the genes 'it is written', it is 'determined' and nothing can be done about it; you might as well give up attempting to teach the child mathematics. This is pernicious rubbish on an almost astrological scale. Genetic causes and environmental causes are in principle no different from each other. Some influences of both types may be hard to reverse; others may be easy to reverse. Some may be usually hard to reverse but easy if the right agent is applied. The important point is that there is no general reason for expecting genetic influences to be any more irreversible than environmental ones.

What did genes do to deserve their sinister, juggernaut-like reputation? Why do we not make a similar bogey out of, say, nursery education or confirmation classes? Why are genes thought to be so much more fixed and inescapable in their effects than television, nuns, or books? Don't blame your mates for sleeping around, ladies, it's not their fault they have been inflamed by pornographic literature! The alleged Jesuit boast, 'Give me the child for his first seven years, and I'll give you the man', may have some truth in it. Educational, or other cultural influences may, in some circumstances, be just as unmodifiable and irreversible as genes and 'stars' are popularly thought to be.

I suppose part of the reason genes have become deterministic bogeys is a confusion resulting from the well-known fact of the non-inheritance of acquired characteristics. Before this century it was widely believed that experience and other acquisitions of an individual's lifetime were somehow imprinted on the hereditary substance and transmitted to the children. The abandoning of this belief, and its replacement by Weismann's doctrine of the continuity of the germ plasm, and its molecular counterpart the 'central dogma', is one of the great achievements of modern biology. If we steep ourselves in the implications of Weismannian orthodoxy, there really does seem to be something juggernaut-like and inexorable about genes. They march through generations, influencing the form and behaviour of a succession of mortal bodies, but, except for rare and non-specific mutagenic effects, they are never influenced by the experience or environment of those bodies. The genes in me came from my four grandparents; they flowed straight through my parents to me, and nothing that my parents achieved, acquired, learned, or experienced had any effect on those genes as they flowed through. Perhaps there is something a little sinister about that. But, however inexorable and undeviating the genes may be as they march down the generations, the nature of their phenotypic effects on the bodies they flow through is by no means inexorable and undeviating. If I am homozygous for a gene G, nothing save mutation can prevent my passing G on to all my children. So much is inexorable. But whether or not I, or my children, show the phenotypic effect normally associated with possession of G may depend very much on how we are brought up, what diet or education we experience, and what other genes we happen to possess. So, of the two effects that genes have on the world—manufacturing copies of themselves, and influencing phenotypes—the first is inflexible apart from the rare possibility of mutation; the second may be exceedingly flexible. I think a confusion between evolution and development is, then, partly responsible for the myth of genetic determinism.

But there is another myth complicating matters, and I have already mentioned it at the beginning of this chapter. The computer myth is almost as deep-seated in the modern mind as the gene myth. Notice that both passages I quoted contain the word 'programmed'. Thus Rose sarcastically absolved promiscuous men from blame because they are genetically *programmed.* Gould says that if we are *programmed* to be what we are then these traits are ineluctable. And it is true that we ordinarily use the word programmed to indicate unthinking inflexibility, the antithesis of freedom of action. Computers and 'robots' are, by repute, notoriously inflexible, carrying out instructions to the letter, even if the consequences are obviously absurd. Why else would they send out those famous million pound bills that everybody's friend's friend's cousin's acquaintance keeps receiving? I had forgotten the great computer myth, as well as the great gene myth, or I would have been more careful when I myself wrote of genes swarming 'inside gigantic lumbering robots . . .', and of ourselves as 'survival machines—robot vehicles blindly programmed to preserve the selfish molecules known as genes' (Dawkins 1976a). These passages have been triumphantly quoted, and requoted apparently from secondary and even tertiary sources, as examples of rabid genetic determinism (e.g. 'Nabi' 1981). I am not apologizing for using the language of robotics. I would use it again without hesitation. But I now realize that it is necessary to give more explanation.

From 13 years' experience of teaching it, I know that a main problem with the 'selfish-gene survival machine' way of looking at natural selection is a particular risk of misunderstanding. The metaphor of the intelligent gene reckoning up how best to ensure its own survival (Hamilton 1972) is a powerful and illuminating one. But it is all too easy to get carried away, and allow hypothetical genes cognitive wisdom and foresight in planning their 'strategy'. At least three out of twelve misunderstandings of kin selection (Dawkins 1979a) are directly attributable to this basic error. Time and again, non-biologists have tried to justify a form of group selection to me by, in effect, imputing foresight to genes: 'The long-term interests of a gene require the continued existence of the species; therefore shouldn't you expect adaptations to prevent species extinction, even at the expense of short-term individual reproductive success?' It was in an attempt to forestall errors like this that I used the language of automation and robotics, and used the word 'blindly' in referring to genetic programming. But it is, of course, the genes that are blind, not the animals they program. Nervous systems, like man-made computers, can be sufficiently complex to show intelligence and foresight.

Symons (1979) makes the computer myth explicit:

> I wish to point out that Dawkins's implication—through the use of words like 'robot' and 'blindly'—that evolutionary theory favors determinism is utterly without foundation . . . A robot is a mindless automaton. Perhaps some animals are robots (we have no way of knowing) ; however, Dawkins is not referring to *some* animals, but to all animals and in this case specifically to human beings. Now, to paraphrase Stebbing, 'robot' can be opposed to 'thinking being' or it can be used figuratively to indicate a person who seems to act mechanically, but there is no common usage of language that provides a meaning for the word 'robot' in which it would make sense to say that all living things are robots [p. 41].

The point of the passage from Stebbing which Symons paraphrased is the reasonable one that X is a useless word unless there are some things that are not X. If everything is a robot, then the word robot doesn't mean anything useful. But the word robot has other associations, and rigid inflexibility was not the association I was thinking of. A robot is a programmed machine, and an important thing about programming is that it is distinct from, and done in advance of, performance of the behaviour itself. A computer is programmed to perform the behaviour of calculating square roots, or playing chess. The relationship between a chess-playing computer and the person who programmed it is not obvious, and is open to misunderstanding. It might be thought that the programmer watches the progress of the game and gives instructions to the computer move by move. In fact, however, the programming is finished before the game begins. The programmer tries to anticipate contingencies, and builds in conditional instructions of great complexity, but once the game begins he has to keep his hands off. He is not allowed to give the computer any new hints during the course of the game. If he did he would not be programming but performing, and his entry would be disqualified from the tournament. In the work criticized by Symons, I made extensive use of the analogy of computer chess in order to explain the point that genes do not control behaviour directly in the sense of interfering in its performance. They only control behaviour in the sense of programming the machine in *advance* of performance. It was this association with the word robot that I wanted to invoke, not the association with mindless inflexibility.

As for the mindless inflexibility association itself, it could have been justified in the days when the acme of automation was the rod and cam control system of a marine engine, and Kipling wrote 'McAndrew's Hymn':

> From coupler-flange to spindle-guide I see Thy Hand, O God—
> Predestination in the stride o' yon connectin'-rod.
> John Calvin might ha' forged the same—

But that was 1893 and the heyday of steam. We are now well embarked on the golden age of electronics. If machines ever had associations with rigid inflexibility—and I accept that they had—it is high time they lived them down. Computer programs have now been written that play chess to International Master standard (Levy 1978), that converse and reason in correct and indefinitely complex grammatical English (Winograd 1972), that create elegant and aesthetically satisfying new proofs of mathematical theorems (Hofstadter 1979), that compose music and diagnose illness; and the pace of progress in the field shows no sign of slowing down (Evans 1979). The advanced programming field known as artificial intelligence is in a buoyant, confident state (Boden 1977). Few who have studied it would now bet against computer programs beating the strongest grand masters at chess within the next 10 years. From being synonymous in the popular mind with a moronically undeviating, jerky-limbed zombie, 'robot' will one day become a byword for flexibility and rapid intelligence.

Unfortunately I jumped the gun a little in the passage quoted. When I wrote it I had just returned from an eye-opening and mind-boggling conference on the state of the art of artificial intelligence programming, and I genuinely and innocently in my enthusiasm forgot that robots are popularly supposed to be inflexible idiots. I also have to apologize for the fact that, without my knowledge, the cover of the German edition of *The Selfish Gene* was given a picture of a human puppet jerking on the end of strings descending from the word gene, and the French edition a picture of little bowler-hatted men with clockwork wind-up keys sticking out of their backs. I have had slides of both covers made up as illustrations of what I was *not* trying to say.

So, the answer to Symons is that of course he was right to criticize what he thought I was saying, but of course I wasn't actually saying it (Ridley 1980). No doubt I was partly to blame for the original misunderstanding, but I can only urge now that we put aside the preconceptions derived from common usage ('most men don't understand computers to even the slightest degree'—Weizenbaum 1976, p. 9), and actually go and read some of the fascinating modern literature on robotics and computer intelligence (e.g. Boden 1977; Evans 1979; Hofstadter 1979).

Once again, of course, philosophers may debate the ultimate determinacy of computers programmed to behave in artificially intelligent ways, but if we are going to get into that level of philosophy many would apply the same arguments to human intelligence (Turing 1950). What is a brain, they would ask, but a computer, and what is education but a form of programming? It is very hard to give a non-supernatural account of the human brain and human emotions, feelings, and apparent free will, *without* regarding the brain as, in some sense, the equivalent of a programmed, cybernetic machine. The astronomer Sir Fred Hoyle (1964) expresses very vividly what, it seems to me, any evolutionist must think about nervous systems:

> Looking back [at evolution] I am overwhelmingly impressed by the way in which chemistry has gradually given way to electronics. It is not unreasonable to describe the first living creatures as entirely chemical in character. Although electrochemical processes are important in plants, organized electronics, in the sense of data processing, does not enter or operate in the plant world. But primitive electronics begins to assume importance as soon as we have a creature that moves around... The first electronic systems possessed by primitive animals were essentially guidance systems, analogous logically to sonar or radar. As we pass to more developed animals we find electronic systems being used not merely for guidance but for directing the animal toward food...
>
> The situation is analogous to a guided missile, the job of which is to intercept and destroy another missile. Just as in our modern world attack and defense become more and more subtle in their methods, so it was the case with animals. And with increasing subtlety, better and better systems of electronics become necessary. What happened in nature has a close parallel with the development of electronics in modern military applications... I find it a sobering thought that but for the tooth-and-claw existence of the jungle we should not possess our intellectual capabilities, we should not be able to inquire into the structure of the Universe, or be able to appreciate a symphony of Beethoven... Viewed in this light, the question that is sometimes asked—can computers think?—is somewhat ironic. Here of course I mean the computers that we ourselves

> make out of inorganic materials. What on earth do those who ask such a question think they themselves are? Simply computers, but vastly more complicated ones than anything we have yet learned to make. Remember that our man-made computer industry is a mere two or three decades old, whereas we ourselves are the products of an evolution that has operated over hundreds of millions of years [pp. 24–6].

Others may disagree with this conclusion, although I suspect that the only alternatives to it are religious ones. Whatever the outcome of that debate, to return to genes and the main point of this chapter, the issue of determinism versus free will is just not affected one way or the other by whether or not you happen to be considering *genes* as causal agents rather than environmental determinants.

But, it will pardonably be said, there is no smoke without fire. Functional ethologists and 'sociobiologists' must have said something to deserve being tarred with the brush of genetic determinism. Or if it is all a misunderstanding there must be some good explanation, because misunderstandings that are so widespread do not come about for no reason, even if abetted by cultural myths as powerful as the gene myth and the computer myth in unholy alliance. Speaking for myself, I think I know the reason. It is an interesting one, and it will occupy the rest of this chapter. The misunderstanding arises from the way we talk about a quite different subject, namely natural selection. Gene selectionism, which is a way of talking about evolution, is mistaken for genetic determinism, which is a point of view about development. People like me are continually postulating genes 'for' this and genes for that. We give the impression of being obsessed with genes and with 'genetically programmed' behaviour. Take this in conjunction with the popular myths of the

Calvinistic determinacy of genes, and of 'programmed' behaviour as the hallmark of jactitating Disneyland puppets, and is it any wonder that we are accused of being genetic determinists?

Why, then, do functional ethologists talk about genes so much? Because we are interested in natural selection, and natural selection is differential survival of genes. If we are to so much as discuss the *possibility* of a behaviour pattern's evolving by natural selection, we have to postulate genetic variation with respect to the tendency or capacity to perform that behaviour pattern. This is not to say that there necessarily *is* such genetic variation for any particular behaviour pattern, only that there must have been genetic variation in the past if we are to treat the behaviour pattern as a Darwinian adaptation. Of course the behaviour pattern may not be a Darwinian adaptation, in which case the argument will not apply.

Incidentally, I should defend my usage of 'Darwinian adaptation' as synonymous with 'adaptation produced by natural selection', for Gould and Lewontin (1979) have recently emphasized, with approval, the 'pluralistic' character of Darwin's own thought. It is indeed true that, especially towards the end of his life, Darwin was driven by criticisms, which we can now see to be erroneous, to make some concessions to 'pluralism': he did not regard natural selection as the only important driving force in evolution. As the historian R. M. Young (1971) has sardonically put it, 'by the sixth edition the book was mistitled and should have read *On the Origin of Species by Means of Natural Selection and All Sorts of Other Things*'. It is, therefore, arguably incorrect to use 'Darwinian evolution' as synonymous with 'evolution by natural selection'. But Darwinian *adaptation* is another matter. Adaptation cannot be produced by random drift, or by any other realistic evolutionary force that we know of save natural selection. It is true that Darwin's pluralism did fleetingly allow for one other driving force that might, in principle, lead to adaptation, but that driving force is inseparably linked with the name of Lamarck, not of Darwin. 'Darwinian adaptation' could not sensibly mean anything other than adaptation produced by natural selection, and I shall use it in this sense. In several other places in this book (e.g. in Chapters 3 and 6), we shall resolve apparent disputes by drawing a distinction between evolution in general and adaptive evolution in particular. The fixation of neutral mutations, for instance, can be regarded as evolution, but it is not adaptive evolution. If a molecular geneticist interested in gene substitutions, or a palaeontologist interested in major trends, argues with an ecologist interested in adaptation, they are likely to find themselves at cross-purposes simply because each of them emphasizes a different aspect of what evolution means.

'Genes for conformity, xenophobia, and aggressiveness are simply postulated for humans because they are needed for the theory, not because any evidence for them exists' (Lewontin 1979b). This is a fair criticism of E. O. Wilson, but not a very damning one. Apart from possible political repercussions which might be unfortunate, there is nothing wrong with cautiously speculating about a possible Darwinian survival value of xenophobia or any other trait. And you cannot begin to speculate, however cautiously, about the survival value of anything unless you postulate a genetic basis for variation in that thing. Of course xenophobia may not vary genetically, and of course xenophobia may not be a Darwinian adaptation, but we can't even discuss the possibility of its being a Darwinian adaptation unless we postulate a genetic basis for it. Lewontin himself has expressed the point as well as anybody: 'In order for a trait to evolve by natural selection it is necessary that there be genetic variation in the population for such a trait' (Lewontin 1979b). And 'genetic variation in the population for' a trait X is exactly what we mean when we talk, for brevity, of 'a gene for' X.

Xenophobia is controversial, so consider a behaviour pattern that nobody would fear to regard as a Darwinian adaptation. Pit-digging in antlions is obviously an adaptation to catch prey. Antlions are insects, neuropteran larvae with the general appearance and demeanour of monsters from outer space. They are 'sit and wait' predators who dig pits in soft sand which trap ants and other small walking insects. The pit is a nearly perfect cone, whose sides slope so steeply that prey cannot climb out once they have fallen in. The antlion sits just under the sand at the bottom of the pit, where it lunges with its horror-film jaws at anything that falls in.

Pit-digging is a complex behaviour pattern. It costs time and energy, and satisfies the most exacting criteria for recognition as an adaptation (Williams 1966; Curio 1973). It must, then, have evolved by natural selection. How might this have happened? The details don't matter for the moral I want to draw. Probably an ancestral antlion existed which did not dig a pit but simply lurked just beneath the sand surface waiting for prey to blunder over it. Indeed some species still do this. Later, behaviour leading to the creation of a shallow depression in the sand probably was favoured by selection because the depression marginally impeded escaping prey. By gradual degrees over many generations the behaviour changed so that what was a shallow depression became deeper and wider. This not only hindered escaping prey but also increased the catchment area over which prey might stumble in the first place. Later still the digging behaviour changed again so that the resulting pit became a steep-sided cone, lined with fine, sliding sand so that prey were unable to climb out.

Nothing in the previous paragraph is contentious or controversial. It will be regarded as legitimate speculation about historical events that we cannot see directly, and it will probably be thought plausible. One reason why it will be accepted as uncontroversial historical speculation is that it makes no mention of genes. But my point is that none of that history, nor any comparable history, could possibly have been true unless there was genetic variation in the behaviour at every step of the evolutionary way. Pit-digging in antlions is only one of the thousands of examples that I could have chosen. Unless natural selection has genetic variation to act upon, it cannot give rise to evolutionary change. It follows that where you find Darwinian adaptation there must have been genetic variation in the character concerned.

Nobody has ever done a genetic study of pit-digging behaviour in antlions (J. Lucas, personal communication). There is no need to do one, if all we want to do is satisfy ourselves of the sometime existence of genetic variation in the behaviour pattern. It is sufficient that we are convinced that it is a Darwinian adaptation (if you are not convinced that pit-digging is such an adaptation, simply substitute any example of which you are convinced).

I spoke of the *sometime* existence of genetic variation. This was because it is quite likely that, were a genetic study to be mounted of antlions today, no genetic variation would be found. It is in general to be expected that, where there is strong selection in favour of some trait, the original variation on which selection acted to guide the evolution of the trait will have become used up. This is the familiar 'paradox' (it is not really very paradoxical when we think about it carefully) that traits under strong selection tend to have low heritability (Falconer 1960); 'evolution by natural selection destroys the genetic variance on which it feeds' (Lewontin 1979b). Functional hypotheses frequently concern phenotypic traits, like possession of eyes, which are all but universal in the population, and therefore without contemporary genetic variation. When we speculate about, or make models of, the evolutionary production of an adaptation, we are necessarily talking about a time when there was appropriate genetic variation.

We are bound, in such discussions, to postulate, implicitly or explicitly, genes 'for' proposed adaptations.

Some may balk at treating 'a genetic contribution to variation in X' as equivalent to 'a gene or genes for X'. But this is a routine genetic practice, and one which close examination shows to be almost inevitable. Other than at the molecular level, where one gene is seen directly to produce one protein chain, geneticists never deal with units of phenotype as such. Rather, they always deal with *differences*. When a geneticist speaks of a gene 'for' red eyes in *Drosophila*, he is not speaking of the cistron which acts as template for the synthesis of the red pigment molecule. He is implicitly saying: there is variation in eye colour in the population; other things being equal, a fly with this gene is more likely to have red eyes than a fly without the gene. That is all that we ever mean by a gene 'for' red eyes. This happens to be a morphological rather than a behavioural example, but exactly the same applies to behaviour. A gene 'for' behaviour X is a gene 'for' whatever morphological and physiological states tend to produce that behaviour.

A related point is that the use of single-locus models is just a conceptual convenience, and this is true of adaptive hypotheses in exactly the same way as it is true of ordinary population genetic models. When we use single-gene language in our adaptive hypotheses, we do not intend to make a point about single-gene models as against multi-gene models. We are usually making a point about *gene* models as against non-gene models, for example as against 'good of the species' models. Since it is difficult enough convincing people that they ought to think in genetic terms *at all* rather than in terms of, say, the good of the species, there is no sense in making things even more difficult by trying to handle the complexities of many loci at the outset. What Lloyd (1979) calls the OGAM (one gene analysis model) is, of course, not the last word in genetic accuracy. *Of course* we shall eventually have to face up to multi-locus complexity. But the OGAM is vastly preferable to modes of adaptive reasoning that forget about genes altogether, and this is the only point I am trying to make at present.

Similarly we may find ourselves aggressively challenged to substantiate our 'claims' of the existence of 'genes for' some adaptation in which we are interested. But this challenge, if it is a real challenge at all, should be directed at the whole of the neo-Darwinian 'modern synthesis' and the whole of population genetics. To phrase a functional hypothesis in terms of genes is to make no strong claims about genes at all: it is simply to make explicit an assumption which is inseparably built into the modern synthesis, albeit it is sometimes implicit rather than explicit.

A few workers have, indeed, flung just such a challenge at the whole neo-Darwinian modern synthesis, and have claimed not to be neo-Darwinians. Goodwin (1979) in a published debate with Deborah Charlesworth and others, said, 'neo-Darwinism has an incoherence in it... we are not given any way of generating phenotypes from genotypes in neo-Darwinism. Therefore the theory is in this respect defective.' Goodwin is, of course, quite right that development is terribly complicated, and we don't yet understand much about how phenotypes are generated. But *that* they are generated, and *that* genes contribute significantly to their variation are incontrovertible facts, and those facts are all we need in order to make neo-Darwinism coherent. Goodwin might just as well say that, before Hodgkin and Huxley worked out how the nerve impulse fired, we were not entitled to believe that nerve impulses controlled behaviour. *Of course* it would be nice to know how phenotypes are made but, while embryologists are busy finding out, the rest of us are entitled by the known facts of genetics to carry on being neo-Darwinians, treating embryonic development as a black box. There is no competing theory that has even a remote claim to be called coherent.

It follows from the fact that geneticists are always concerned with phenotypic *differences* that we need not be afraid of postulating genes with indefinitely complex phenotypic effects, and with phenotypic effects that show themselves only in highly complex developmental conditions. Together with Professor John Maynard Smith, I recently took part in a public debate with two radical critics of 'sociobiology', before an audience of students. At one time in the discussion we were trying to establish that to talk of a gene 'for X' is to make no outlandish claim, even where X is a complex, learned behaviour pattern. Maynard Smith reached for a hypothetical example and came up with a 'gene for skill in tying shoelaces'. Pandemonium broke loose at this rampant genetic determinism! The air was thick with the unmistakable sound of worst suspicions being gleefully confirmed. Delightedly sceptical cries drowned the quiet and patient explanation of just what a *modest* claim is being made whenever one postulates a gene for, say, skill in tying shoelaces. Let me explain the point with the aid of an even more radical-sounding yet truly innocuous thought experiment (Dawkins 1981).

Reading is a learned skill of prodigious complexity, but this provides no reason in itself for scepticism about the possible existence of a gene for reading. All we would need in order to establish the existence of a gene for reading is to discover a gene for not reading, say a gene which induced a brain lesion causing specific dyslexia. Such a dyslexic person might be normal and intelligent in all respects except that he could not read. No geneticist would be particularly surprised if this type of dyslexia turned out to breed true in some Mendelian fashion. Obviously, in this event, the gene would only exhibit its effect in an environment which included normal education. In a prehistoric environment it might have had no detectable effect, or it might have had some different effect and have been known to cave-dwelling geneticists as, say, a gene for inability to read animal footprints. In our educated environment it would properly be called a gene 'for' dyslexia, since dyslexia would be its most salient consequence. Similarly, a gene which caused total blindness would also prevent reading, but it would not usefully be regarded as a gene for not reading. This is simply because preventing reading would not be its most obvious or debilitating phenotypic effect.

Returning to our gene for specific dyslexia, it follows from the ordinary conventions of genetic terminology that the wild-type gene at the same locus, the gene that the rest of the population has in double dose, would properly be called a gene 'for reading'. If you object to that, you must also object to our speaking of a gene for tallness in Mendel's peas, because the logic of the terminology is identical in the two cases. In both cases the character of interest is a *difference*, and in both cases the difference only shows itself in some specified environment. The reason why something so simple as a one gene difference can have such a complex effect as to determine whether or not a person can learn to read, or how good he is at tying shoelaces, is basically as follows. However complex a given state of the world may be, the *difference* between that state of the world and some alternative state of the world may be caused by something extremely simple.

The point I made using antlions is a general one. I could have used any real or purported Darwinian adaptation whatsoever. For further emphasis I shall use one more example. Tinbergen *et al.* (1962) investigated the adaptive significance of a particular behaviour pattern in black-headed gulls (*Larus ridibundus*), eggshell removal. Shortly after a chick hatches, the parent bird grasps the empty eggshell in the bill and removes it from the vicinity of the nest. Tinbergen and his colleagues considered a number of possible hypotheses about the survival value of this behaviour pattern. For instance they suggested that the empty eggshells might serve as breeding grounds for harmful bacteria, or the sharp edges might cut the chicks. But the hypothesis for which they ended up finding evidence was that the empty eggshell serves as a conspicuous visual beacon summoning crows and other predators of chicks or eggs to the nest. They did ingenious experiments, laying out artificial nests with and without empty eggshells, and showed that eggs accompanied by empty eggshells were, indeed, more likely to be attacked by crows than eggs without empty eggshells by their side. They concluded that natural selection had favoured eggshell removal behaviour of adult gulls, because past adults who did not do it reared fewer children.

As in the case of antlion digging, nobody has ever done a genetic study of eggshell removal behaviour in black-headed gulls. There is no direct evidence that variation in tendency to remove empty eggshells breeds true. Yet clearly the assumption that it does, or once did, is essential for the Tinbergen hypothesis. The Tinbergen hypothesis, as normally phrased in gene-free language, is not particularly controversial. Yet it, like all the rival functional hypotheses that Tinbergen rejected, rests fundamentally upon the assumption that once upon a time there must have been gulls with a genetic tendency to remove eggshells, and other gulls with a genetic tendency not to remove them, or to be less likely to remove them. There must have been genes for removing eggshells.

Here I must enter a note of caution. Suppose we actually did a study of the genetics of eggshell removal behaviour in modern gulls. It would be a behaviour-geneticist's dream to find a simple Mendelian mutation which radically altered the behaviour pattern, perhaps abolished the behaviour altogether. By the argument given above, this mutant would truly be a gene 'for' not removing eggshells, and, by definition, its wild-type allele would have to be called a gene for eggshell removal. But now comes the note of caution. It most definitely does not follow that this particular locus 'for' eggshell removal was one of the ones upon which natural selection worked during the evolution of the adaptation. On the contrary, it seems much more probable that a complex behaviour pattern like eggshell removal must have been built up by selection on a large number of loci, each having a small effect in interaction with the others. Once the behaviour complex had been built up, it is easy to imagine a single major mutation arising, whose effect is to destroy it. Geneticists perforce must exploit the genetic variation available for them to study. They also believe that natural selection must have worked on similar genetic variation in wreaking evolutionary change. But there is no reason for them to believe that the loci controlling modern variation in an adaptation were the very same loci at which selection acted in building up the adaptation in the first place.

Consider the most famous example of single gene control of complex behaviour, the case of Rothenbuhler's (1964) hygienic bees. The point of using this example is that it illustrates well how a highly complex behaviour difference can be due to a single gene difference. The hygienic behaviour of the Brown strain of honeybees involves the whole neuromuscular system, but the fact that they perform the behaviour whereas Van Scoy bees do not is, according to Rothenbuhler's model, due to differences at two loci only. One locus determines the uncapping of cells containing diseased brood, the other locus determines the removing of diseased brood after uncapping. It would be possible, therefore, to imagine a natural selection in favour of uncapping behaviour and a natural selection in favour of removing behaviour, meaning selection of the two genes versus their respective alleles. But the point I am making here is that, although that could happen, it is not likely to be very interesting evolutionarily. The modern uncapping gene and the modern removing gene may very well not have been involved in the original natural selection process that steered the evolutionary putting together of the behaviour.

Rothenbuhler observed that even Van Scoy bees sometimes perform hygienic behaviour. They are just quantitatively much less likely to do so than are Brown bees. It is likely, therefore, that both Brown and Van Scoy bees have hygienic ancestors, and both have in their nervous systems the machinery of uncapping and removing behaviour: it is just that Van Scoy bees have genes that prevent them from turning the machinery on. Presumably if we went back even further in time we should find an ancestor of all modern bees which not only was not hygienic itself but had never had a hygienic ancestor. There must have been an evolutionary progression building up the uncapping and removing behaviour from nothing, and this evolutionary progression involved the selection of many genes which are now fixed in both the Brown and the Van Scoy strains. So, although the uncapping and the removing genes of the Brown strain really are rightly called genes for uncapping and removing, they are defined as such only because they happen to have alleles whose effect is to prevent the behaviour from being performed. The mode of action of these alleles could be boringly destructive. They might simply cut some vital link in the neural machinery. I am reminded of Gregory's (1961) vivid illustration of the perils of making inferences from ablation experiments on the brain: 'the removal of any of several widely spaced resistors may cause a radio set to emit howls, but it does not follow that howls are immediately associated with these resistors, or indeed that the causal relation is anything but the most indirect. In particular, we should not say that the function of the resistors in the normal circuit is to inhibit howling. Neurophysiologists, when faced with a comparable situation, have postulated "suppressor regions".'

This consideration seems to me to be a reason for caution, not a reason for rejecting the whole genetic theory of natural selection! Never mind if living geneticists are debarred from studying the particular loci at which selection in the past gave rise to the original evolution of interesting adaptations. It is too bad if geneticists usually are forced to concentrate on loci that are convenient rather than evolutionarily important. It is *still* true that the evolutionary putting together of complex and interesting adaptation consisted in the replacement of genes by their alleles.

This argument can contribute tangentially to the resolution of a fashionable contemporary dispute, by helping to put the issue in perspective. It is now highly, indeed passionately, controversial whether there is significant genetic variation in human mental abilities. Are some of us genetically brainier than others? What we mean by 'brainy' is also highly contentious, and rightly so. But I suggest that, by any meaning of the term, the following propositions cannot be denied. (1) There was a time when our ancestors were less brainy than we are. (2) Therefore there has been an increase in braininess in our ancestral lineage. (3) That increase came about through evolution, probably propelled by natural selection. (4) Whether propelled by selection or not, at least part of the evolutionary change in phenotype reflected an underlying genetic change: allele replacement took place and consequently mean mental ability increased over generations. (5) By definition therefore, at least in the past, there must have been significant genetic variation in braininess within the human population. Some people were genetically clever in comparison with their contemporaries, others were genetically relatively stupid.

The last sentence may engender a *frisson* of ideological disquiet, yet none of my five propositions could be seriously doubted, nor could their logical sequence. The argument works for brain size, but it equally works for any behavioural measure of cleverness we care to dream up. It does not depend on simplistic views of human intelligence as being a one-dimensional scalar quantity. The fact that intelligence is not a simple scalar quantity, important as that fact is, is simply irrelevant. So is the difficulty of measuring intelligence in practice. The conclusion of the previous paragraph is inevitable, provided only that we are evolutionists who agree to the proposition that once upon a time our ancestors were less clever (by whatever criterion) then we are. Yet in spite of all that, it still does not follow that there is any genetic variation in mental abilities left in the human population today: the genetic variance might all have been used up by selection. On the other hand it might not, and my thought experiment shows at least the inadvisability of dogmatic and hysterical opposition to the very possibility of genetic variation in human mental abilities. My own opinion, for what it is worth, is that even if there is such genetic variation in modern human populations, to base any policy on it would be illogical and wicked.

The existence of a Darwinian adaptation, then, implies the sometime existence of genes for producing the adaptation. This is not always made explicit. It is always possible to talk about the natural selection of a behaviour pattern in two ways. We can either talk about individuals with a tendency to perform the behaviour pattern being 'fitter' than individuals with a less strongly developed tendency. This is the now fashionable phraseology, within the paradigm of the 'selfish organism' and the 'central theorem of sociobiology'. Alternatively, and equivalently, we can talk directly of genes for performing the behaviour pattern surviving better than their alleles. It is always legitimate to postulate genes in any discussion of Darwinian adaptation, and it will be one of my central points in this book that it is often positively beneficial to do so. Objections, such as I have heard made, to the 'unnecessary geneticizing' of the language of functional ethology, betray a fundamental failure to face up to the reality of what Darwinian selection is all about.

Let me illustrate this failure by another anecdote. I recently attended a research seminar given by an anthropologist. He was trying to interpret the incidence among various human tribes of a particular mating system (it happened to be polyandry) in terms of a theory of kin selection. A kin selection theorist can make models to predict the conditions under which we would expect to find polyandry. Thus, on one model applied to Tasmanian native hens (Maynard Smith & Ridpath 1972), the population sex ratio would need to be male-biased, and partners would need to be close kin, before a biologist would predict polyandry. The anthropologist sought to show that his polyandrous human tribes lived under such conditions, and, by implication, that other tribes showing the more normal patterns of monogamy or polygyny lived under different conditions.

Though fascinated by the information he presented, I tried to warn him of some difficulties in his hypothesis. I pointed out that the theory of kin selection is fundamentally a genetic theory, and that kin-selected adaptations to local conditions had to come about through the replacement of alleles by other alleles, over generations. Had his polyandrous tribes been living, I asked, under their current peculiar conditions for long enough—enough generations—for the necessary genetic replacement to have taken place? Was there, indeed, any reason to believe that variations in human mating systems are under genetic control at all?

The speaker, supported by many of his anthropological colleagues in the seminar, objected to my dragging genes into the discussion. He was not talking about genes, he said, but about a social behaviour pattern. Some of his colleagues seemed uncomfortable with the very mention of the four-letter word 'gene'. I tried to persuade him that it was *he* who had 'dragged genes in' to the discussion although, to be sure, he had not mentioned the word gene in his talk. That is exactly the point I am trying to make. You cannot talk about kin selection, or any other form of Darwinian selection, *without* dragging genes in, whether you do so explicitly or not. By even speculating about kin selection as an explanation of differences in tribal mating systems, my anthropologist friend was implicitly dragging genes into the discussion. It is a pity he did not make it *explicit*, because he would then have realized what formidable difficulties lay in the path of his kin selection hypothesis: either his polyandrous tribes had to have been living, in partial genetic isolation, under their peculiar conditions for a large number of centuries, or natural selection had to have favoured the universal occurrence of genes programming some complex 'conditional strategy'. The irony is that, of all the participants in that seminar on polyandry, it was I who was advancing the least 'genetically deterministic' view of the behaviour under discussion. Yet because I insisted on making the genetic nature of the kin selection hypothesis explicit, I expect I appeared to be characteristically obsessed with genes, a 'typical genetic determinist'. The story illustrates well the main message of this chapter, that frankly facing up to the fundamental genetic nature of Darwinian *selection* is all too easily mistaken for an unhealthy preoccupation with hereditarian interpretations of ontogenetic *development*.

The same prejudice against explicit mention of genes where one can get away with an individual-level circumlocution is common among biologists. The statement 'genes for performing behaviour X are favoured over genes for not performing X' has a vaguely naive and unprofessional ring to it. What evidence is there for such genes? How dare you conjure up *ad hoc* genes simply to satisfy your hypothetical convenience! To say 'individuals that perform X are fitter than individuals that do not perform X' sounds much more respectable. Even if it is not known to be true, it will probably be accepted as a permissible speculation. But the two sentences are exactly equivalent in meaning. The second one says nothing that the first does not say more clearly. Yet if we recognize this equivalence and talk explicitly about genes 'for' adaptations, we run the risk of being accused of 'genetic determinism'. I hope I have succeeded in showing that this risk results from nothing more than misunderstanding. A sensible and unexceptionable way of thinking about natural selection—'gene selectionism'—is mistaken for a strong belief about development—'genetic determinism'. Anyone who thinks clearly about the details of how adaptations come into being is almost bound to think, implicitly if not explicitly, about genes, albeit they may be hypothetical genes. As I shall show in this book, there is much to be said for making the genetic basis of Darwinian functional speculations explicit rather than implicit. It is a good way of avoiding certain tempting errors of reasoning (Lloyd 1979). In doing this we may give the impression, entirely for the wrong reason, of being obsessed with genes and all the mythic baggage that genes carry in the contemporary journalistic consciousness. But determinism, in the sense of an inflexible, tramline-following ontogeny, is, or should be, a thousand miles from our thoughts. Of course, individual sociobiologists may or may not be genetic determinists. They may be Rastafarians, Shakers, or Marxists. But their private opinions on genetic determinism, like their private opinions on religion, have nothing to do with the fact that they use the language of 'genes for behaviour' when talking about natural selection.

A large part of this chapter has been based on the assumption that a biologist might wish to speculate on the Darwinian 'function' of behaviour patterns. This is not to say that all behaviour patterns necessarily have a Darwinian function. It may be that there is a large class of behaviour patterns which are selectively neutral or deleterious to their performers, and cannot usefully be regarded as the products of natural selection. If so, the arguments of this chapter do not apply to them. But it is legitimate to say 'I am interested in adaptation. I don't necessarily think all behaviour patterns are adaptations, but I want to study those behaviour patterns that are adaptations.' Similarly, to express a preference for studying vertebrates rather than invertebrates does not commit us to the belief that all animals are vertebrates. Given that our field of interest is adaptive behaviour, we cannot talk about the Darwinian evolution of the objects of interest without postulating a genetic basis for them. And to use 'a gene for X' as a convenient way of talking about 'the genetic basis of X', has been standard practice in population genetics for over half a century.

The question of how large is the class of behaviour patterns that we can consider to be adaptations is an entirely separate question. It is the subject of the next chapter.

3 CONSTRAINTS ON PERFECTION

In one way or another, this book is largely preoccupied with the logic of Darwinian explanations of function. Bitter experience warns that a biologist who shows a strong interest in functional explanation is likely to be accused, sometimes with a passion that startles those more accustomed to scientific than ideological debate (Lewontin 1977), of believing that all animals are perfectly optimal—accused of being an 'adaptationist' (Lewontin 1979a,b; Gould & Lewontin 1979). Adaptationism is defined as 'that approach to evolutionary studies which assumes without further proof that all aspects of the morphology, physiology, and behavior of organisms are adaptive optimal solutions to problems' (Lewontin 1979b). In the first draft of this chapter I expressed doubts that anyone was truly an adaptationist in the extreme sense, but I have recently found the following quotation from, ironically enough, Lewontin himself: 'That is the one point which I think all evolutionists are agreed upon, that it is virtually impossible to do a better job than an organism is doing in its own environment' (Lewontin 1967). Lewontin has since, it seems, travelled his road to Damascus, so it would be unfair to use him as my adaptationist spokesman. Indeed together with Gould he has, in recent years, been one of the most articulate and forceful critics of adaptationism. As my representative adaptationist I take A. J. Cain, who has remained (Cain 1979) consistently true to the views expressed in his trenchant and elegant paper on 'The perfection of animals'.

Writing as a taxonomist, Cain (1964) is concerned to attack the traditional dichotomy between 'functional' characters, which by implication are not reliable taxonomic indicators, and 'ancestral' characters, which are. Cain argues forcefully that ancient 'ground-plan' characters, like the pentadactyl limb of tetrapods and the aquatic phase of amphibians, are there because they are functionally useful, rather than because they are inescapable historical legacies as is often implied. If one of two groups 'is in any way more primitive than the other, then its primitiveness must in itself be an adaptation to some less specialized mode of life which it can pursue successfully; it cannot be merely a sign of inefficiency' (p. 57). Cain makes a similar point about so-called trivial characters, criticizing Darwin for being too ready, under the at first sight surprising influence of Richard Owen, to concede functionlessness: 'No one will suppose that the stripes on the whelp of a lion, or the spots on the young blackbird, are of any use to these animals'. Darwin's remark must sound foolhardy today even to the most extreme critic of adaptationism. Indeed, history seems to be on the side of the adaptationists, in the sense that in particular instances they have confounded the scoffers again and again. Cain's own celebrated work, with Sheppard and their school, on the selection pressures maintaining the banding polymorphism in the snail *Cepaea nemoralis* may have been partly provoked by the fact that 'it had been confidently asserted that it could not matter to a snail whether it had one band on its shell or two' (Cain, p. 48). 'But perhaps the most remarkable functional interpretation of a "trivial" character is given by Manton's work on the diplopod *Polyxenus*, in which she has shown that a character formerly described as an "ornament" (and what could sound more useless?) is almost literally the pivot of the animal's life' (Cain, p. 51).

Adaptationism as a working hypothesis, almost as a faith, has undoubtedly been the inspiration for some outstanding discoveries. Von Frisch (1967), in defiance of the prestigious orthodoxy of von Hess, conclusively demonstrated colour vision in fish and in honeybees by controlled experiments. He was driven to undertake those experiments by his refusal to believe that, for example, the colours of flowers were there for no reason, or simply to delight men's eyes. This is, of course, not evidence for the validity of adaptationist faith. Each question must be tackled afresh, on its merits.

Wenner (1971) performed a valuable service in questioning von Frisch's dance language hypothesis, since he provoked J. L. Gould's (1976) brilliant confirmation of von Frisch's theory. If Wenner had been more of an adaptationist Gould's research might never have been done, but Wenner would also not have allowed himself to be so blithely wrong. Any adaptationist, while perhaps conceding that Wenner had usefully exposed lacunae in von Frisch's original experimental design, would instantly have jumped, with Lindauer (1971), on the fundamental question of why bees dance at all. Wenner never denied that they dance, nor that the dance contained all the information about the direction and distance of food that von Frisch claimed. All he denied was that other bees used the dance information. An adaptationist could not have rested happy with the idea of animals performing such a time-consuming, and above all complex and statistically improbable, activity for nothing. Adaptationism cuts both ways, however. I am now delighted that Gould did his clinching experiments, and it is entirely to my discredit that, even in the unlikely event of my having been ingenious enough to think of them,

I would have been too adaptationist to have bothered. I just *knew* Wenner was wrong (Dawkins 1969)!

Adaptationist thinking, if not blind conviction, has been a valuable stimulator of testable hypotheses in physiology. Barlow's (1961) recognition of the overwhelming functional need in sensory systems to reduce redundancy in input led him to a uniquely coherent understanding of a variety of facts about sensory physiology. Analogous functional reasoning can be applied to the motor system, and to hierarchical systems of organization generally (Dawkins 1976b; Hailman 1977). Adaptationist conviction cannot tell us about physiological mechanism. Only physiological experiment can do that. But cautious adaptationist reasoning can suggest which of many possible physiological hypotheses are most promising and should be tested first.

I have tried to show that adaptationism can have virtues as well as faults. But this chapter's main purpose is to list and classify constraints on perfection, to list the main reasons why the student of adaptation should proceed with caution. Before coming to my list of six constraints on perfection, I should deal with three others that have been proposed, but which I find less persuasive. Taking, first, the modern controversy among biochemical geneticists about 'neutral mutations', repeatedly cited in critiques of adaptationism, it is simply irrelevant. If there are neutral mutations in the biochemists' sense, what this means is that any change in polypeptide structure which they induce has no effect on the enzymatic activity of the protein. This means that the neutral mutation will not change the course of embryonic development, will have no phenotypic effect *at all*, as a whole-organism biologist would understand phenotypic effect. The biochemical controversy over neutralism is concerned with the interesting and important question of whether all gene substitutions have phenotypic effects. The adaptationism controversy is quite different. It is concerned with whether, *given* that we are dealing with a phenotypic effect big enough to see and ask questions about, we should assume that it is the product of natural selection. The biochemist's 'neutral mutations' are more than neutral. As far as those of us who look at gross morphology, physiology, and behaviour are concerned, they are not mutations at all. It was in this spirit that Maynard Smith (1976b) wrote: 'I interpret "rate of evolution" as a rate of adaptive change. In this sense, the substitution of a neutral allele would not constitute evolution'. If a whole-organism biologist sees a genetically determined difference among phenotypes, he already knows he cannot be dealing with neutrality in the sense of the modern controversy among biochemical geneticists.

He might, nevertheless, be dealing with a neutral character in the sense of an earlier controversy (Fisher & Ford 1950; Wright 1951). A genetic difference could show itself at the phenotypic level, yet still be selectively neutral. But mathematical calculations such as those of Fisher (1930b) and Haldane (1932a) show how unreliable human subjective judgement can be on the 'obviously trivial' nature of some biological characters. Haldane, for example, showed that, with plausible assumptions about a typical population, a selection pressure as weak as 1 in 1000 would take only a few thousand generations to push an initially rare mutation to fixation, a small time by geological standards. It appears that, in the controversy referred to above, Wright was misunderstood (see below). Wright (1980) was embarrassed at finding the idea of evolution of nonadaptive characters by genetic drift labelled the 'Sewall Wright effect', 'not only because others had previously advanced the same idea, but because I myself had strongly rejected it from the first (1929), stating that pure random drift leads "inevitably to degeneration and extinction"'. I have attributed apparent nonadaptive taxonomic differences to pleiotropy, where not merely ignorance of an adaptive significance.' Wright was in fact showing how a subtle mixture of drift and selection can produce adaptations *superior* to the products of selection alone (see pp. 60–1).

A second suggested constraint on perfection concerns allometry (Huxley 1932): 'In cervine deer, antler size increases more than proportionately to body size...so that larger deer have more than proportionately large antlers. It is then unnecessary to give a specifically adaptive reason for the extremely large antlers of large deer' (Lewontin 1979b). Well, Lewontin has a point here, but I would prefer to rephrase it. As it stands it suggests that the allometric constant is constant in a God-given immutable sense. But constants on one time-scale can be variables on another. The allometric constant is a parameter of embryonic development. Like any other such parameter it may be subject to genetic variation and therefore it may change over evolutionary time (Clutton-Brock & Harvey 1979). Lewontin's remark turns out to be analogous to the following: all primates have teeth; this is just a plain fact about primates, and it is therefore unnecessary to give a specifically adaptive reason for the presence of teeth in primates. What he probably meant to say is something like the following.

Deer have evolved a developmental mechanism such that growth of antlers relative to body size is allometric with a particular constant of allometry. Very probably the evolution of this allometric system of development occurred under the influence of selection pressures having nothing to do with the social function of antlers: probably it was conveniently compatible with pre-existing developmental processes in a way which we shall not understand until we know more about the biochemical and cellular details of embryology. Maybe ethological consequences of the extra large antlers of large deer exert a selective effect, but this selection pressure is likely to be swamped in importance by other selection pressures concerned with concealed internal embryological details.

Williams (1966, p. 16) invoked allometry in the service of a speculation about the selection pressures leading to increased brain size in man. He suggested that the prime focus of selection was on early teachability, at an elementary level, of children. 'The resulting selection for acquiring verbal facility as early as possible might have produced, as an allometric effect on cerebral development, populations in which an occasional Leonardo might arise.' Williams, however, did not see allometry as a weapon against the use of adaptive explanations. One feels that he was rightly less loyal to his particular theory of cerebral hypertrophy than to the general principle enunciated in his concluding rhetorical question: 'Is it not reasonable to anticipate that our understanding of the human mind would be aided greatly by knowing the purpose for which it was designed?'

What has been said of allometry applies also to pleiotropy, the possession by one gene of more than one phenotypic effect. This is the third of the suggested constraints on perfection that I want to get out of the way before embarking on my main list. It has already been mentioned in my quotation from Wright. A possible source of confusion here is that pleiotropy has been used as a weapon by both sides in this debate, if indeed it is a real debate. Fisher (1930b) reasoned that it was unlikely that any one of a gene's phenotypic effects was neutral, so how much more unlikely was it that *all* of a gene's pleiotropic effects could be neutral. Lewontin (1979b), on the other hand, remarked that 'many changes in characters are the result of pleiotropic gene action, rather than the direct result of selection on the character itself. The yellow color of the Malpighian tubules of an insect cannot itself be the subject of natural selection since that color can never be seen by any organism. Rather it is the pleiotropic consequence of red eye pigment metabolism, which may be adaptive.' There is no real disagreement here. Fisher was talking of the selective effects on a genetic mutation, Lewontin of selective effects on a phenotypic character; it is the same distinction, indeed, as I was making in discussing neutrality in the biochemical geneticists' sense.

Lewontin's point about pleiotropy is related to another one which I shall come on to below, about the problem of defining what he calls the natural 'suture lines', the 'phenotypic units' of evolution. Sometimes the dual effects of a gene are in principle inseparable; they are different views of the same thing, just as Everest used to have two names depending on which side it was seen from. What a biochemist sees as an oxygen-carrying molecule may be seen by an ethologist as red coloration. But there is a more interesting kind of pleiotropy in which the two phenotypic effects of a mutation are separable. The phenotypic effect of any gene (versus its alleles) is not a property of the gene alone, but also of the embryological context in which it acts. This allows abundant opportunities for the phenotypic effects of one mutation to be modified by others, and is the basis of such respected ideas as Fisher's (1930a) theory of the evolution of dominance, the Medawar (1952) and Williams (1957) theories of senescence, and Hamilton's (1967) theory of Y-chromosome inertness. In the present connection, if a mutation has one beneficial effect and one harmful one, there is no reason why selection should not favour modifier genes that detach the two phenotypic effects, or that reduce the harmful effect while enhancing the beneficial one. As in the case of allometry, Lewontin took too static a view of gene action, treating pleiotropy as if it was a property of the gene rather than of the interaction between the gene and its (modifiable) embryological context.

This brings me to my own critique of naive adaptationism, my own list of constraints on perfection, a list which has much in common with those of Lewontin and Cain, and those of Maynard Smith (1978b), Oster and Wilson (1978), Williams (1966), Curio (1973), and others. There is, indeed, much more agreement than the polemical tone of recent critiques would suggest. I shall not be concerned with particular cases, except as examples. As Cain and Lewontin both stress, it is not of general interest to challenge our ingenuity in dreaming up possible advantages of particular strange things that animals do. Here we are interested in the more general question of what the theory of natural selection entitles us to expect. My first constraint on perfection is an obvious one, mentioned by most writers on adaptation.

Time lags

The animal we are looking at is very probably out of date, built under the influence of genes that were selected in some earlier era when conditions were different. Maynard Smith (1976b) gives a quantitative measure of this effect, the 'lag load'. He (Maynard Smith 1978b) cites Nelson's demonstration that gannets, who normally lay only one egg, are quite capable of successfully incubating and rearing two if an extra one is experimentally added. Obviously an awkward case for the Lack hypothesis on optimal clutch size, and Lack himself (1966) was not slow to use the 'time-lag' escape route. He suggested, entirely plausibly, that the gannet clutch size of one egg evolved during a time when food was less plentiful, and that there had not yet been time for them to evolve to meet the changed conditions.

Such *post hoc* rescuing of a hypothesis in trouble is apt to provoke accusations of the sin of unfalsifiability, but I find such accusations rather unconstructive, almost nihilistic. We are not in Parliament or a court of law, with advocates of Darwinism scoring debating points against opponents, and vice versa. With the exception of a few genuine opponents of Darwinism, who are unlikely to be reading this, we are all in this together, all Darwinians who substantially agree on how we interpret what is, after all, the only workable theory we have to explain the organized complexity of life. We should all sincerely want to *know* why gannets lay only one egg when they could lay two, rather than treating the fact as a debating point. Lack's invoking of the 'time-lag' hypothesis may have been *post hoc*, but it is still thoroughly plausible, and it is testable. No doubt there are other possibilities which, with luck, may also be testable. Maynard Smith is surely right that we should leave aside the 'defeatist' (Tinbergen 1965) and untestable 'natural selection has bungled again' explanation as a last resort, as a matter of simple research strategy if nothing else. Lewontin (1978) says much the same: 'In a sense, then, biologists are forced to the extreme adaptationist program because the alternatives, although they are undoubtedly operative in many cases, are untestable in particular cases.'

Returning to the time-lag effect itself, since modern man has drastically changed the environment of many animals and plants over a time-scale that is negligible by ordinary evolutionary standards, we can expect to see anachronistic adaptations rather often. The hedgehog antipredator response of rolling up into a ball is sadly inadequate against motor cars.

Lay critics frequently bring up some apparently maladaptive feature of modern human behaviour—adoption, say, or contraception—and fling down a challenge to 'explain that if you can with your selfish genes'. Obviously, as Lewontin, Gould, and others have rightly stressed, it would be possible, depending on one's ingenuity, to pull a 'sociobiological' explanation out of a hat, a 'just-so story', but I agree with them and Cain that the answering of such challenges is a trivial exercise; indeed it is likely to be positively harmful. Adoption and contraception, like reading, mathematics, and stress-induced illness, are products of an animal that is living in an environment radically different from the one in which its genes were naturally selected. The question, about the adaptive significance of behaviour in an artificial world, should never have been put; and although a silly question may deserve a silly answer, it is wiser to give no answer at all and to explain why.

A useful analogy here is one that I heard from R. D. Alexander. Moths fly into candle flames, and this does nothing to help their inclusive fitness. In the world before candles were invented, small sources of bright light in darkness would either have been celestial bodies at optical infinity, or they might have been escape holes from caves or other enclosed spaces. The latter case immediately suggests a survival value for approaching light sources. The former case also suggests one, but in a more indirect sense (Fraenkel & Gunn 1940). Many insects use celestial bodies as compasses. Since these are at optical infinity, rays from them are parallel, and an insect that maintains a fixed orientation of, say, 30° to them will go in a straight line. But if the rays do not come from infinity they will not be parallel, and an insect that behaves in this way will spiral in to the light source (if steering an acute-angled course) or spiral away (if steering an obtuse-angled course) or orbit the source (if steering a course of exactly 90° to the rays). Self-immolation by insects in candle flames, then, has no survival value in itself: according to this hypothesis, it is a byproduct of the useful habit of steering by means of sources of light which are 'assumed' to be at infinity. That assumption was once safe. It now is safe no longer, and it may be that selection is even now working to modify the insects' behaviour. (Not necessarily, however. The overhead costs of making the necessary improvements may outweigh the benefits they might bring: moths that pay the costs of discriminating candles from stars may be less successful, on average, than moths that do not attempt the costly discrimination and accept the low risk of self-immolation—see next chapter.)

But now we have reached a problem which is more subtle than the simple time-lag hypothesis itself. This is the problem, already mentioned, about what characteristics of animals we choose to recognize as units which require explanation. As Lewontin (1979b) puts it, 'What are the "natural" suture lines for evolutionary dynamics? What is the topology of phenotype in evolution? What are the phenotypic units of evolution?' The candle flame paradox arose only because of the way in which we chose to characterize the moth's behaviour. We asked 'Why do moths fly into candle flames?' and were puzzled. If we had characterized the behaviour differently and asked 'Why do moths maintain a fixed angle to light rays (a habit which incidentally causes them to spiral into the light source if the rays happen not to be parallel)?', we should not have been so puzzled.

Consider human male homosexuality as a more serious example. On the face of it, the existence of a substantial minority of men who prefer sexual relations with their own sex rather than with the opposite sex constitutes a problem for any simple Darwinian theory. The rather discursive title of a privately circulated homosexualist pamphlet, which the author was kind enough to send me, summarizes the problem: 'Why are there "gays" at all? Why hasn't evolution eliminated "gayness" millions of years ago?' The author, incidentally, thinks the problem so important that it seriously undermines the whole Darwinian view of life. Trivers (1974), Wilson (1975, 1978), and especially Weinrich (1976) have considered various versions of the possibility that homosexuals may, at some time in history, have been functionally equivalent to sterile workers, foregoing personal reproduction the better to care for other relatives. I do not find this idea particularly plausible (Ridley & Dawkins, 1981), certainly no more so than a 'sneaky male' hypothesis. According to this latter idea, homosexuality represents an 'alternative male tactic' for obtaining matings with females. In a society with harem defence by dominant males, a male who is known to be homosexual is more likely to be tolerated by a dominant male than a known heterosexual male, and an otherwise subordinate male may be able, by virtue of this, to obtain clandestine copulations with females.

But I raise the 'sneaky male' hypothesis not as a plausible possibility so much as a way of dramatizing how easy and inconclusive it is to dream up explanations of this kind (Lewontin, 1979b, used the same didactic trick in discussing apparent homosexuality in *Drosophila*). The main point I wish to make is quite different and much more important. It is again the point about how we characterize the phenotypic feature that we are trying to explain.

Homosexuality is, of course, a problem for Darwinians only if there is a genetic component to the difference between homosexual and heterosexual individuals. While the evidence is controversial (Weinrich 1976), let us assume for the sake of argument that this is the case. Now the question arises, what does it *mean* to say there is a genetic component to the difference, in common parlance that there is a gene (or genes) 'for' homosexuality? It is a fundamental truism, of logic more than of genetics, that the phenotypic 'effect' of a gene is a concept that has meaning only if the context of environmental influences is specified, environment being understood to include all the other genes in the genome. A gene 'for' A in environment X may well turn out to be a gene for B in environment Y. It is simply meaningless to speak of an absolute, context-free, phenotypic effect of a given gene.

Even if there are genes which, in today's environment, produce a homosexual phenotype, this does not mean that in another environment, say that of our Pleistocene ancestors, they would have had the same phenotypic effect. A gene for homosexuality in our modern environment might have been a gene for something utterly different in the Pleistocene. So, we have the possibility of a special kind of 'time-lag effect' here. It may be that the phenotype which we are trying to explain did not even exist in some earlier environment, even though the gene did then exist. The ordinary time-lag effect which we discussed at the beginning of this section was concerned with changes in the environment as manifested in changed selection pressures. We have now added the more subtle point that changes in the environment may change the very nature of the phenotypic character we set out to explain.

Historical constraints

The jet engine superseded the propeller engine because, for most purposes, it was superior. The designers of the first jet engine started with a clean drawing board. Imagine what they would have produced if they had been constrained to 'evolve' the first jet engine from an existing propeller engine, changing one component at a time, nut by nut, screw by screw, rivet by rivet. A jet engine so assembled would be a weird contraption indeed. It is hard to imagine that an aeroplane designed in that evolutionary way would ever get off the ground. Yet in order to complete the biological analogy we have to add yet another constraint. Not only must the end product get off the ground; so must every intermediate along the way, and each intermediate must be superior to its predecessor. When looked at in this light, far from expecting animals to be perfect we may wonder that anything about them works at all.

Examples of the Heath Robinson (or Rube Goldberg—Gould 1978) character of animals are harder to be confident of than the previous paragraph might lead us to expect. A favourite example, suggested to me by Professor J. D. Currey, is the recurrent laryngeal nerve. The shortest distance from the brain to the larynx in a mammal, especially a giraffe, is emphatically not via the posterior side of the aorta, yet that is the route taken by the recurrent laryngeal. Presumably there once was a time in the remote ancestry of the mammals when the straight line from origin to end organ of the nerve did run posterior to the aorta. When, in due course, the neck began to lengthen, the nerve lengthened its detour posterior to the aorta, but the marginal cost of each step in the lengthening of the detour was not great. A major mutation might have re-routed the nerve completely, but only at a cost of great upheaval in early embryonic processes. Perhaps a prophetic, God-like designer back in the Devonian could have foreseen the giraffe and designed the original embryonic routing of the nerve differently, but natural selection has no foresight. As Sydney Brenner has remarked, natural selection could not be expected to have favoured some useless mutation in the Cambrian simply because 'it might come in handy in the Cretaceous'.

The Picasso-like face of a flatfish such as a sole, grotesquely twisted to bring both eyes round to the same side of the head, is another striking demonstration of a historical constraint on perfection. The evolutionary history of these fish is so clearly written into their anatomy that the example is a good one to thrust down the throats of religious fundamentalists. Much the same could be said of the curious fact that the retina of the vertebrate eye appears to be installed backwards. The light-sensitive 'photocells' are at the back of the retina, and light has to pass through the connecting circuitry, with some inevitable attenuation, before it reaches them. Presumably it would be possible to write down a very long sequence of mutations which would eventually lead to the production of an eye whose retina was 'the right way round' as it is in cephalopods, and this might be, in the end, slightly more efficient. But the cost in embryological upheaval would be so great that the intermediate stages would be heavily disfavoured by natural selection in comparison with the rival, patched-up job which does, after all, work pretty well. Pittendrigh (1958) has well said of adaptive organization that it is 'a patchwork of makeshifts pieced together, as it were, from what was available when opportunity knocked, and accepted in the hindsight, not the foresight, of natural selection' (see also Jacob, 1977, on 'tinkering').

Sewall Wright's (1932) metaphor, which has become known under the name of the 'adaptive landscape', conveys the same idea that selection in favour of local optima prevents evolution in the direction of ultimately superior, more global optima. His somewhat misunderstood (Wright 1980) emphasis on the role of genetic drift in allowing lineages to escape from the pull of local optima, and thereby attain a closer approximation to what a human might recognize as 'the' optimal solution, contrasts interestingly with Lewontin's (1979b) invoking of drift as an 'alternative to adaptation'. As in the case of pleiotropy, there is no paradox here. Lewontin is right that 'the finiteness of real populations results in random changes in gene frequency so that, with a certain probability, genetic combinations with lower reproductive fitness will be fixed in a population'. But on the other hand it is also true that, to the extent that local optima constitute a limitation on the attainment of design perfection, drift will tend to provide an escape (Lande 1976). Ironically, then, a *weakness* in natural selection can theoretically *enhance* the likelihood of a lineage attaining optimal design! Because it has no foresight, unalloyed natural selection is in a sense an *anti*-perfection mechanism, hugging, as it will, the tops of the low foot-hills of Wright's landscape. A mixture of strong selection interspersed with periods of relaxation of selection and drift may be the formula for crossing the valleys to the high uplands. Clearly if 'adaptationism' is to become an issue where debating points are scored, there is scope for both sides to have it both ways!

My own feeling is that somewhere here may lie the solution to the real paradox of this section on historical constraints. The jet engine analogy suggested that animals ought to be risible monstrosities of lashed-up improvisation, top-heavy with grotesque relics of patched-over antiquity. How can we reconcile this reasonable expectation with the formidable grace of the hunting cheetah, the aerodynamic beauty of the swift, the scrupulous attention to deceptive detail of the leaf insect? Even more impressive is the detailed agreement between different convergent solutions to common problems, for instance the multiple parallels that exist between the mammal radiations of Australia, South America, and the Old World. Cain (1964) remarks that, 'Up to now it has usually been assumed, by Darwin and others, that convergence will never be so good as to mislead us' but he goes on to give examples where competent taxonomists have been fooled. More and more groups which had hitherto been regarded as decently monophyletic are now being suspected of polyphyletic origin.

The citation of example and counter-example is mere idle fact-dropping. What we need is constructive work on the relation between local and global optima in an evolutionary context. Our understanding of natural selection itself needs to be supplemented by a study of 'escapes from specialization', to use Hardy's (1954) phrase. Hardy himself was suggesting neoteny as an escape from specialization, while in this chapter, following Wright, I have emphasized drift in this role.

Müllerian mimicry in butterflies may prove to be a useful case-study here. Turner (1977) remarks that 'among the long-winged butterflies of the tropical American rainforests (ithomiids, heliconids, danaids, pierids, pericopids) there are six distinct warning patterns, and although all the warningly colored species belong to one of these mimicry "rings" the rings themselves coexist in the same habitats through most of the American tropics and remain very distinct.... Once the difference between two patterns is too great to be jumped by a single mutation, convergence becomes virtually impossible, and the mimicry rings will coexist indefinitely.' This is one of the only cases where 'historical constraints' may be close to being understood in full genetic detail. It may provide a worthwhile opportunity also for the study of the genetic details of 'valley-crossing', which in the present case would consist in the detachment of a type of butterfly from the orbit of one mimicry ring, and its eventual 'capture' by the 'pull' of another mimicry ring. Though he does not invoke drift as an explanation in this case, Turner tantalizingly indicates that 'In southern Europe *Amata phegea*...has...captured *Zygenea ephialtes* from the Müllerian mimicry ring of zygaenids, homopterans, etc. to which it still belongs outside the range of *A. phegea* in northern Europe'.

At a more general theoretical level, Lewontin (1978) notes that 'there may often be several alternative stable equilibriums of genetic composition even when the force of natural selection remains the same. Which of these adaptive peaks in the space of genetic composition is eventually reached by a population depends entirely on chance events at the beginning of the selective process.... For example, the Indian rhinoceros has one horn and the African rhinoceros has two. Horns are an adaptation for protection against predators but it is not true that one horn is specifically adaptive under Indian conditions as opposed to two horns on the African plains. Beginning with two somewhat different developmental systems, the two species responded to the same selective forces in slightly different ways.' The point is basically a good one, although it is worth adding that Lewontin's uncharacteristically 'adaptationist' blunder about the functional significance of rhinoceros horns is not trivial. If horns really *were* an adaptation against predators it would indeed be hard to imagine how a single horn could be more useful against Asian predators while two horns were of more help against African predators. However if, as seems much more likely, rhinoceros horns are an adaptation for intraspecific combat and intimidation, it could well be the case that a one-horned rhino would be at a disadvantage in one continent while a two-horned rhino would suffer in the other. Whenever the name of the game is intimidation (or sexual attraction as Fisher taught us long ago), mere conformity to the majority style, whatever that majority style may happen to be, can have advantages. The details of a threat display and its associated organs may be arbitrary, but woe betide any mutant individual that departs from established custom (Maynard Smith & Parker 1976).

Available genetic variation

No matter how strong a potential selection pressure may be, no evolution will result unless there is genetic variation for it to work on. 'Thus, although I might argue that the possession of wings in addition to arms and legs might be advantageous to some vertebrates, none has ever evolved a third pair of appendages, presumably because the genetic variation has never been available' (Lewontin 1979b). One could reasonably dissent from this opinion. It may be that the only reason pigs have no wings is that selection has never favoured their evolution. Certainly we must be careful before we assume, on human-centred common-sense grounds, that it would obviously be handy for any animal to have a pair of wings even if it didn't use them very often, and that therefore the absence of wings in a given lineage must be due to lack of available mutations. Female ants can sprout wings if they happen to be nurtured as queens, but if nurtured as workers they do not express their capacity to do so. More strikingly, the queens in many species use their wings only once, for their nuptial flight, and then take the drastic step of biting or breaking them off at the roots in preparation for the rest of their life underground. Evidently wings have costs as well as benefits.

One of the most impressive demonstrations of the subtlety of Charles Darwin's mind is given by his discussion of winglessness and the costs of having wings in the insects of oceanic islands. For present purposes, the relevant point is that winged insects may risk being blown out to sea, and Darwin (1859, p. 177) suggested that this is why many island insects have reduced wings. But he also noted that some island insects are far from wingless; they have extra large wings.

> This is quite compatible with the action of natural selection. For when a new insect first arrived on the island, the tendency of natural selection to enlarge or to reduce the wings, would depend on whether a greater number of individuals were saved by successfully battling with the winds, or by giving up the attempt and rarely or never flying. As with mariners shipwrecked near a coast, it would have been better for the good swimmers if they had been able to swim still further, whereas it would have been better for the bad swimmers if they had not been able to swim at all and had stuck to the wreck.

A neater piece of evolutionary reasoning would be hard to find, although one can almost hear the baying chorus of 'Unfalsifiable! Tautological! Just-so story!'

Returning to the question of whether pigs ever could develop wings, Lewontin is undoubtedly right that biologists interested in adaptation cannot afford to ignore the question of the availability of mutational variation. It is certainly true that many of us, with Maynard Smith (1978a) though without his and Lewontin's authoritative knowledge of genetics, tend to assume 'that genetic variance of an appropriate kind will usually exist'. Maynard Smith's grounds are that 'with rare exceptions, artificial selection has always proved effective, whatever the organism or the selected character'. A notorious case, fully conceded by Maynard Smith (1978b), where the genetic variation necessary to an optimality theory often seems to be lacking, is that of Fisher's (1930a) sex-ratio theory. Cattle breeders have had no trouble in breeding for high milk yield, high beef production, large size, small size, hornlessness, resistance to various diseases, and fierceness in fighting bulls. It would obviously be of immense benefit to the dairy industry if cattle could be bred with a bias towards producing heifer calves rather than bull calves. All attempts to do this have singularly failed, apparently because the necessary genetic variation does not exist. It may be the measure of how misled is my own biological intuition that I find this fact rather astonishing, indeed worrying. I would like to think that it is an exceptional case, but Lewontin is certainly right that we need to pay more attention to the problem of the limitations of available genetic variation. From this point of view, a compilation of the amenability or resistance to artificial selection of a wide variety of characters would be of great interest.

Meanwhile, there are certain common-sense things that can be said. Firstly, it may make sense to invoke lack of available mutation to explain why animals do not have some adaptation which we think reasonable, but it is harder to apply the argument the other way round. For instance, we might indeed think that pigs would be better off with wings and suggest that they lack them only because their ancestors never produced the necessary mutations. But if we see an animal with a complex organ, or a complex and time-consuming behaviour pattern, we would seem to be on strong grounds in guessing that it must have been put together by natural selection. Habits such as dancing in bees as already discussed, 'anting' in birds, 'rocking' in stick insects, and egg-shell removal in gulls are positively time-consuming, energy-consuming, and complex. The working hypothesis that they must have a Darwinian survival value is overwhelmingly strong. In a few cases it has proved possible to find out what that survival value is (Tinbergen 1963).

The second common-sense point is that the hypothesis of 'no available mutations' loses some of its force if a related species, or the same species in other contexts, has shown itself capable of producing the necessary variation. I shall mention below a case where the known capabilities of the digger wasp *Ammophila campestris* were used to illuminate the lack of similar capabilities in the related species *Sphex ichneumoneus.* A more subtle version of the same argument can be applied within any one species. For instance, Maynard Smith (1977, see also Daly 1979) concludes a paper with an up-beat question: 'Why do male mammals not lactate?' We need not go into the details of why he thought they ought to; he may have been wrong, his model may have been wrongly set up, and the real answer to his question may be that it would not pay male mammals to lactate. The point here is that this is a slightly different kind of question from 'Why don't pigs have wings?' We know that male mammals contain the genes necessary for lactation, because all the genes in a female mammal have passed through male ancestors and may be handed on to male descendants. Genetic male mammals treated with hormones, indeed, can develop as lactating females. This all makes it less plausible that the reason male mammals don't lactate is simply that they haven't 'thought of it' mutationally speaking. (Indeed, I bet I could breed a race of spontaneously lactating males by selecting for increased sensitivity to progressively reduced dosages of injected hormone, an interesting practical application of the Baldwin/Waddington Effect.)

The third common-sense point is that if the variation that is being postulated consists in a simple quantitative extension of already existing variation it is more plausible than a radical qualitative innovation. It may be implausible to postulate a mutant pig with wing rudiments, but it is not implausible to postulate a mutant pig with a curlier tail than existing pigs. I have elaborated this point elsewhere (Dawkins 1980).

In any case, we need a more subtle approach to the question of what is the evolutionary impact of differing degrees of mutability. It is not good enough to ask, in an all or none way, whether there is or is not genetic variation available to respond to a given selection pressure. As Lewontin (1979a) rightly says, 'Not only is the qualitative possibility of adaptive evolution constrained by available genetic variation, but the relative rates of evolution of different characters are proportional to the amount of genetic variance for each.' I think this opens up an important line of thought when combined with the notion of historical constraints treated in the previous section. The point can be illustrated with a fanciful example.

Birds fly with wings made of feathers, bats with wings consisting of flaps of skin. Why do they not both have wings made in the same way, whichever way is 'superior'? A confirmed adaptationist might reply that birds must be better off with feathers and bats better off with skin flaps. An extreme anti-adaptationist might say that very probably feathers would actually be better than skin-flaps for both birds and bats, but bats never had the good fortune to produce the right mutations. But there is an intermediate position, one which I find more persuasive than either extreme. Let us concede to the adaptationist that, given enough time, the ancestors of bats probably could have produced the sequence of mutations necessary for them to sprout feathers. The operative phrase is 'given enough time'. We are not making an all-or-none distinction between impossible and possible mutational variation, but simply stating the undeniable fact that some mutations are quantitatively more probable than others. In this case, ancestral mammals might have produced both mutants with rudimentary feathers and mutants with rudimentary skin flaps. But the proto-feather mutants (they might have had to go through an intermediate stage of small scales) were so slow in making their appearance in comparison with the skin-flap mutants, that skin-flap wings had long ago appeared and led to the evolution of passably efficient wings.

The general point is akin to the one already made about adaptive landscapes. There we were concerned with selection preventing lineages from escaping the clutches of local optima. Here we have a lineage faced with two alternative routes of evolution, one leading to, say, feathered wings, the other to skin-flap wings. The feathered design may be not only a global optimum but the present local optimum as well. The lineage, in other words, may be sitting exactly at the foot of the slope leading to the feathered peak of the Sewall Wright landscape. If only the necessary mutations were available it would climb easily up the hill. Eventually, according to this fanciful parable, those mutations might have come, but—and this is the important point—they were too late. Skin-flap mutations had come before them, and the lineage had already climbed too far up the slopes of the skin-flap adaptive hill to turn back. As a river takes the line of least resistance downhill, thereby meandering in a route that is far from the most direct one to the sea, so a lineage will evolve according to the effects of selection on the variation available at any given moment. Once a lineage has begun to evolve in a given direction, this may in itself close options that were formerly available, sealing off access to a global optimum. My point is that lack of available variation does not have to be absolute in order to become a significant constraint on perfection. It need only be a quantitative brake to have dramatic qualitative effects. In spirit, then, I agree with Gould and Calloway (1980) when they say, citing Vermeij's (1973) stimulating paper on the mathematics of morphological versatility, that 'Some morphologies can be twisted, bent and altered in a variety of ways, and others cannot.' But I would prefer to soften 'cannot', to make it a quantitative constraint, not an absolute barrier.

McCleery (1978), in an agreeably comprehensible introduction to the McFarland school of ethological optimality theory, mentions H. A. Simon's concept of 'satisficing' as an alternative to optimizing. If optimizing systems are concerned with maximizing something, satisficing systems get away with doing just enough. In this case, doing enough means doing enough to stay alive. McCleery contents himself with complaining that such 'adequacy' concepts have not generated much experimental work. I think evolutionary theory entitles us to be a bit more negative *a priori*. Living things are not selected for their capacity simply to stay alive; they are staying alive in competition with other such living things. The trouble with satisficing as a concept is that it completely leaves out the competitive element which is fundamental to all life. In Gore Vidal's words: 'It is not enough to succeed. Others must fail.'

On the other hand 'optimizing' is also an unfortunate word because it suggests the attainment of what an engineer would recognize as the best design in a global sense. It tends to overlook the constraints on perfection which are the subject of this chapter. In many ways the word 'meliorizing' expresses a sensible middle way between optimizing and satisficing. Where *optimus* means best, *melior* means better. The points we have been considering about historical constraints, about Wright's adaptive landscapes, and about rivers following the line of immediate least resistance are all related to the fact that natural selection chooses the better of present available alternatives. Nature does not have the foresight to put together a sequence of mutations which, for all that they may entail temporary disadvantage, set a lineage on the road to ultimate global superiority. It cannot refrain from favouring slightly advantageous available mutations now, so as to take better advantage of superior mutations which may arrive later. Like a river, natural selection blindly meliorizes its way down successive lines of immediately available least resistance. The animal that results is not the most perfect design conceivable, nor is it merely good enough to scrape by. It is the product of a historical sequence of changes, each one of which represented, at best, the *better* of the alternatives that happened to be around at the time.

Constraints of costs and materials

'If there were no constraints on what is possible, the best phenotype would live for ever, would be impregnable to predators, would lay eggs at an infinite rate, and so on' (Maynard Smith 1978b). 'An engineer, given carte blanche on his drawing board could design an "ideal" wing for a bird, but he would demand to know the constraints under which he must work. Is he constrained to use feathers and bones, or may he design the skeleton in titanium alloy? How much is he allowed to spend on the wings, and how much of the available economic investment must be diverted into, say, egg production?' (Dawkins & Brockmann 1980). In practice, an engineer will normally be given a specification of minimum performance such as 'The bridge must bear a load of ten tons.... The aeroplane wing must not break until it receives a stress three times what would be expected in worst-case turbulent conditions; now go ahead and build it as cheaply as you can.' The best design is the one that satisfies ('satisfices') the criterion specification at the least cost. Any design that achieves 'better' than the specified criterion performance is likely to be rejected, because presumably the criterion could be achieved more cheaply.

The particular criterion specification is an arbitrary working rule. There is nothing magic about a safety margin of three times the expected worst-case conditions. Military aircraft may be designed with more risky safety margins than civilian ones. In effect, the engineer's optimization instructions amount to a monetary evaluation of human safety, speed, convenience, pollution of the atmosphere, etc. The price put on each of these is a matter of judgement, and is often a matter of controversy.

In the evolutionary design of animals and plants, judgement does not enter into it, nor does controversy except among the human spectators of the show. In some way, however, natural selection must provide the equivalent of such judgement: risks of predation must be evaluated against risks of starving and benefits of mating with an extra female. For a bird, resources spent on making breast muscles for powering wings are resources that could have been spent on making eggs. An enlarged brain would permit a finer tuning of behaviour to environmental details, past and present, but at a cost of an enlarged head, which means extra weight at the front end of the body, which in turn necessitates a larger tail for aerodynamic stability, which in turn.... Winged aphids are less fecund than wingless ones of the same species

(J. S. Kennedy, personal communication). That every evolutionary adaptation must cost something, costs being measured in lost opportunities to do other things, is as true as that gem of traditional economic wisdom, 'There is no such thing as a free lunch'.

Of course the mathematics of biological currency-conversion, of evaluating the costs of wing muscle, singing time, predator-vigilance time, etc., in some common currency such as 'gonad equivalents', are likely to be very complex. Whereas the engineer is allowed to simplify his mathematics by working to an arbitrarily chosen minimum threshold of performance, the biologist is granted no such luxury. Our sympathy and admiration must go out to those few biologists who have attempted to grapple with these problems in detail (e.g. Oster & Wilson 1978; McFarland & Houston 1981).

On the other hand, although the mathematics may be formidable, we don't need mathematics to deduce the most important point, which is that any view of biological optimization that denies the existence of costs and trade-offs is doomed. An adaptationist who looks at one aspect of an animal's body or behaviour, say the aerodynamic performance of its wings, while forgetting that efficiency in the wings can only be bought at a cost which will be felt somewhere else in the animal's economy, would deserve all the criticism he gets. It has to be admitted that too many of us, while never actually denying the importance of costs, forget to mention them, perhaps even forget to think about them, when we discuss biological function. This has probably provoked some of the criticism that has come our way. In an earlier section I quoted Pittendrigh's remark that adaptive organization was a 'patchwork of makeshifts'. We must also not forget that it is a tangle of compromises (Tinbergen 1965).

In principle, it would seem a valuable heuristic procedure to *assume* that an animal is optimizing something under a given set of constraints, and to try to work out what those constraints are. This is a restricted version of what McFarland and his colleagues call the 'reverse optimality' approach (e.g. McCleery 1978). As a case study I shall take some work with which I happen to be familiar.

Dawkins and Brockmann (1980) found that the digger wasps (*Sphex ichneumoneus*) studied by Brockmann behaved in a way that a naive human economist might have criticized as maladaptive. Individual wasps appeared to commit the 'Concorde Fallacy' of valuing a resource according to how much they had already spent on it, rather than according to how much they could get out of it in the future. Very briefly, the evidence is as follows. Solitary females provision burrows with stung and paralysed katydids which are to serve as food for their larvae (see Chapter 7). Occasionally two females find themselves provisioning the same burrow, and they usually end up fighting over it. Each fight goes on until one wasp, thereby defined as the loser, flees from the area, leaving the winner in control of the burrow and all the katydids caught by both wasps. We measured the 'real value' of a burrow as the number of katydids which it contained. The 'prior investment' by each wasp in the burrow was measured as the number of katydids which she, as an individual, had put into it. The evidence suggested that each wasp fought for a time proportional to her own investment, rather than proportional to the 'true value' of the burrow.

Such a policy has great human psychological appeal. We too tend to fight tenaciously for property which we have put great effort into acquiring. The fallacy gets its name from the fact that, at a time when sober economic judgement of future prospects counselled abandoning the developing of the Concorde airliner, one of the arguments in favour of continuing with the half completed project was retrospective: 'We have already spent so much on it that we cannot back out now.' A popular argument for prolonging wars gave rise to the other name for the fallacy, the 'Our boys shall not have died in vain' fallacy.

When Dr Brockmann and I first realized that digger wasps behaved in like manner, I was, it has to be confessed, a little disconcerted, possibly because of my own past investment of effort (Dawkins & Carlisle 1976; Dawkins 1976a) in persuading my colleagues that the psychologically appealing Concorde Fallacy was, indeed, a fallacy! But then we started to think more seriously about cost constraints. Could it be that what appeared to be maladaptive was better interpreted as an optimum, *given certain constraints?* The question then became: Is there a constraint such that the wasps' Concordian behaviour is the best they can achieve under it?

In fact the question was more complicated than that, because it was necessary to substitute Maynard Smith's (1974) concept of evolutionary stability ('ESS'—see Chapter 7) for that of simple optimality, but the principle remains that a reverse optimality approach might be heuristically valuable. If we can show that an animal's behaviour is what would be produced by an optimizing system working under constraint X, maybe we can use the approach to learn something of the constraints under which animals actually do work.

In the present case it seemed that the relevant constraint might be one of sensory capacity. If the wasps, for some reason, cannot count katydids in the burrow, but can metre some aspect of their own hunting efforts, there is an asymmetry of information possessed by the two combatants. Each one 'knows' that the burrow contains at least *b* katydids, where *b* is the number she herself has caught. She may 'estimate' that the true number in the burrow is larger than *b*, but she does not know how much larger. Under such conditions Grafen (in preparation) has shown that the expected ESS is approximately the one originally calculated by Bishop and Cannings (1978) for the so-called 'generalized war of attrition'. The mathematical details can be left aside; for present purposes what matters is that the behaviour expected by the extended war of attrition model would look very like the Concordian behaviour actually shown by the wasps.

If we were interested in testing the general hypothesis that animals optimize, this kind of *post hoc* rationalization would be suspect. By *post hoc* modification of the details of the hypothesis, one is bound to find a version which fits the facts. Maynard Smith's (1978b) reply to this kind of criticism is very relevant: 'in testing a model we are *not* testing the general proposition that nature optimizes, but the specific hypotheses about constraints, optimization criteria, and heredity'. In the present case we are making a general assumption that nature does optimize within constraints, and testing particular models of what those constraints might be.

The particular constraint suggested—inability of the wasp's sensory system to assess the contents of a burrow—is in accordance with independent evidence from the same population of wasps (Brockmann, Grafen & Dawkins 1979; Brockmann & Dawkins 1979). There is no reason to regard it as an irrevocably binding limitation for all time. Probably the wasps could evolve the capacity to assess nest contents, but only at a cost. Digger wasps of the related species *Ammophila campestris* have long been known to make an assessment of the contents of each of their nests every day (Baerends 1941). Unlike *Sphex*, which provisions one burrow at a time, lays an egg, then fills the burrow in with soil and leaves the larva to eat the provision on its own, *Ammophila campestris* is a progressive provisioner of several burrows concurrently. A female tends two or three growing larvae, each in a separate burrow, at the same time. The ages of her various larvae are staggered, and their food needs are different. Every morning she assesses the current contents of each burrow on a special early morning 'inspection round'. By experimentally changing the contents of burrows, Baerends showed that the female adjusts her whole day's provisioning of each burrow according to what it contained at the time of her morning inspection. The contents of the burrow at any other time of day have no effect on her behaviour, even though she is provisioning it all day. She appears, therefore, to use her assessment faculty sparingly, switching it off for the rest of the day after the morning inspection, almost as though it was a costly, power-consuming instrument. Fanciful as that analogy may be, it surely suggests that the assessment faculty, whatever it is, may have overhead running costs, even if (G. P. Baerends, personal communication) these consist only in the time consumed.

Sphex ichneumoneus, not being a progressive provisioner, and tending only one burrow at a time, presumably has less need than *Ammophila* for a burrow-assessment faculty. By not attempting to count prey in the burrow, it can save itself not only the running expenditure that *Ammophila* seems so careful to ration; it can also save itself the initial manufacturing costs of the necessary neural and sensory apparatus. Probably it could benefit slightly from having an ability to assess burrow contents, but only on the comparatively rare occasions when it finds itself competing for a burrow with another wasp. It is easy to believe that the costs outweigh the benefits, and that selection has therefore never favoured the evolution of assessment apparatus. I think this is a more constructive and interesting hypothesis than the alternative hypothesis that the necessary mutational variation has never arisen. Of course we have to admit that the latter might be right, but I would prefer to keep it as a hypothesis of last resort.

Imperfections at one level due to selection at another level

One of the main topics to be tackled in this book is that of the level at which natural selection acts. The kind of adaptations we should see if selection acted at the level of the group would be quite different from the adaptations we should expect if selection acts at the level of the individual. It follows that a group selectionist might well see as imperfections features which an individual selectionist would see as adaptations. This is the main reason why I regard as unfair Gould and Lewontin's (1979) equating of modern adaptationism with the naive perfectionism that Haldane named after Voltaire's Dr Pangloss. With reservations due to the various constraints on perfection, an adaptationist may believe that all aspects of organisms are 'adaptive optimal solutions to problems', or that 'it is virtually impossible to do a better job than an organism is doing in its given environment'. Yet the same adaptationist may be extremely fussy about the kind of meaning he allows to words like 'optimal' and 'better'. There are many kinds of adaptive, indeed Panglossian, explanations, for example most group selectionist ones, which would be utterly ruled out by the modern adaptationist.

For the Panglossian the demonstration that something is 'beneficial' (to whom or to what is often not specified) is a sufficient explanation for its existence. The neo-Darwinian adaptationist, on the other hand, insists upon knowing the exact nature of the selective process that has led to the evolution of the putative adaptation. In particular, he insists on precise language about the level at which natural selection is supposed to have acted. The Panglossian looks at a one-to-one sex ratio and sees that it is good: does it not minimize the wastage of the population's resources? The neo-Darwinian adaptationist considers in detail the fates of genes acting on parents to bias the sex ratio of their offspring, and calculates the evolutionarily stable state of the population (Fisher 1930a). The Panglossian is disconcerted by 1:1 sex ratios in polygynous species, in which a minority of males hold harems and the rest sit about in bachelor herds consuming almost half the population's food resources yet contributing not at all to the population's reproduction. The neo-Darwinian adaptationist takes this in his stride. The system may be hideously uneconomical from the population's point of view, but, from the point of view of the genes influencing the trait concerned, there is no mutant that could do better. My point is that neo-Darwinian adaptationism is not a catch-all, blanket faith in all being for the best. It rules out of court most of the adaptive explanations that readily occur to the Panglossian.

Some years ago, a colleague received an application from a prospective graduate student wishing to work on adaptation, who was brought up a religious fundamentalist and did not believe in evolution. He believed in adaptations, but thought they were designed by God, designed for the benefit of... ah, but that is just the problem! It might be thought that it did not matter whether the student believed adaptations were produced by natural selection or by God. Adaptations are 'beneficial' whether because of natural selection or because of beneficient design, and could not a fundamentalist student be usefully employed in uncovering the detailed ways in which they were beneficial? My point is that this argument will not do, because what is beneficial to one entity in the hierarchy of life is harmful to another, and creationism gives us no grounds for supposing that one entity's welfare will be preferred to another's. In passing, the fundamentalist student might pause to wonder at a God who goes to great trouble to provide predators with beautiful adaptations to catch prey, while with the other hand giving prey beautiful adaptations to thwart them. Perhaps He enjoys the spectator sport. Returning to the main point, if adaptations were designed by God, He might have designed them to benefit the individual animal (its survival or—not the same thing—its inclusive fitness), the species, some other species such as mankind (the usual view of religious fundamentalists), the 'balance of nature', or some other inscrutable purpose known only to Him. These are frequently incompatible alternatives. It really *matters* for whose benefit adaptations are designed. Facts such as the sex ratio in harem-forming mammals are inexplicable on certain hypotheses and easily explicable on others. The adaptationist working within the framework of a proper understanding of the genetical theory of natural selection countenances only a very restricted set of the possible functional hypotheses which the Panglossian might admit.

One of the main messages of this book is that, for many purposes, it is better to regard the level at which selection acts as neither the organism nor the group or any larger unit, but the gene or small genetic fragment. This difficult topic will be debated in later chapters. For the present, it is sufficient to note that selection at the level of the gene can give rise to apparent imperfections at the level of the individual. I shall discuss 'meiotic drive' and related phenomena in Chapter 8, but the classic example is the case of heterozygous advantage. A gene may be positively selected because of its beneficial effects when heterozygous, even though it has harmful effects when homozygous. As a consequence of this, a predictable proportion of the individual organisms in the population will have defects. The general point is this. The genome of an individual organism in a sexual population is the product of a more or less random shuffling of the genes in the population. Genes are selected over their alleles because of their phenotypic effects, averaged over all the individual bodies in which they are distributed, over the whole population, and through many generations. The effects that a given gene has will usually depend upon the other genes with which it shares a body: heterozygous advantage is just a special case of this. A certain proportion of bad bodies seems an almost inevitable consequence of selection for good genes, where good refers to the average effects of a gene on a statistical sample of bodies in which it finds itself permuted with other genes.

Inevitable, that is, as long as we accept the Mendelian shuffle as given and inescapable. Williams (1979), disappointed at finding no evidence for adaptive fine-adjustment of the sex ratio, makes the perceptive point that

> Sex is only one of many offspring characters that would seem adaptive for a parent to control. For instance, in human populations affected by sickle-cell anaemia, it would be advantageous for a heterozygous woman to have her *A* eggs fertilized only by *a*-bearing sperm, and vice versa, or even to abort all homozygous embryos. Yet if mated to another heterozygote she will reliably submit to the Mendelian lottery, even though this means markedly lowered fitness for half her children.... The really fundamental questions in evolution may be answerable only by regarding each gene as ultimately in conflict with every other gene, even those at other loci in the same cell. A really valid theory of natural selection must be based ultimately on selfish replicators, genes and all other entities capable of the biased accumulation of different variant forms.

Amen!

Mistakes due to environmental unpredictability or 'malevolence'

However well adapted an animal may be to environmental conditions, those conditions must be regarded as a statistical average. It will usually be impossible to cater for every conceivable contingency of detail, and any given animal will therefore frequently be observed to make 'mistakes', mistakes which can easily be fatal. This is not the same point as the time-lag problem already mentioned. The time-lag problem arises because of non-stationarities in the statistical properties of the environment: average conditions now are different from the average conditions experienced by the animal's ancestors. The present point is more inescapable. The modern animal may be living under identical *average* conditions to those of an ancestor, yet the detailed moment to moment occurrences facing either of them are not the same from day to day, and are too complex for precise prediction to be possible.

It is particularly in behaviour that such mistakes are seen. The more static attributes of an animal, its anatomical structure for instance, are obviously adapted only to long-term average conditions. An individual is either big or small, it cannot change size from minute to minute as the need arises. Behaviour, rapid muscular movement, is that part of an animal's adaptive repertoire which is specifically concerned with high speed adjustment. The animal can be now here, now there, now up a tree, now underground, rapidly accommodating to environmental contingencies. The number of such *possible* contingencies, when defined in all their detail, is like the number of possible chess positions, virtually infinite. Just as chess-playing computers (and chess-playing people) learn to classify chess positions into a manageable number of generalized classes, so the best that an adaptationist can hope for is that an animal will have been programmed to behave in ways appropriate to a manageable number of general contingency classes. Actual contingencies will fit these general classes only approximately, and apparent mistakes are therefore bound to be made.

The animal that we see up a tree may come from a long line of tree-dwelling ancestors. The trees in which the ancestors underwent natural selection were, in general, much the same as the trees of today. General rules of behaviour which worked then, such as 'Never go out on a limb that is too thin', still work. But the details of any one tree are inevitably different from the details of another. The leaves are in slightly different places, the breaking strain of the branches is only approximately predictable from their diameter, and so on. However strongly adaptationist our beliefs may be, we can only expect animals to be average statistical optimizers, never perfect anticipators of every detail.

So far we have considered the environment as statistically complex and therefore hard to predict. We have not reckoned on its being actively malevolent from our animal's point of view. Tree boughs surely do not deliberately snap out of spite when monkeys venture on to them. But a 'tree bough' may turn out to be a camouflaged python, and our monkey's last mistake is then no accident but is, in a sense, deliberately engineered. Part of a monkey's environment is non-living or at least indifferent to the monkey's existence, and the monkey's mistakes can be put down to statistical unpredictability. But other parts of the monkey's environment consist of living things that are themselves adapted to profit at the expense of monkeys. This portion of the monkey's environment may be called malevolent.

Malevolent environmental influences may themselves be hard to predict for the same reasons as indifferent ones, but they introduce an added hazard; an added opportunity for the victim to make 'mistakes'. The mistake made by a robin in feeding a cuckoo in its nest is presumably in some sense a maladaptive blunder. This is not an isolated, unpredictable occurrence such as arises because of the statistical unpredictability of the non-malevolent part of the environment. It is a recurrent blunder, afflicting generation after generation of robins, even the same robin several times in its life. Examples of this kind always make us wonder at the compliance, in evolutionary time, of the organisms that are manipulated against their best interests. Why doesn't selection simply eliminate the susceptibility of robins to the deception of cuckoos? This kind of problem is one of many which I believe will one day become the stock in trade of a new subdiscipline of biology—the study of manipulation, arms races, and the extended phenotype. Manipulation and arms races form the subject of the next chapter, which in some ways can be regarded as an expansion of the theme of the final section of this chapter.

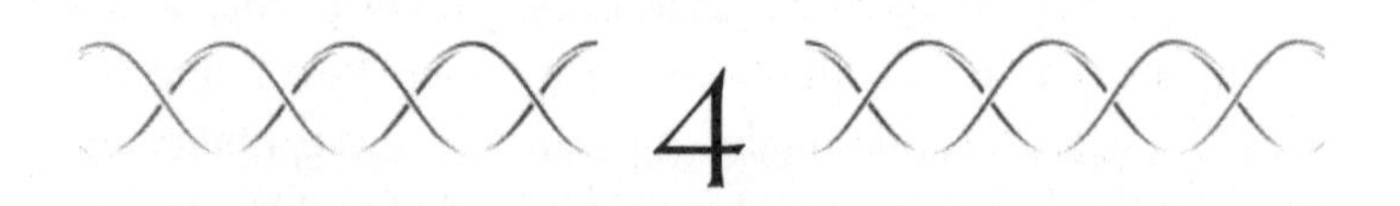

4

ARMS RACES AND MANIPULATION

One of my purposes in this book is to question the 'central theorem' that it is useful to expect individual organisms to behave in such a way as to maximize their own inclusive fitness, or in other words to maximize the survival of copies of the genes inside them. The end of the previous chapter suggests one way in which the central theorem might be violated. Organisms might consistently work in the interests of other organisms rather than of themselves. That is, they might be 'manipulated'.

The fact that animals frequently cause other animals to perform some action that is against their own best interests is, of course, well known. Obviously it happens every time an angler fish catches prey, every time a cuckoo is fed by its foster mother. I shall make use of both these examples in this chapter, but I shall also emphasize two points that have not always been stressed. Firstly, it is natural to assume that even if a manipulator gets away with it temporarily, it is only a matter of evolutionary time before the lineage of manipulated organisms comes up with a counter-adaptation. In other words, we tend to assume that manipulation only works because of the 'time-lag' constraint on perfection. In this chapter I shall point out that, on the contrary, there are conditions under which we should expect manipulators to succeed consistently and for indefinite lengths of evolutionary time. I shall discuss this later under the catch phrase of 'arms races'.

Secondly, until the last decade or so, most of us have paid insufficient attention to the likelihood of intraspecific manipulation, especially exploitative manipulation within the family. I attribute this deficiency to a residuum of group selectionist intuition which often lurks in the depths of the biologist's mind even after group selection has been rejected at the surface level of reason. I think that a minor revolution has taken place in the way we think about social relationships. 'Genteel' (Lloyd 1979) ideas of vaguely benevolent mutual coöperation are replaced by an expectation of stark, ruthless, opportunistic mutual exploitation (e.g. Hamilton 1964a,b, 1970, 1971a; Williams 1966; Trivers 1972, 1974; Ghiselin 1974a; Alexander 1974). This revolution is popularly associated with the name 'sociobiology', although the association is somewhat ironic since, as I have suggested before, Wilson's (1975) great book of that name is in many respects pre-revolutionary in attitude: not the new synthesis, but the last and greatest synthesis of the old, benevolent regime (e.g. his Chapter 5).

I can exemplify the changed view by quoting from one recent paper, Lloyd's (1979) entertaining review of sexual dirty tricks in insects.

> Selection for haste in males, and coyness in females, results in what amounts to competition *between* the sexes. Males may be selected to bypass any choice that the females attempt to exercise, and then females selected to maintain their options, to not be misled or to have their choices subverted. If males subdue and seduce females with true aphrodisiacs [Lloyd gives evidence elsewhere in the paper], females may be expected to escape sooner or later in evolutionary time. And after sperm has been placed in a female, she should manipulate it: store, transfer (from

> chamber to chamber), use, eat, or dissolve it, as she makes additional observations on males. Females may accept and store sperm from a male for insurance that they will get a mate, and then become choosy.... It is possible for sperm to be manipulated in the female (e.g. sex determination in Hymenoptera). Female reproductive morphology often includes sacs, valves, and tubes that could have evolved in this context. In fact, it is possible that some reported examples of sperm competition are actually cases of sperm manipulation.... Given that females, to one extent or another, subvert male interests by the internal manipulation of ejaculate, it is not inconceivable that males will have evolved little openers, snippers, levers and syringes that put sperm in the places females have evolved ('intended') for sperm with priority usage—collectively, a veritable Swiss Army Knife of gadgetry!

This kind of unsentimental, dog eat dog, language would not have come easily to biologists a few years ago, but nowadays I am glad to say it dominates the textbooks (e.g. Alcock 1979).

Such dirty tricks, as often as not, involve direct action, the muscles of one individual moving to molest the body of another. The manipulation which is the subject of this chapter is more indirect and more subtle. An individual induces the effectors of another individual to work against its own best interests, and in favour of the interests of the manipulator. Alexander (1974) was one of the first to emphasize the importance of such manipulation. He generalized his concept of queen domination in the evolution of social insect worker behaviour, to produce a wide-ranging theory of 'parental manipulation' (see also Ghiselin 1974a). He suggested that parents are in such a commanding position over their offspring that offspring may be forced to work in the interests of their parents' genetic fitness, even where this conflicts with their own. West-Eberhard (1975) follows

Alexander in dignifying parental manipulation as one of three general ways in which individual 'altruism' can evolve, the others being kin selection and reciprocal altruism. Ridley and I make the same point, but do not restrict ourselves to *parental* manipulation (Ridley & Dawkins 1981).

The argument is as follows. Biologists define behaviour as altruistic if it favours other individuals at the expense of the altruist himself. A problem arises, incidentally, over how benefit and expense are to be defined. If they are defined in terms of individual survival, acts of altruism are expected to be very common, and will be taken to include parental care. If they are defined in terms of individual reproductive success, parental care no longer counts as altruism, but altruistic acts towards other relatives are predicted by neo-Darwinian theory. If benefit and expense are defined in terms of individual inclusive fitness, neither parental care nor care for other genetic relatives counts as altruism, and indeed a naive version of the theory expects that altruism should not really exist. Any of the three definitions can be justified, although if we must talk of altruism at all I prefer the first definition, the one that allows parental care as altruistic. But my point here is that, whichever definition we favour, it is met if the 'altruist' is forced—manipulated—by the beneficiary into donating something to him. For instance, the definition (any of the three) allows us no choice but to regard the feeding of a cuckoo nestling by its foster parent as altruistic behaviour. Maybe this means we need a new kind of definition, but that is another issue. Krebs and I take the argument to its logical conclusion, and interpret all of animal communication as manipulation of signal-receiver by signal-sender (Dawkins & Krebs 1978).

Manipulation is, indeed, pivotal to the view of life expounded in this book, and it is a trifle ironic that I am one of those who has criticized Alexander's concept of parental manipulation

(Dawkins 1976a, pp. 145–8; Blick 1977; Parker & Macnair 1978; Krebs & Davies 1978; Stamps & Metcalf 1980), and, in turn, been criticized for doing so (Sherman 1978; Harpending 1979; Daly 1980). Notwithstanding these defenders, Alexander (1980, pp. 38–9) himself has conceded that his critics were right.

Clearly some clarification is necessary. Neither I nor any of the later critics mentioned doubted that selection would favour parents who succeeded in manipulating their offspring over parents who did not. Nor did we doubt that parents would indeed, in many cases, 'win the arms race' against offspring. All we were objecting to was the logic of suggesting that parents enjoy a built-in advantage over their children, simply because all children aspire to become parents. That is no more true than that children enjoy a built-in advantage simply because all parents were once children. Alexander had suggested that tendencies to selfishness in children, tendencies to act against the interests of their parents, could not spread because, when the child grew up, its own children's inherited selfishness against *it* would detract from its own reproductive success. For Alexander this sprang from his conviction that 'the entire parent–offspring interaction has evolved because it benefited one of the two individuals—the parent. No organism can evolve parental behaviour, or extend its parental care, unless its own reproduction is thereby enhanced' (Alexander 1974, p. 340). Alexander was, therefore, thinking firmly within the paradigm of the selfish organism. He upheld the central theorem that animals act in the interests of their own inclusive fitness, and he understood this to preclude the possibility of offspring acting against their parent's interests. But the lesson I prefer to learn from Alexander is that of the central importance of manipulation itself, which I regard as a *violation* of the central theorem.

I believe animals exert strong power over other animals, and that frequently an animal's actions are most usefully interpreted as working in the interests of *another* individual's inclusive fitness, rather than its own. Later in this book we shall dispense with the use of the concept of inclusive fitness altogether, and the principles of manipulation will be subsumed under the umbrella of the extended phenotype. But for the rest of this chapter it will be convenient to discuss manipulation at the level of the individual organism.

There is an inevitable overlap here with a paper which I wrote jointly with J. R. Krebs on animal signals as manipulation (Dawkins & Krebs 1978). Before proceeding, I feel obliged to acknowledge that this paper has been severely criticized by Hinde (1981). Some of his criticisms, which do not affect the parts of the paper I wish to use here, are answered by Caryl (1982), who had also been criticized by Hinde. Hinde took us to task for unfairness in quoting apparently group selectionist statements by Tinbergen (1964) and others whom we labelled, for want of a better title, classical ethologists. I have sympathy with this historical criticism. The 'good of the species' passage which we quoted from Tinbergen was a genuine quotation, but I agree that it was not typical of Tinbergen's (e.g. 1965) thinking at that time. The following much earlier quotation from Tinbergen would, perhaps, have been a fairer one for us to have chosen: 'just as the functioning of hormones and the nervous system implies not only the sending out of signals but also a specific responsiveness in the reacting organ, so the releaser system involves a specific responsiveness to particular releasers in the reacting individual as well as a specific tendency to send out the signals in the initiator. The releaser system ties individuals into units of a super-individual order and renders them higher units subject to natural selection' (Tinbergen 1954). Even if outright group selectionism was strongly opposed by Tinbergen and most of his pupils in the early 1960s, I still think that nearly all of us thought of animal signals in terms of a vague notion of 'mutual benefit': if signals were not actually 'for the good of the species' (as in the unrepresentative quotation from Tinbergen), they were 'for the mutual benefit of both signaller and receiver'. The evolution of ritualized signals was regarded as mutual evolution: enhanced signalling power on one side was accompanied by increased sensitivity to the signals on the other side.

Today we would recognize that if receiver sensitivity increases, signal strength does not need to increase but, instead, is more likely to decrease owing to the attendant costs of conspicuous or loud signals. This might be called the Sir Adrian Boult principle. Once, in rehearsal, Sir Adrian turned to the violas and told them to play out more. 'But Sir Adrian', protested the principal viola, 'you were indicating less and less with your baton.' 'The idea', retorted the maestro, 'is that I should do less and less, and you should do more and more!' In those cases where animal signals really are of mutual benefit, they will tend to sink to the level of a conspiratorial whisper: indeed this may often have happened, the resulting signals being too inconspicuous for us to have noticed them. If signal strength increases over the generations this suggests, on the other hand, that there has been increasing sales resistance on the side of the receiver (Williams 1966).

As mentioned earlier, an animal will not necessarily submit passively to being manipulated, and an evolutionary 'arms race' is expected to develop. Arms races were the subject of a second joint paper (Dawkins & Krebs 1979). We were not, of course, the first to make the points that follow in this chapter, but nevertheless it will be convenient to quote from our two joint papers. With permission from Dr Krebs, I shall do so without breaking the flow with repeated bibliographic citations and quotation marks.

An animal often needs to manipulate objects in the world around it. A pigeon carries twigs to its nest. A cuttlefish blows sand from the sea bottom to expose prey. A beaver fells trees and, by means of its dam, manipulates the entire landscape for miles around its lodge. When the object an animal seeks to manipulate is non-living, or at least when it is not self-mobile, the animal has no choice but to shift it by brute force. A dung beetle can move a ball of dung only by forcibly pushing it. But sometimes an animal may benefit by moving an 'object' which happens, itself, to be another living animal. The object has muscles and limbs of its own, controlled by a nervous system and sense organs. While it may still be possible to shift such an 'object' by brute force, the goal may often be more economically engineered by subtler means. The object's internal chain of command—sense organs, nervous system, muscles—may be infiltrated and subverted. A male cricket does not physically roll a female along the ground and into his burrow. He sits and sings, and the female comes to him under her own power. From his point of view this *communication* is energetically more efficient than trying to take her by force.

A question immediately arises. Why should the female stand for it? Since she is in control of her own muscles and limbs, why would she approach the male unless it is in her genetic interests to do so? Surely the word manipulation is appropriate only if the victim is unwilling? Surely the male cricket is simply informing the female of a fact that is useful to her, that over here is a ready and willing male of her own species. Having given her this information, doesn't he then leave it up to her to approach him or not as she pleases, or as natural selection has programmed her?

Well and good when males and females happen to have identical interests, but examine the premise of the last paragraph. What entitles us to assert that the female is 'in control of her own muscles and limbs'? Doesn't this beg the very question we are interested in? By advancing a manipulation hypothesis we are, in effect, suggesting that the female may *not* be in control of her own muscles and limbs, and that the male may be. This example could, of course, be reversed, and the female be said to manipulate the male. The point being made has no specific connection with sexuality. I could have used the example of plants, lacking muscles of their own, using insect muscles as effector organs to transport their pollen, and fuelling those muscles with nectar (Heinrich 1979). The general point is that an organism's limbs may be manipulated to work in the interests of the genetic fitness of another organism. This statement cannot be made convincing until later in the book when we have introduced the idea of the extended phenotype. For this chapter we are still working within the paradigm of the selfish organism, albeit we are starting to stretch it so that it creaks ominously at the edges.

The example of male and female cricket may have been poorly chosen, because, as I said above, many of us have only recently become used to the idea of sexual relations as a battle. Many of us have yet to absorb into our consciousness the fact that 'selection can act in opposition on the two sexes. Commonly, for a given type of encounter, males will be favoured if they do mate and females if they don't' (Parker 1979; see also West-Eberhard 1979). I shall return to this, but for the moment let us use a starker example of manipulation. As starkly ruthless as any battle in nature is that between predator and prey. There are various techniques a predator may use to catch his prey. He may run after them and attempt to outpace, outstay, or outflank them. He may sit in one place and ambush or trap them. Or he may do as angler fish and 'femmes fatales' fireflies (Lloyd 1975, 1981) do, and manipulate the prey's own nervous system so that it actively approaches its own doom. An angler fish sits on the sea bottom and is highly camouflaged except for a long rod projecting from the top of the head, on the end of which is the 'lure', a flexible piece of tissue which resembles some appetizing morsel such as a worm. Small fish, prey of the angler, are attracted by the lure which resembles their own prey. When they approach it the angler 'plays' them down into the vicinity of his mouth, then suddenly opens his jaws and the prey are engulfed in the inrush of water. Instead of using massive body and tail muscles in active pursuit of prey, the angler uses the small economical muscles controlling his rod, to titillate the prey's nervous system via its eyes. Finally it is the prey fish's own muscles that the angler uses to close the gap between them. Krebs and I informally characterized animal 'communication' as a means by which one animal makes use of another animal's muscle power. This is roughly synonymous with manipulation.

The same question arises as before. Why does the victim of manipulation stand for it? Why does the prey fish rush literally into the jaws of death? Because it 'thinks' it is really rushing to get a meal itself. More formally, natural selection has acted on its ancestors, favouring tendencies to approach small wriggling objects, because small wriggling objects are usually worms. Since they are not always worms but sometimes angler fish lures, there may well be some selection on prey fish to be cautious, or to sharpen up their powers of discrimination. To the extent that lures are good mimics of worms, we may surmise that selection has acted on the angler fishes' ancestors to improve them, in the face of improved discrimination by their prey. Some prey are caught, and angler fish do make a living, so some manipulation is going on successfully.

It is convenient to use the metaphor of an arms race whenever we have progressive improvements in adaptations in one lineage, as an evolutionary response to progressive counter-improvements in an enemy lineage. It is important to realize who are the parties that are 'racing' against one another. They are not individuals but lineages. To be sure, it is individuals who attack and defend, individuals who kill or resist being killed. But the arms race takes place on the evolutionary time-scale, and individuals do not evolve. It is lineages that evolve, and lineages that exhibit progressive trends in response to the selection pressures set up by the progressive improvements in other lineages.

One lineage will tend to evolve adaptations to manipulate the behaviour of another lineage, then the second lineage will evolve counter-adaptations. We must obviously be interested in any general rules governing whether one or the other lineage can 'win', or have a built-in advantage. It was just such a built-in advantage that Alexander attributed to parents over their children. Apart from his major theoretical argument which, as we have seen, is now not favoured, he also plausibly suggested various practical advantages of parents over offspring: 'the parent is bigger and stronger than the offspring, hence in a better position to impose its will' (Alexander 1974). This is true, but we must not forget the lesson of the previous paragraph, that an arms race is run in evolutionary time. In any one generation the muscles of a parent are stronger than those of its offspring, and whoever controls those muscles must have the upper hand. But the question at issue is, *who* controls the parents' muscles? As Trivers (1974) says, 'An offspring cannot fling its mother to the ground at will and nurse...the offspring is expected to employ psychological rather than physical tactics.'

Krebs and I suggested that animal signals could be thought of as employing psychological tactics in rather the same way as human advertisements. Advertisements are not there to inform, or to misinform, they are there to *persuade*. The advertiser uses his knowledge of human psychology, of the hopes, fears, and secret motives of his targets, and he designs an advertisement which is effective in manipulating their behaviour. Packard's (1957) exposé of the deep psychological techniques of commercial advertisers makes fascinating reading for the ethologist. A supermarket manager is quoted as saying, 'People like to see a lot of merchandise. When there are only three of four cans of an item on a shelf, they just won't move.' The obvious analogy with lek birds does not lose its value merely because the physiological mechanism of the effect will probably prove to be different in the two cases. Hidden cine-cameras recording the eye-blinking rate of housewives in a supermaket indicated that in some cases the effect of the multiplicity of bright-coloured packages was to induce a mild hypnoidal trance.

K. Nelson once gave a talk at a conference, entitled 'Is bird song music? Well, then, is it language? Well, then, what is it?' Perhaps bird song is more akin to hypnotic persuasion, or to a form of drugging. The nightingale's song induced in John Keats a drowsy numbness 'as though of hemlock I had drunk'. Might it not have an even more powerful effect on the nervous system of another nightingale? If nervous systems are susceptible to drug-like influences via the normal sense organs, should we not positively expect that natural selection would have favoured the exploitation of such possibilities, would have favoured the development of visual, olfactory, or auditory 'drugs'?

A neurophysiologist, presented with an animal with a complex nervous system and asked to manipulate the animal's behaviour, may insert electrodes in sensitive points of the brain and stimulate electrically, or make pinpoint lesions. An animal normally does not have direct access to the brain of another animal, although I shall mention one example, the so-called brainworm, in Chapter 12. But the eyes and ears are also entry ports to the nervous system, and there might be patterns of light or sound which, if properly deployed, could be as effective as direct electrical stimulation. Grey Walter (1953) vividly illustrates the power of flashing lights tuned to the frequency of human EEG rhythms: in one case a man felt 'an irresistible urge to strangle the person next to him'.

If we were asked to imagine what an 'auditory drug' would sound like, say for making the soundtrack of a science fiction film, what would we say? The incessant rhythm of an African drum; the eerie trilling of the tree cricket *Oecanthus*, of which it has been said that if moonlight could be heard that is how it would sound (quoted in Bennet-Clark 1971); or the song of the nightingale? All three strike me as worthy candidates, and I believe all three have been in some sense designed for the same purpose: the manipulation of one nervous system by another, in a way that is not in principle different from manipulation of a nervous system by a neurophysiologist's electrodes. Of course if an animal sound powerfully affects a human nervous system this is probably incidental. The hypothesis being advanced is that selection has shaped animal sounds to manipulate some nervous system, not necessarily a human one. The snort of a pig-frog *Rana grylio* may affect another pig-frog as the nightingale affected Keats, or the skylark Shelley. The nightingale happened to be a better example for me to choose, because human nervous systems can be moved to deep emotion by nightingale song, whereas they are usually moved to laugh at pig-frog snorts.

Consider that other celebrated songster, the canary, since much happens to be known about the physiology of its reproductive behaviour (Hinde & Steel 1978). If a physiologist wants to bring a female canary into reproductive condition, increase the size of her functional ovary, and cause her to start nest-building and other reproductive behaviour patterns, there are various things he can do. He can inject her with gonadotropins or oestrogens. He can use electric light to increase the day-length that she experiences. Or, most interestingly from our point of view, he can play her a tape recording of male canary song. It apparently has to be canary song; budgerigar song will not do, although budgerigar song has a similar effect on female budgerigars.

Now suppose a male canary wanted to bring a female into reproductive condition, what could *he* do? He does not have a syringe to inject hormones. He cannot switch on artificial lights in the female's environment. Of course what he does is sing. The particular pattern of sounds that he makes enters the female's head through her ears, is translated into nerve impulses, and bores insidiously into her pituitary. The male does not have to synthesize and inject gonadotropins; he makes the female's pituitary work to synthesize them for him. He stimulates her pituitary by means of nerve impulses. They are not 'his' nerve impulses, in the sense that they all occur within the female's own nerve cells. But they are his in another sense. It is his particular sounds which are subtly fashioned to make the female's nerves work on her pituitary. Where a physiologist might pump gonadotropins into the female's breast muscle, or electric current into her brain, the male canary pours song into her ear. And the effect is the same. Schleidt (1973) discusses other examples of such 'tonic' effects of signals on the physiology of the receiver.

'Bores insidiously into her pituitary' may be too much for some readers. Certainly it begs important questions. The obvious point that will be made is that the female may benefit from the transaction as much as the male, in which case 'insidiously' and 'manipulation' are inappropriate words. Alternatively, if there is any insidious manipulation it may be by the female on the male. Maybe the female 'insists' upon an exhausting performance of song by her mate before she will come into reproductive condition, thereby selecting only the most robust male for a mate. I think something along those lines is quite probably the explanation of the extraordinarily high rate of copulation observed in large cats (Eaton 1978). Schaller (1972) followed a sample male lion for 55 hours, during which he copulated 157 times, with an average inter-copulation interval of 21 minutes. Ovulation in cats is induced by copulation. It seems plausible that the prodigious copulation rates shown by male lions are the end product of a runaway arms race, in which females insisted on progressively more copulations before ovulating, and males were selected for ever-increased sexual stamina. Lion copulation stamina might be regarded as the behavioural equivalent of a peacock's tail. A version of this hypothesis is compatible with that of Bertram (1978), that females have devalued the currency of copulation as a means of reducing the occurrence of disruptive male fights.

To continue the quotation from Trivers on psychological manipulation tactics that children might use against their parents:

> Since an offspring will often have better knowledge of its real needs than will its parents, selection should favour parental attentiveness to signals from its offspring that apprize the parent of the offspring's condition.... But once such a system has evolved, the offspring can begin to employ it out of context. The offspring can cry not only when it is famished but also when it merely wants more food than the parent is selected to give.... Selection will then of course favor parental ability to discriminate the two uses of the signals, but still subtler mimicry and deception by the offspring are always possible [Trivers 1974].

We have arrived again at the main question raised by arms races. Are there any generalizations we can make about which side is likely to 'win' an arms race?

First, what does it *mean* to speak of one side or the other 'winning'? Does the 'loser' eventually go extinct? At times this may happen. A bizarre possibility is suggested by Lloyd and Dybas (1966) for periodical cicadas (Simon, 1979, gives an entertaining recent account). Periodical cicadas are three species of the genus *Magicicada*. All three species have two varieties, a 17-year and a 13-year variety. Each individual spends 17 (or 13) years feeding as an underground nymph, before emerging for a few weeks of adult reproductive life, after which it dies. In any particular area all life cycles are synchronized, with the result that each area experiences a cicada plague every 17 (or 13) years. The functional significance of this is presumably that would-be predators or parasites are swamped in plague years and starved in intervening years. The risk to an individual who remains synchronized is therefore less than the risk to one who breaks synchrony and emerges, say, a year early. But given the admitted advantage of synchrony, why didn't the cicadas settle on a shorter life cycle than 17 (or 13) years, thereby reducing the unfortunate delay before reproduction? Lloyd and Dybas's suggestion is that the long life cycle is the end result of an 'evolutionary race through time' with a now extinct predator (or parasitoid). 'This hypothetical parasitoid presumably had a life cycle that was almost synchronized with, and nearly equal in length to (but always slightly less than), the ancestral protoperiodical cicada. As the theory goes, the cicadas finally outran their parasitoid pursuer and the poor specialized beast went extinct' (Simon 1979). Ingeniously, Lloyd and Dybas go further and suggest that the arms race culminated in life cycles lasting a prime number of years (13 or 17) because otherwise a predator with a shorter life cycle might synchronize with the cicadas every second or every third time around!

There is really no need for the 'loser' lineage of an arms race to go extinct, as suggested for the cicada parasitoid. It may be that the 'winner' is such a rare species that it constitutes a relatively negligible risk to individuals of the 'loser' species. The winner only wins in the sense that its adaptations against the loser are not effectively countered. This is good for individuals of the winner lineage, but it may not be very bad for individuals of the loser lineage who, after all, are running other races simultaneously against other lineages, possibly very successfully.

This 'rare-enemy effect' is an important example of an asymmetry in selection pressures acting on the two sides of an arms race. Although we gave no formal models, Krebs and I considered one such asymmetry qualitatively under the catchphrase heading of the 'life/dinner principle', named after a fable of Aesop which was called to our attention by M. Slatkin: 'The rabbit runs faster than the fox, because the rabbit is running for his life while the fox is only running for his dinner.' The general point here is that for an animal on one side of the arms race the penalty of failure is more severe than for an animal on the other side of the arms race. Mutations that make foxes run more slowly than rabbits might therefore survive in the fox gene-pool longer than mutations that cause rabbits to run slowly can expect to survive in the rabbit gene-pool. A fox may reproduce after being outrun by a rabbit. No rabbit has ever reproduced after being outrun by a fox. Foxes may therefore be better able than rabbits to divert resources away from adaptations for running fast and into other adaptations; rabbits therefore appear to 'win' the arms race as far as running speed is concerned. The point is really one about asymmetries in strengths of selection pressure.

Probably the simplest asymmetry arises from what I have just called the rare-enemy effect, where one side in the arms race is rare enough to exert a relatively negligible influence on any given individual on the other side. This can be illustrated by the angler fish example again, and the example has the additional merit of showing that there is no necessary reason why prey ('rabbits') should come out 'ahead' of predators ('foxes'). Assume that angler fish are rather rare, so however unpleasant it may be for an individual prey to be caught by an angler fish, the risk of this happening to a randomly designated individual is not high. Any adaptation costs something. For a small fish to discriminate angler fish lures from real worms it needs sophisticated visual processing apparatus. To make this apparatus may cost it resources which could otherwise have been put into, say, gonads (see Chapter 3). Then again, using the apparatus takes time, which could have been spent on courting females or defending a territory, or indeed chasing the prey if it is judged to be a real worm. Finally, a fish that is very cautious about approaching worm-like objects may cut its risk of being eaten, but it also increases its risk of starving. This is because it may avoid many perfectly good worms because they *might* have been angler fish lures. It may well be that the balance of costs and benefits favours complete recklessness on the part of some prey animals. Individuals that *always* try to eat small wriggling objects, and damn the consequences, may, on average, do better than individuals that pay the costs of attempting to discriminate real worms from angler fish lures. William James made much the same point in 1910: 'There are more worms unattached to hooks than impaled upon them; therefore, on the whole, says Nature to her fishy children, bite at every worm and take your chances' (quoted by Staddon 1981).

Now look at it from the point of view of the angler fish. It, too, needs to spend resources on apparatus to outwit its opponents in the arms race. The resources that go into making a lure could have been put into gonads. The time spent sitting motionless waiting patiently for prey could have been spent in actively searching for a mate. But the angler fish absolutely depends on the success of its lure for survival. An angler fish that is not well equipped to lure prey will starve. A prey fish that is not well equipped to avoid being lured may run only a small risk of being eaten, and may more than compensate for the risk by saving the cost of making and using the equipment.

In this example the angler wins the arms race against its prey, simply because its side of the arms race constitutes a relatively negligible threat to any given individual of the other side, through being rare. This does *not*, by the way, mean that selection will favour predators that take steps to be rare, or to constitute a small threat to their prey! Within the angler fish population selection will favour those individuals who are the most efficient killers and the most prolific contributors to reducing the rarity of their species. But the angler fish lineage may 'keep ahead' in the arms race against their prey, simply as an incidental consequence of being rare for some other reason, rare in spite of all their efforts.

In addition, there are likely to be frequency-dependent effects for various reasons (Slatkin & Maynard Smith 1979). For instance, as individuals on one side in an arms race, say angler fish, become rarer, they will constitute a steadily weaker threat to individuals on the other side of the arms race. Therefore individuals on the other side will be selected to shift precious resources away from adaptations concerned with this particular arms race. In the present example, prey fish are selected to shift resources away from anti-angler adaptations and into other things such as gonads, possibly to the extent of total recklessness as suggested above. This will make life easier for individual anglers, who will therefore become more numerous. Anglers will then come to constitute a severer threat to the prey fish, who will then be selected to shift resources back into anti-angler adaptations. As usual in such arguments, we do not have to imagine an oscillation. There may, instead, be an evolutionarily stable endpoint to the arms race; stable, that is, until an environmental change shifts the relevant costs and benefits.

Obviously, until we know a great deal more about costs and benefits in the field, we cannot predict the outcome of particular arms races. For present purposes this does not matter. All we need is to satisfy ourselves that in any particular arms race individuals on one side may have more to lose than individuals on the other side. The rabbit has his life to lose, the fox only has his dinner. The incompetent angler fish dies; the incompetent avoider of angler fish only runs a tiny risk of dying, and may, by saving costs, end up better off than the 'competent' prey fish.

We only have to accept the general plausibility of such asymmetries in order to answer the question raised by our discussion of manipulation. We agreed that if one organism could get away with manipulating the nervous system of another, and exploiting its muscle power, selection would favour such manipulation. But we were brought up short by the reflection that selection would also favour resistance to being manipulated. Should we, then, really expect to see effective manipulation in nature? The life/dinner principle, and other such principles as the rare-enemy effect, provide us with an answer. If the individual manipulator has more to lose by failing to manipulate than the individual victim has to lose by failing to resist manipulation, we should expect to see successful manipulation in nature. We should expect to see animals working in the interests of other animals' genes.

Brood parasitism provides what may be the most striking example. A parent bird such as a reed warbler sweeps a copious flow of food from a large catchment area into the narrow funnel of its nest. There is a living to be made by any creature that evolves the necessary adaptations to insert itself in the funnel and intercept the flow. That is what cuckoos and other brood parasites have done. But the reed warbler is not an uncomplaining cornucopia of free food. It is an active, complex machine, with sense organs, muscles, and a brain. The brood parasite must not only have its body inserted in the host's nest. It must also infiltrate the defences of the host's nervous system, and its ports of entry are the host's sense organs. The cuckoo uses key stimuli to unlock the host's machinery of parental care and subvert it.

The advantages of the brood-parasite way of life are so manifest that today we are taken aback to find Hamilton and Orians needing, in 1965, to defend the proposition that it has been favoured by natural selection, against theories of 'degenerative breakdown' of normal breeding behaviour. Hamilton and Orians went on to provide a satisfying discussion of the probable evolutionary origins of brood parasitism, the preadaptations which preceded its evolution, and the adaptations which have accompanied its evolution.

One of these adaptations is egg mimicry. The perfection of egg mimicry, in at least some 'gentes' of cuckoo, shows that foster parents are potentially capable of keen-eyed discrimination against interlopers. This only underlines the mystery of why cuckoo hosts seem to be so poor at discriminating against cuckoo nestlings. Hamilton and Orians (1965) express the problem vividly: 'Young brown-headed cowbirds and European Cuckoos, as they reach their maximum dimension, dwarf their foster parents. Consider the ludicrous sight of a tiny Garden Warbler... [Latin name leaves exact species intended unclear] standing atop a cuckoo to reach the mouth of the gaping parasite. Why does not the Garden Warbler take the adaptive measure of abandoning the nestling prematurely, especially when to the human observer it is so clearly identifiable?' When a parent is a small fraction of the size of the nestling it is feeding, the most rudimentary eyesight should suffice to show that something has gone seriously wrong with the normal parental process. Yet, at the same time, the existence of egg mimicry shows that hosts are capable of fastidious discrimination, making use of keen eyesight. How are we to explain this paradox (see also Zahavi 1979)?

One fact that helps to reduce the mystery is that there must be stronger selection pressure on hosts to discriminate against cuckoos at the egg stage than against cuckoos at the nestling stage, simply because eggs occur earlier. The benefit of detecting a cuckoo egg is the potential gaining of an entire breeding cycle in the future. The benefit of detecting a nearly fledged cuckoo is the saving of only a few days, at a time when it may be too late to breed again anyway. Another mitigating circumstance in the case of *Cuculus canorus* (Lack 1968) is that the host's own young is usually not there for simultaneous comparison, having been tipped out by the baby cuckoo. It is well known that discrimination is easier if there is a model actually present for comparison.

Various authors have invoked the 'supernormal stimulus', in one form or another. Thus Lack remarks (p. 88) that 'the young cuckoo, with its huge gape and loud begging call, has evidently evolved in exaggerated form the stimuli which elicit the feeding response of parent passerine birds. So much is this so that there are many records of adult passerine birds feeding a fledged young *C. canorus* raised by a different host species; this, like lipstick in the courtship of mankind, demonstrates successful exploitation by means of a "super-stimulus".' Wickler (1968) makes a similar point, quoting Heinroth as having referred to foster parents as behaving like 'addicts', and to the cuckoo nestling as a 'vice of its foster parents'. As it stands, this kind of suggestion will strike many critics as unsatisfying, because it immediately prompts a question at least as big as the one it answers. Why doesn't selection eliminate from the host species the tendency to be 'addicted' to 'supernormal stimuli'?

This, of course, is where the arms race concept comes in again. When a human behaves in a way that is manifestly bad for him, for instance when he continually takes poison, we may explain his behaviour in at least two ways. He may not realize that the substance that he is drinking is poison, so closely does it resemble a genuinely nutritous substance. This corresponds to the host bird's being fooled by the cuckoo's egg mimicry. Or he may be unable to save himself because of some direct subverting influence of the poison on his nervous system. Such is the case of the heroin addict who knows the drug is killing him, but who cannot stop taking it because the drug itself controls his nervous system. We have already seen that the cuckoo nestling's lipstick-like gape is regarded as a supernormal stimulus, and that foster parents have been described as apparently 'addicted' to the supernormal stimulus. Could it be that the host can no more resist the supernormal manipulative power of the cuckoo nestling than the junkie can resist his fix, or than the brainwashed prisoner can resist the orders of his captor, however much it would benefit him to do so? Perhaps cuckoos have put their adaptive emphasis on mimetic deception at the egg stage, but on positive manipulation of the host's nervous system at the late nestling stage.

Any nervous system can be subverted if treated in the right way. Any evolutionary adaptation of the host nervous system to resist manipulation by cuckoo nestlings lays itself open to counter-adaptation by the cuckoos. Selection acting on cuckoos will work to find whatever chinks there may be in the hosts' newly evolved psychological armour. Host birds may be very good at resisting psychological manipulation, but cuckoos might become even better at manipulating. All we need to postulate is that, for some reason such as that suggested by the life/dinner principle or the rare-enemy effect, cuckoos have won the arms race: a cuckoo in the nest has *got* to manipulate its host successfully or it will surely die; its individual foster parent will benefit somewhat if it resists manipulation, but it still has a good chance of future reproductive success in other years even if it fails to resist this particular cuckoo. Moreover, cuckoos might be sufficiently rare that the risk of an individual of the foster species being parasitized is low; conversely the 'risk' of an individual cuckoo's being a parasite is 100 per cent, no matter how common or rare either party to the arms race may be. The cuckoo is descended from a line of ancestors, every single one of whom has successfully fooled a host. The host is descended from a line of ancestors, many of whom may never have encountered a cuckoo in their lives, or may have reproduced successfully after being parasitized by a cuckoo. The arms race concept completes the classical supernormal stimulus explanation, by providing a functional account of the host's maladaptive behaviour, instead of leaving it as an unexplained limitation of the nervous system.

In one respect my treatment of cuckoos as manipulators may be found unsatisfying. The cuckoo is, after all, only diverting the normal parental behaviour of its host. It has not succeeded in building into the host's behavioural repertoire a whole new behaviour pattern that was not there, in some form, before. Some might find analogies with drugs, hypnotism, and electrical stimulation of the brain more persuasive if an example of this more extreme kind of manipulation could be found. A possible case is the 'preening invitation' display of another brood parasite, the brown-headed cowbird *Molothrus ater* (Rothstein 1980). Allopreening, the preening of one individual by another, is not uncommon within various species of birds. It is not particularly surprising, therefore, that cowbirds should succeed in getting other birds of several species to preen them. Again, this can be seen as a simple diversion of intraspecific allopreening, with the cowbird providing a supernormal exaggeration of the normal eliciting stimuli for allopreening. What *is* rather surprising is that cowbirds manage to get themselves preened by species that never engage in intraspecific allopreening.

The drug analogy is especially apt for insect 'cuckoos' that use chemical means to coerce their hosts into acts that are profoundly damaging to their own inclusive fitness. Several species of ant have no workers of their own. The queens invade nests of other species, dispose of the host queen, and use the host workers to bring up their own reproductive young. The method of disposing of the host queen varies. In some species, such as the descriptively named *Bothriomyrmex regicidus* and *B. decapitans*, the parasite queen rides about on the back of the host queen and then, in Wilson's (1971) delightful description, 'begins the one act for which she is uniquely specialized: slowly cutting off the head of her victim' (p. 363).

Monomorium santschii achieves the same result by more subtle means. The host workers have weapons wielded by strong muscles, and nerves attached to the muscles; why should the parasite queen exert her own jaws if she can subvert the nervous systems controlling the numerous jaws of the host workers? It does not seem to be known how she achieves it, but she does: the host workers kill their own mother and adopt the usurper. A chemical secreted by the parasite queen seems the likely weapon, in which case it might be labelled a pheromone, but it is probably more illuminating to think of it as a formidably powerful drug. In line with this interpretation, Wilson (1971, p. 413) writes of symphylic substances as being 'more than just elementary nutritive substances or even analogues of the natural host pheromones. Several authors have spoken of a narcotizing effect of symphylic substances.' Wilson also uses the word 'intoxicant' and quotes a case in which worker ants under the influence of such a substance became temporarily disoriented and less sure of their footing.

Those who have never been brainwashed or addicted to a drug find it hard to understand their fellow men who are driven by such compulsions. In the same naive way we cannot understand a host bird's being compelled to feed an absurdly oversized cuckoo, or worker ants wantonly murdering the only being in the whole world that is vital to their genetic success. But such subjective feelings are misleading, even where the relatively crude achievements of human pharmacology are concerned. With natural selection working on the problem, who would be so presumptious as to guess what feats of mind control might not be achieved? Do not expect to see animals always behaving in such a way as to maximize their own inclusive fitness. Losers in an arms race may behave in some very odd ways indeed. If they appear disoriented and unsure of their footing, this may be only the beginning.

Let me stress again what a feat of mind-control the *Monomorium santschii* queen achieves. To a sterile worker ant, her mother is a kind of genetic goldmine. For a worker ant to kill her own mother is an act of genetic madness. Why do the workers do it? I am sorry I can do no more than, once again, vaguely talk about arms races. Any nervous system is vulnerable to manipulation by a clever-enough pharmacologist. There is no difficulty in believing that natural selection acting on *M. santschii* would seek out the weak points in the host workers' nervous system, and insert a pharmacological key in the lock. Selection on the host species would soon have plugged those weak points, whereupon selection on the parasite would improve the drug, and the arms race was under way. If *M. santschii* is sufficiently rare, it is easy to see that it might 'win' the arms race, even though regicide is such a disastrous act for each host colony whose workers succumb to it.

The overall risk of parasitization by M. *santschii* could be very low even though the marginal cost of regicide, given that an M. *santschii* queen has entered, is disastrously high. Each individual M. *santschii* queen is descended from a line of ancestors every one of whom has succeeded in manipulating host workers into regicide. Each host worker is descended from a line of ancestors whose colonies may seldom have been within 10 miles of an M. *santschii* queen. The costs of 'bothering' to be equipped to resist manipulation by an occasional M. *santschii* queen may outweigh the benefits. Reflections such as this lead me to believe that the hosts might well lose the arms race.

Other species of parasitic ants use a different system. Instead of sending out queens to implant their eggs in host nests and using host labour there, they transport host labour back to their own nests. These are the so-called slavemaking ants. Slavemaking species have workers, but these workers devote part, or in some cases all, of their energy to going on slaving expeditions. They raid nests of other species and carry off larvae and pupae. These subsequently hatch in the slavers' nest, where they work normally, foraging and tending brood, not 'realizing' that they are, in effect, slaves. The advantage of the slavemaking way of life is presumably that most of the cost of feeding the workforce in the larval stage is saved. That cost is borne by the home colony from which the slave pupae were taken.

The slavemaking habit is interesting from the present point of view, because it raises an unusual arms race asymmetry. Presumably there is an arms race between slavemaking species and slave species. Adaptations to counter slavery, for instance enlarged soldier jaws for driving off slave-raiders, should be expected in species that are victims of slave raids. But surely the more obvious countermeasure the slaves could take would simply be to withhold their labour in the slavers' nest, or to kill slavemaker brood instead of feeding them? It *seems* the obvious countermeasure, but there are formidable obstacles to its evolution. Consider an adaptation to 'go on strike', to refuse to work in the slavemakers' nest. The slave workers would of course have to have some means of recognizing that they had hatched in a foreign nest, but that should not be difficult in principle. The problem arises when we think in detail of how the adaptations would be passed on.

Since workers don't reproduce, all worker adaptations, in any social insect species, have to be passed on by reproductive relatives of the workers. This normally presents no insuperable problems, because workers directly assist their own reproductive relatives, so genes giving rise to worker adaptations directly assist copies of themselves in reproductives. But take the example of a mutant gene causing slave workers to go on strike. It may very effectively sabotage the slavemakers' nest, possibly wipe it out altogether. And to what effect? The area now contains one less slavemaking nest, presumably a good thing for *all* potential victim nests in the area, not just the nest from which the rebel slaves came, but nests containing non-striking genes as well. The same kind of problem arises in the general case of the spread of 'spiteful' behaviour (Hamilton 1970; Knowlton & Parker 1979).

The only easy way the genes for striking can be preferentially passed on is for striking to benefit, selectively, the strikers' own home nest, the nest they left behind and in which their own reproductive relatives are being reared. This could happen if slavemakers habitually returned to make repeat raids on the same nest, but otherwise we must conclude that anti-slavery adaptations must be confined to the period *before* the slave pupae have left their home nest. Once the slaves have arrived in the slavemakers' nest they effectively drop out of the arms race since they no longer have any power to influence the success of their reproductive relatives. The slavemakers can develop manipulative adaptations of any degree of sophistication, physical or chemical, pheromones or powerful drugs, and the slaves cannot evolve countermeasures.

Actually, the very fact that the slaves cannot evolve countermeasures will tend to reduce the likelihood that the manipulative techniques evolved by the slavemakers will be very sophisticated: the fact that the slaves cannot retaliate, in an evolutionary sense, means that the slavemakers do not need to spend costly resources on elaborate and sophisticated manipulation adaptations, because simple and cheap ones will do. The example of slavery in ants is rather a special one, but it illustrates a particularly interesting sense in which one side in an arms race can be said to lose completely.

A case could be made for drawing an analogy here with the hybrid frog *Rana esculenta* (White 1978). This common European frog, the edible frog of French restaurants, is not a species in the normal sense of the word. Individuals of the 'species' are really various kinds of hybrids between two other species, *Rana ridibunda* and *R. lessonae.* There are two different diploid forms and two different triploid forms of *R. esculenta.* For simplicity I shall consider only one of the diploid forms, but the argument holds for all the varieties. These frogs coexist with *R. lessonae.* Their diploid karyotype consists of one set of *lessonae* chromosomes and one set of *ridibunda* chromosomes. At meiosis they discard the *lessonae* chromosomes and produce pure *ridibunda* gametes. They mate with *lessonae* individuals, thereby restoring the hybrid genotype in the next generation. In this race of *Rana esculenta* bodies, therefore, *ridibunda* genes are germ-line replicators, *lessonae* genes dead-end ones. Dead-end replicators can exert phenotypic effects. They can even be naturally selected. But the consequences of that natural selection are irrelevant to evolution (see Chapter 5). To make the next paragraph easier to follow, I shall call *R. esculenta* H (for hybrid), *R. ridibunda* G (for germ-line), and *R. lessonae* D (for dead-end, although it should be remembered that 'D' genes are dead-end replicators only when in H frogs; when in 'D' frogs they are normal germ-line replicators).

Now, consider a gene in the D gene-pool which exerts an effect on D bodies to make them refuse to mate with H. Such a gene should be favoured by natural selection over its H-tolerating allele, since the latter will tend to end up in H bodies of the next generation, and will be discarded at meiosis. G genes are not discarded at meiosis, and they will tend to be selected if they influence H bodies so that they overcome the reluctance of D individuals to mate with them. We should therefore see an arms race between G genes acting on H bodies, and D genes acting on D bodies. In those respective bodies, both sets of genes are germ-line replicators. But what about D genes acting on H bodies? They should have just as powerful an influence over H phenotypes as G genes, since they constitute exactly half of the H genome. Naively, we might expect them to carry their arms race against G genes over into the H bodies which they share. But in H bodies, those D genes are in the same position as ants that have been taken as slaves. Any adaptation that they mediate in the H body cannot be passed on to the next generation; for the gametes produced by the H individual, regardless of how his developing phenotype, and indeed his survival, may have been influenced by D genes, are strictly G gametes. Just as would-be slave ants can be selected to resist being taken into captivity while they are still in their home nest, but cannot be selected to subvert the slavemaking nest once they are in it, so D genes can be selected to influence D bodies so that they resist being incorporated in H genomes in the first place, but once so incorporated they are no longer under selection, even though they can still have phenotypic effects. They lose the arms race because they are a dead-end. A similar argument could be made for fish of the hybrid 'species' *Poeciliopsis monacha-occidentalis* (Maynard Smith 1978a).

The inability of slaves to evolve counter-adaptations was originally invoked by Trivers and Hare (1976), in their theory of the sex-ratio arms race in social Hymenoptera. This is one of the best known of recent discussions of a particular arms race, and it is worth considering further. Elaborating on ideas of Fisher (1930a) and Hamilton (1972), Trivers and Hare reasoned that the evolutionarily stable sex ratio in ant species with one singly mated queen per nest cannot be simply predicted. If the queen is assumed to have all power over the sex of reproductive offspring (young queens and males), the stable ratio of economic investment in male and female reproductives is 1:1. If, on the other hand, non-laying workers are assumed to hold all power over investment in young, the stable ratio will be 3:1 in favour of females, ultimately because of the haplodiploid genetic system. There is, therefore, a potential conflict between queen and workers. Trivers and Hare reviewed the, admittedly imperfect, available data, and reported a good average fit to the 3:1 prediction, from which they concluded that they had found evidence for worker power winning the battle against queen power. It was a clever attempt to use real data to test a hypothesis of a kind that is often criticized as untestable, but like other innovative first attempts it is easy to find fault with it. Alexander and Sherman (1977) complained about Trivers and Hare's handling of the data, and also suggested an alternative explanation for the female-biased sex ratio common in ants. Their explanation ('local mate competition'), like that of Trivers and Hare, was originally derived from Hamilton, in this case his paper on extraordinary sex ratios (1967).

This controversy has had the good effect of stimulating further work. Especially illuminating in the context of arms races and manipulation is the paper by Charnov (1978), which is concerned with the origins of eusociality, and which introduces a potentially important version of the life/dinner principle. His argument works for diploid as well as haplodiploid organisms, and I shall consider the diploid case first. Consider a mother whose elder children have still not left the nest when the next brood hatches. When the time comes for them to leave the nest and begin their own reproduction, the young have the option, instead, of staying behind and helping to rear their young siblings. As is now well known, all other things being equal such a fledgling should be genetically indifferent between rearing offspring and rearing full siblings (Hamilton 1964a,b). But suppose the old mother could exert any manipulative power over the decision of her elder children: would she 'prefer' that they leave and rear families of their own, or stay and rear her next brood? Obviously that they should stay and rear her next brood, since grandchildren are half as valuable to her as children. (The argument as it stands is incomplete. If she manipulated *all* her children for the whole of her life into rearing yet more non-producing child labourers, her germ-line would peter out. We must assume that she manipulates some offspring of the same genetic type into developing into reproductives and others into developing as workers.) Selection will, then, favour such manipulative tendencies in parents.

Normally, when we postulate selection in favour of manipulation we dutifully pay lip service to counter-selection on the victim to resist manipulation. The beauty of Charnov's point is that in this case there will be no counter-selection. The 'arms race' is a walk-over because one side, so to speak, doesn't even try. The offspring being manipulated are, as we have already seen, indifferent to whether they rear young siblings or offspring of their own (again assuming all other things equal). Therefore, although we may *postulate* reverse manipulation by offspring of parents, this is bound, at least in the simple example visualized by Charnov, to be outweighed by parental manipulation of offspring. This is an asymmetry to be added to the list of parental advantages offered by Alexander (1974), but I find it more generally convincing than any others on the list.

At first sight it might appear that Charnov's argument does not apply to haplodiploid animals, and that would be a pity since most social insects are haplodiploid. But this view is mistaken. Charnov himself shows this for the special assumption that the population has an unbiased sex ratio, in which case even in haplodiploid species females are indifferent between rearing siblings (r = the average of $\frac{3}{4}$ and $\frac{1}{4}$) and rearing offspring ($r = \frac{1}{2}$). But Craig (1980) and Grafen (in preparation) independently show that Charnov did not even need to assume an unbiased sex ratio. The potential worker is *still* indifferent between rearing siblings and rearing offspring at any conceivable population sex ratio. Thus suppose the population sex ratio is female-biased, even suppose it conforms to Trivers and Hare's predicted 3:1. Since the worker is more closely related to her sister than to her brother or her offspring of either sex, it might seem that she would 'prefer' to rear siblings over offspring given such a female-biased sex ratio: is she not gaining mostly valuable sisters (plus only a few relatively worthless brothers) when she opts for siblings? But this reasoning neglects the relatively great reproductive value of males in such a population as a consequence of their rarity. The worker may not be closely related to each of her brothers, but if males are rare in the population as a whole each one of those brothers is correspondingly highly likely to be an ancestor of future generations.

The mathematics confirm that Charnov's conclusion is even more general than he suggested. In both diploid and haplodiploid species, at any population sex ratio, an individual female is theoretically indifferent whether she herself rears offspring or younger siblings. She is not, however, indifferent whether her offspring rear their own children or their siblings: she prefers them to rear their siblings (her offspring) over their offspring (her grand-children). Therefore if there is any question of manipulation in this situation, parental manipulation of offspring is more likely than offspring manipulation of parents.

It might appear that Charnov's, Craig's and Grafen's conclusions radically contradict those of Trivers and Hare on sex ratios in social Hymenoptera. The statement that, at any sex ratio, a female hymenopteran is indifferent between rearing siblings and rearing offspring, sounds tantamount to saying that she is also indifferent to what the sex ratio in her nest is. But this is not so. It is still true that, given the assumption of worker control over investment in male and female reproductives, the resulting evolutionarily stable sex ratio will not be necessarily the same as the evolutionarily stable sex ratio given queen control. In this sense a worker is not indifferent to the sex ratio: she may well work to shift the sex ratio away from what the queen is 'trying' to achieve.

Trivers and Hare's analysis of the exact nature of the conflict between queen and workers over the sex ratio can be extended in ways that further illuminate the concept of manipulation (e.g. Oster & Wilson 1978). The following account is derived from Grafen (in preparation). I shall not anticipate his conclusions in detail, but wish to emphasize one principle which is explicit in his analysis as well as implicit in that of Trivers and Hare. The question is not 'Has the "best" sex ratio been successfully achieved?' On the contrary, we make a working assumption that natural selection has produced a result, given some constraints, and then ask what those constraints are (see Chapter 3). In the present case we follow Trivers and Hare in recognizing that the evolutionarily stable sex ratio depends crucially upon which parties to the arms race have practical power, but we recognize a wider range of possible dispositions of power than they did. In effect, Trivers and Hare deduced the consequences of two alternative assumptions about practical power; firstly the assumption that the queen exerts all the power, and secondly the assumption that the workers exert all the power. But many other possible assumptions could be made, and each gives rise to a different prediction of the evolutionarily stable sex ratio. In other parts of their paper, indeed, Trivers and Hare consider some of these, for instance the assumption that workers are able to lay their own male eggs.

Grafen, like Bulmer and Taylor (in preparation), has explored the consequences of assuming that power is *divided* as follows: the queen has absolute power over the sex of the eggs that she lays; the workers have absolute power over feeding the larvae. The workers can thus determine how many of the available female eggs shall develop into queens and how many into workers. They have the power to starve the young of one sex or the other, but they have to work within the constraint of what the queen gives them in the way of eggs. Queens have the power to lay eggs in any sex ratio they choose, including withholding, totally, eggs of one sex or the other. But, once laid, those eggs are at the mercy of the workers. A queen might, for instance, play the strategy (in the game theory sense) of laying only male eggs in a given year. Reluctant as we might expect them to be, the workers have no option but to rear their brothers. The queen, in this case, can preempt certain worker strategies, such as 'preferentially feed sisters', simply because she 'plays' first. But there are other things workers can do.

Using game theory, Grafen shows that only certain queen strategies are evolutionarily stable replies to particular worker strategies, and only certain worker strategies are evolutionarily stable replies to particular queen strategies. The interesting question is 'What are the evolutionarily stable combinations of worker and queen strategies?' It turns out that there is more than one answer, and there can be as many as three evolutionarily stable states for a given set of parameters. Grafen's particular conclusions are not my concern here, although I will remark that they are interestingly 'counter-intuitive'. What is my concern is that the evolutionarily stable state of the model population depends upon the assumptions we make about *power*. Trivers and Hare contrasted two possible absolute assumptions (absolute worker power versus absolute queen power). Grafen investigated one plausible *division* of power (queens have power over eggs, workers over larval feeding). But, as I have already noted, numerous other assumptions about power could be made. Each assumption generates different predictions about evolutionarily stable sex ratios, and tests of the predictions can therefore be regarded as providing evidence about the disposition of power in the nest.

For instance, we might focus our research attention on the exact moment when a queen 'decides' whether to fertilize a given egg or not. It is plausible to assume that, since the event takes place within the queen's own body, that particular decision is likely to have been selected to benefit the queen's genes. Plausible it may be, but it is precisely this kind of assumption that the doctrine of the extended phenotype is going to call in question. For the moment, we simply note the possibility that workers might manipulate the queen's nervous system, by pheromonal or other means, so as to subvert her behaviour in their genetic interests. Similarly, it is worker nerves and muscles that are immediately responsible for feeding the larvae, but we are not, therefore, necessarily entitled to assume that worker limbs move only in the interests of worker genes. As is well known, there is massive pheromonal traffic flow from queen to workers, and it is easy to imagine powerful manipulation of worker behaviour by queens. The point is that each assumption about power which we might make yields a testable prediction about sex ratios, and it is for this insight that we have to thank Trivers and Hare, not for the particular model whose predictions they happened to test.

It is even conceivable, in some Hymenoptera, that males might exert power. Brockmann (1980) is making an intensive study of mud-daubing wasps *Trypoxylon politum*. These are 'solitary' (as opposed to truly social) wasps, but they are not always totally alone. As in other sphecids, each female builds her own nest (in this case out of mud), provisions it with paralysed prey (spiders), lays one egg on the prey, then seals up the nest and begins the cycle again. In many Hymenoptera, the female carries a lifetime's supply of sperm from one brief period of insemination early in life. *T. politum* females, however, copulate frequently throughout adult life. Males haunt female nests, losing no opportunity to copulate with the female on each of her returns to the nest. A male may spend hours at a time sitting passively in the nest, probably helping to guard it against parasites, and fighting with other males who attempt to enter. Unlike most male Hymenoptera then, the male *T. politum* is present at the scene of the action. Might he not, therefore, be potentially in a position to influence the sex ratio, in the same kind of way as has been postulated for worker ants?

If males did exert power, what would we expect the consequences to be? Since a male passes all his genes on to his daughters, and none to his mate's sons, genes tending to make males favour daughters over sons would be favoured. If males exerted total power, completely determining the sex ratio of their mates' offspring, the consequence would be odd. No males would be born in the first generation of male power. As a result, in the following year all eggs laid would be unfertilized and therefore male. The population would therefore oscillate violently and then go extinct (Hamilton 1967). If males exerted a limited amount of power, less drastic consequences would follow, the situation being formally analogous to that of the 'driving X chromosome' in the normal diploid genetic system (Chapter 8). In any case a male hymenopteran, if he found himself in a position to influence the sex ratio of his mate's children, would be expected to try to do so in a female direction. He might do this by trying to influence his mate's decision whether to release sperm from her spermatheca. It is not obvious how he might actually do this, but it is known that honeybee queens take longer over laying a female egg than a male one, perhaps using the extra time to achieve fertilization. It would be interesting to try experimentally interrupting a queen in the middle of egg-laying, to see if the delay increased the chance of a female egg emerging.

Do male *T. politum* show any behaviour that we might suspect of being an attempt at such manipulation, for example do they behave as if trying to prolong egg-laying? Brockmann describes a curious behaviour pattern called 'holding'. This is seen alternating with copulation during the final minutes before the female lays her egg. In addition to brief copulations throughout the provisioning phase of the nest, the final egg-laying and sealing up of the nest is heralded by a prolonged bout of repeated copulations, which lasts many minutes. The female goes head first into the vertical, organ pipe-shaped mud nest, and pushes her head up into the cluster of paralysed spiders lodged in the top of the nest. Her abdomen is facing the entrance at the bottom of the nest, and in this position the male copulates with her. The female then turns round so that she is head downwards facing out of the nest, and probes the spiders with the tip of her abdomen, as if about to lay an egg. The male meanwhile 'holds' her head in his forelegs for about half a minute, grabs her antennae and pulls her downwards away from the spiders. She then turns around and they copulate again. She again turns to probe the spiders with her abdomen, the male again holds her head and drags her down. The whole cycle repeats some half dozen times. Finally, after one especially long bout of head-holding, the female lays her egg.

Once the egg is laid, its sex is determined. We have already considered the hypothetical possibility of worker ants manipulating their mother's nervous system, forcing her to change her fertilizing decision in their genetic interest. Brockmann's suggestion is that male *T. politum* might attempt similar subversion, and that the head-holding and dragging behaviour may be a manifestation of their manipulation technique. When the male seizes the female's antennae and drags her away from the spiders which she is probing with her abdomen, is he forcing a postponement of egg-laying as a means of increasing the chance of the egg's being fertilized in the oviduct? The plausibility of this suggestion might depend on exactly where the egg is in the female's body during the time of holding. Or is he, as Dr W. D. Hamilton has suggested to us, blackmailing the female by, in effect, threatening to bite her head off unless she postpones egg-laying until after further copulation? Perhaps he gains by repeated copulations, simply by flooding the female's internal passages with his sperm, thereby raising the chance that the egg will encounter a sperm without one being deliberately released from the spermatheca by the female. Clearly these are just suggestions for further research, and Brockmann, together with

Grafen and others, is following them up. Preliminary indications, I understand, do not support the hypothesis that males actually succeed in exerting power over the sex ratio.

This chapter is intended to begin the process of undermining the reader's confidence in the central theorem of the selfish organism. That theorem states that individual animals are expected to work for the good of their own inclusive fitness, for the good of copies of their own genes. The chapter has shown that animals are quite likely to work hard and vigorously for the good of some other individual's genes, and to the detriment of their own. This is not necessarily just a temporary departure from the central theorem, a brief interlude of manipulative exploitation before counterselection on the victim lineage redresses the balance. I have suggested that fundamental asymmetries such as the life/dinner principle, and the rare-enemy effect, will see to it that many arms races reach a stable state in which animals on one side permanently work for the benefit of animals on the other side, and to their own detriment; work hard, energetically, wantonly against their own genetic interests. When we see the members of a species consistently behaving in a certain way, 'anting' in birds or whatever it is, we are apt to scratch our heads and wonder how the behaviour benefits the animals' inclusive fitness. How does it benefit a bird to allow ants to run all through its feathers? Is it using the ants to clean it of parasites, or what? The conclusion of this chapter is that we might instead ask *whose* inclusive fitness the behaviour is benefiting! Is it the animal's own, or that of some manipulator lurking behind the scenes? In the case of 'anting' it does seem reasonable to speculate about advantages to the bird, but perhaps we should give at least a sideways glance at the possibility that it is an adaptation for the good of the ants!

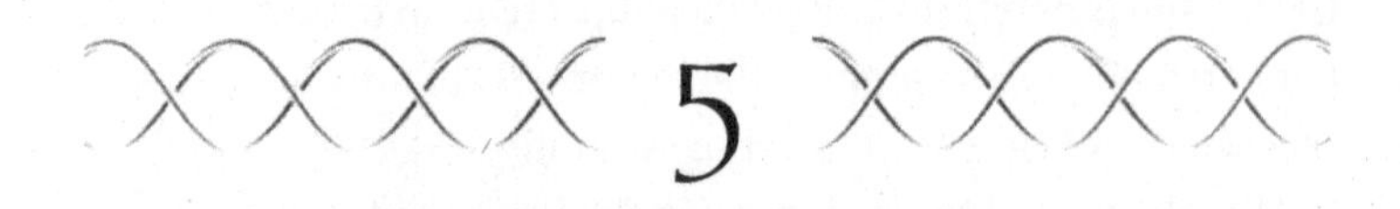

5
THE ACTIVE GERM-LINE REPLICATOR

In 1957, Benzer argued that 'the gene' could no longer continue as a single, unitary concept. He split it into three: the muton was the minimum unit of mutational change; the recon was the minimum unit of recombination; and the cistron was defined in a way that was directly applicable only to microorganisms, but it was effectively equivalent to the unit responsible for synthesizing one polypeptide chain. I have suggested adding a fourth unit, the *optimon*, the unit of natural selection (Dawkins 1978b). Independently, E. Mayr (personal communication) coined the term 'selecton' to serve the same purpose. The optimon (or selecton) is the 'something' to which we refer when we speak of an adaptation as being 'for the good of' something. The question is, what is that something; what is the optimon?

The question of what is the 'unit of selection' has been debated from time to time in the literature both of biology (Wynne-Edwards 1962; Williams 1966; Lewontin 1970a; Leigh 1977; Dawkins 1978a; Alexander & Borgia 1978; Wright 1980) and of philosophy (Hull 1980a,b; Wimsatt in preparation). At first sight it seems a rather uselessly theological argument. Hull, indeed, explicitly regards it as 'metaphysical' (though none the worse for that). I must justify my interest in it. Why does it matter what we consider to be the unit of selection? There are various reasons, but I shall give only one. I agree with Williams (1966), Curio (1973), and others that there is a need to develop a serious science of adaptation—teleonomy as Pittendrigh (1958) called it. The central theoretical problem of teleonomy will be that of the nature of the entity for whose benefit adaptations may be said to exist. Are they for the benefit of the individual organism, for the benefit of the group or species of which it is a member, or for the benefit of some smaller unit inside the individual organism? As already emphasized in Chapter 3, this really matters. Adaptations for the good of a group will look quite different from adaptations for the good of an individual.

Gould (1977b) sets out what, at first sight, appears to be the issue:

> The identification of individuals as the unit of selection is a central theme in Darwin's thought.... Individuals are the unit of selection; the 'struggle for existence' is a matter among individuals.... In the last fifteen years challenges to Darwin's focus on individuals have come from above and from below. From above, Scottish biologist V. C. Wynne-Edwards raised orthodox hackles fifteen years ago by arguing that groups, not individuals, are units of selection, at least for the evolution of social behaviour. From below, English biologist Richard Dawkins has recently raised my hackles with his claim that genes themselves are units of selection, and individuals merely their temporary receptacles.

Gould is invoking the idea of a hierarchy of levels in the organization of life. He sees himself as perched on an intermediate rung of a ladder, with group selectionists above and gene selectionists below. The present chapter and the next will show that this kind of analysis is false. There is, of course, a hierarchy of levels of biological organization (see next chapter), but Gould is applying it incorrectly. The conventional dispute between group selection and individual selection is different in category from the apparent dispute between individual selection and gene selection. It is wrong to think of the three as arranged on a single-dimensional ladder, such that words like 'above' and 'below' have transitive meaning. I shall show that the well-aired dispute between group and individual is concerned with what I shall call 'vehicle selection' and can be regarded as a factual biological dispute about units of natural selection. The attack 'from below', on the other hand, is really an argument about what we ought to *mean* when we talk about a unit of natural selection.

To anticipate the conclusion of these two chapters, there are two ways in which we can characterize natural selection. Both are correct; they simply focus on different aspects of the same process. Evolution is the external and visible manifestation of the differential survival of alternative *replicators* (Dawkins 1978a). Genes are replicators; organisms and groups of organisms are best not regarded as replicators; they are *vehicles* in which replicators travel about. Replicator selection is the process by which some replicators survive at the expense of other replicators. Vehicle selection is the process by which some vehicles are more successful than other vehicles in ensuring the survival of their replicators. The controversy about group selection versus individual selection is a controversy about the rival claims of two suggested kinds of vehicle. The controversy about gene selection versus individual (or group) selection is a controversy about whether, when we talk about a unit of selection, we ought to mean a vehicle at all, or a replicator. Much the same point has been realized by the philosopher D. L. Hull (1980a,b), but after some thought I prefer to persist with my own terminology rather than adopt his 'interactors' and 'evolvors'.

I define a *replicator* as anything in the universe of which copies are made. Examples are a DNA molecule, and a sheet of paper that is xeroxed. Replicators may be classified in two ways. They may be 'active' or 'passive', and, cutting across this classification, they may be 'germ-line' or 'dead-end' replicators.

An *active replicator* is any replicator whose nature has some influence over its probability of being copied. For example a DNA molecule, via protein synthesis, exerts phenotypic effects which influence whether it is copied: this is what natural selection is all about. A *passive replicator* is a replicator whose nature has no influence over its probability of being copied. A xeroxed sheet of paper at first sight seems to be an example, but some might argue that its nature *does* influence whether it is copied, and therefore that it is active: humans are more likely to xerox some sheets of paper than others, because of what is written on them, and these copies are, in their turn, relatively likely to be copied again. A section of DNA that is never transcribed might be a genuine example of a passive replicator (but see Chapter 9 on 'selfish DNA').

A *germ-line replicator* (which may be active or passive) is a replicator that is potentially the ancestor of an indefinitely long line of descendant replicators. A gene in a gamete is a germ-line replicator. So is a gene in one of the germ-line cells of a body, a direct mitotic ancestor of a gamete. So is any gene in *Amoeba proteus*. So is an RNA molecule in one of Orgel's (1979) test-tubes. A *dead-end replicator* (which also may be active or passive) is a replicator which may be copied a finite number of times, giving rise to a short chain of descendants, but which is definitely not the potential ancestor of an indefinitely long line of descendants. Most of the DNA molecules in our bodies are dead-end replicators. They may be the ancestors of a few dozen generations of mitotic replication, but they will definitely not be long-term ancestors.

A DNA molecule in the germ-line of an individual who happens to die young, or who otherwise fails to reproduce, should not be called a dead-end replicator. Such germ-lines are, as it turns out, terminal. They fail in what may metaphorically be called their aspiration to immortality. Differential failure of this kind is what we mean by natural selection. But whether it succeeds in practice or not, any germ-line replicator is *potentially* immortal. It 'aspires' to immortality but in practice is in danger of failing. All the DNA molecules in fully sterile social insect workers, however, are true dead-end replicators. They do not even aspire to replicate indefinitely. Workers lack a germ-line, not as a matter of misfortune but as a matter of design. In this respect they resemble human liver cells rather than the spermatogonia of a human who happens to be celibate. There may be awkward intermediate cases, for instance the 'sterile' worker who becomes facultatively fertile if her mother dies, and the *Streptocarpus* leaf which is not expected to propagate a new plant but which can do so if planted as a cutting. But this is getting theological: let us not worry *precisely* how many angels can dance on a pin's head.

As I said, the active/passive distinction cuts across the germ-line/dead-end distinction. All four combinations are conceivable. Particular interest attaches to one of the four, the active germ-line replicator, for it is, I suggest, the 'optimon', the unit for whose benefit adaptations exist. The reason active germ-line replicators are important units is that, wherever in the universe they may be found, they are likely to become the basis for natural selection and hence evolution. If replicators exist that are active, variants of them with certain phenotypic effects tend to out-replicate those with other phenotypic effects. If they are also germ-line replicators, these changes in relative frequency can have long-term, evolutionary impact. The world tends automatically to become populated by germ-line replicators whose active phenotypic effects are such as to ensure their successful replication. It is these phenotypic effects that we see as adaptations to survival. When we ask *whose* survival they are adapted to ensure, the fundamental answer has to be not the group, nor the individual organism, but the relevant replicators themselves.

I have previously summed up the qualities of a successful replicator 'in a slogan reminiscent of the French Revolution: Longevity, Fecundity, Fidelity' (Dawkins 1978a). Hull (1980b) explains the point clearly.

> Replicators need not last forever. They need only last long enough to produce additional replicators [fecundity] that retain their structure largely intact [fidelity]. The relevant longevity concerns the retention of structure through descent. Some entities, though structurally similar, are not copies because they are not related by descent. For example, although atoms of gold are structurally similar, they are not copies of one another because atoms of gold do not give rise to other atoms of gold. Conversely, a large molecule can break down into successively smaller molecules as its quaternary, tertiary, and secondary bonds are severed. Although descent is present, these successively smaller molecules cannot count as copies because they lack the requisite structural similarity.

A replicator may be said to 'benefit' from anything that increases the number of its descendant ('germ-line') copies. To the extent that active germ-line replicators benefit from the survival of the bodies in which they sit, we may expect to see adaptations that can be interpreted as for bodily survival. A large number of adaptations are of this type. To the extent that active germ-line replicators benefit from the survival of bodies *other* than those in which they sit, we may expect to see 'altruism', parental care, etc. To the extent that active germ-line replicators benefit from the survival of the group of individuals in which they sit, over and above the two effects just mentioned, we may expect to see adaptations for the preservation of the group. But all these adaptations will exist, fundamentally, through differential replicator survival. The basic beneficiary of any adaptation is the active germ-line replicator, the optimon.

It is important not to forget the 'germ-line' proviso in the specification of the optimon. This is the point of Hull's gold atom analogy. Krebs (1977) and I (Dawkins 1979a) have previously criticized Barash (1977) for suggesting that sterile worker insects care for *other workers* because they share genes with them. I would not harp on this again, had the error not been repeated twice recently in print (Barash 1978; Kirk 1980). It would be more correct to say that workers care for their *reproductive* siblings who carry germ-line copies of the caring genes. If they care for other workers, it is because those other workers are likely to work on behalf of the same reproductives (to whom they also are kin), *not* because the workers are kin to each other. Worker genes may be active, but they are dead-end, not germ-line replicators.

No copying process is infallible. It is no part of the definition of a replicator that its copies must all be perfect. It is fundamental to the idea of a replicator that when a mistake or 'mutation' does occur it is passed on to future copies: the mutation brings into existence a new kind of replicator which 'breeds true' until there is a further mutation. When a sheet of paper is xeroxed, a blemish may appear on the copy which was not present on the original. If the xerox copy itself is now copied, the blemish is incorporated into the second copy (which may also introduce a new blemish of its own). The important principle is that in a chain of replicators errors are cumulative.

I have previously used the word 'gene' in the same sense as I would now use 'genetic replicator', to refer to a genetic fragment which, for all that it serves as a unit of selection, does not have rigidly fixed boundaries. This has not met with uniform approval. The eminent molecular biologist Gunther Stent (1977) wrote that

'One of the great triumphs of 20th century biology was the eventual unambiguous identification of the Mendelian hereditary factor, or gene... as that unit of genetic material... in which the amino acid sequence of a particular protein is encoded.' Stent therefore violently objected to my adopting the equivalent of Williams's (1966) definition of a gene as 'that which segregates and recombines with appreciable frequency', describing it as a 'heinous terminological sin'.

Such motherly protectiveness towards a rather recently usurped technical term is not universal among molecular biologists, for one of the greatest of them has recently written that 'the theory of the "selfish gene" will have to be extended to any stretch of DNA' (Crick 1979). And, as we saw at the beginning of this chapter, another molecular biologist of the first rank, Seymour Benzer (1957), recognized the shortcomings of the traditional gene concept, but rather than grab the traditional word gene itself for one particular molecular usage, he chose the more modest course of coining a useful set of new terms—muton, recon, and cistron, to which we may add the optimon. Benzer recognized that all three of his units had claims to be regarded as equivalent to the gene of the earlier literature. Stent's uncompromising elevation of the cistron to the honour is arbitrary, though admittedly it is quite common. A more balanced view was given by the lamented W. T. Keeton (1980): 'It may seem strange that geneticists continue to employ different definitions of the gene for different purposes. The fact is that, at the present stage of knowledge, one definition is more useful in one context and another in another context; a rigid terminology would only hamper the formulation of current ideas and research aims.' Lewontin (1970b), too, gets it right when he says that 'it is only the chromosomes that obey Mendel's Law of Independent Assortment, and only the nucleotide base that is indivisible. The codons and the genes [cistrons] lie in between, being neither unitary nor independent in their behavior at meiosis.'

But let us not become worked up over terminology. Meanings of words are important, but not important enough to justify the ill-feeling they sometimes provoke, as in the present case of Stent (and also as in Stent's passionate and apparently sincere denunciation of my following the standard modern fashion of redefining 'selfishness' and 'altruism' in non-subjective senses—see Dawkins, 1981, for a reply to a similar criticism). I am happy to replace 'gene' with 'genetic replicator' where there is any doubt.

Heinous terminological sins aside, Stent makes the more important point that my unit is not precisely delimited in the way that the cistron is. Well, perhaps I should say 'in the way that the cistron once seemed to be', for the recent discovery of 'embedded' cistrons in virus ΦX174, and of 'exons' surrounding 'introns' must be causing a little discomfort to anyone who likes his units rigid. Crick (1979) expresses the sense of novelty well: 'In the last 2 years there has been a mini-revolution in molecular genetics. When I came to California, in September 1976, I had no idea that a typical gene might be split into several pieces and I doubt if anyone else had.' Crick significantly adds a footnote to the word gene: 'Throughout this article I have deliberately used the word "gene" in a loose sense since at this time any precise definition would be premature.' My unit of selection, whether I called it gene (Dawkins 1976a) or replicator (1978a) never had any pretensions to unitariness anyway. For the purposes for which it was defined, unitariness is not an important consideration, although I readily see that it might be important for other purposes.

The word replicator is purposely defined in a general way, so that it does not even have to refer to DNA. I am, indeed, quite sympathetic towards the idea that human culture provides a new milieu in which an entirely different kind of replicator selection can go on. In the next chapter we shall look briefly at this matter, and also at the claims of species gene-pools to be regarded as replicators in a large-scale selection process governing 'macroevolutionary' trends. But in the rest of this chapter we shall be concerned only with genetic fragments, and 'replicator' will be used as an abbreviation for 'genetic replicator'.

In principle, we may consider any portion of chromosome as a potential candidate for the title of replicator. Natural selection may usually be safely regarded as the differential survival of replicators relative to their alleles. The word allele is nowadays customarily used of cistrons but it is clearly easy, and in the spirit of this chapter, to generalize it to any portion of chromosome. If we look at a portion of chromosome five cistrons long, its alleles are the alternative sets of five cistrons that exist at the homologous loci of all the chromosomes in the population. An allele of an arbitrary sequence of twenty-six codons is an alternative homologous sequence of twenty-six codons somewhere in the population. Any stretch of DNA, beginning and ending at arbitrarily chosen points on the chromosome, can be considered to be competing with allelomorphic stretches for the region of chromosome concerned. It further follows that we can generalize the terms homozygous and heterozygous. Having picked out an arbitrary length of chromosome as our candidate replicator, we look at the homologous chromosome in the same diploid individual. If the two chromosomes are identical over the whole length of the replicator, the individual is homozygous for that replicator, otherwise it is heterozygous.

When I said 'arbitrarily chosen portion of chromosome', I really meant arbitrarily. The twenty-six codons that I chose might well span the border between two cistrons. The sequence still potentially fits the definition of a replicator, it is still possible to think of it as having alleles, and it still may be thought of as homozygous or heterozygous to the corresponding portion of the homologous chromosome in a diploid genotype. This, then, is our candidate replicator. But a candidate should be regarded as an *actual* replicator only if it possesses some minimum degree of longevity/fecundity/fidelity (there may be trade-offs among the three). Other things being equal, it is clear that larger candidates will have lower longevity/fecundity/fidelity than smaller ones because they are more vulnerable to being broken by recombination events. So, *how* large and how small a portion of chromosome is it useful to treat as a replicator?

This depends on the answer to another question: 'Useful for what?' The reason a replicator is interesting to Darwinians is that it is potentially immortal, or at least very long-lived in the form of copies. A successful replicator is one that succeeds in lasting, in the form of copies, for a very long time measured in generations, and succeeds in propagating many copies of itself. An unsuccessful replicator is one that potentially might have been long-lived, but in fact failed to survive, say because it caused the successive bodies in which it found itself to be sexually unattractive. We may apply the terms 'successful' and 'unsuccessful' to any arbitrarily defined portion of chromosome. Its success is measured relative to its alleles, and, if there is heterozygosity at the replicator locus in the population, natural selection will change the relative frequencies of the allelomorphic replicators in the population. But if the arbitrarily chosen portion of chromosome is very long it is not even *potentially* long-lived in its present form, for it is likely to be split apart by crossing-over in any given generation, regardless of how successful it may be in making a body survive and reproduce. To go to an extreme, if the potential replicator we consider is a whole chromosome, the difference between a 'successful' and an unsuccessful chromosome is of no significance, since both are almost bound to be split by crossing-over before the next generation in any case: their 'fidelity' is zero.

This can be put in another way. An arbitrarily defined length of chromosome, or potential replicator, may be said to have an expected half-life, measured in generations. Two kinds of factor will affect this half-life. Firstly, replicators whose phenotypic effects render them successful at their business of propagating themselves will tend to have a long half-life. Replicators with longer half-lives than their alleles will come to predominate in the population, and this is the familiar process of natural selection. But if we set selection pressures on one side, we can say something about the half-life of a replicator on the basis of its length alone. If the stretch of chromosome we choose to define as our replicator of interest is long, it will tend to have a shorter half-life than a shorter replicator, simply because it is more likely to be broken by crossing-over. A very long portion of chromosome ceases to deserve the title of replicator at all.

A corollary is that a long portion of chromosome, even if successful in terms of its phenotypic effects, will not be represented in many copies in the population. The Y chromosome excepted, and depending upon crossover rates, it seems unlikely that I share any whole chromosomes with any other individual. I assuredly share many small portions of chromosomes with others, and if we choose our portions small enough their likelihood of being shared becomes very high indeed. It is not normally useful to speak of inter-chromosome selection, therefore, since each chromosome is probably unique. Natural selection is the process by which replicators change in frequency in the population relative to their alleles. If the replicator under consideration is so large that it is probably unique, it cannot be said to have a 'frequency' to change. We must choose our arbitrary portion of chromosome so that it is small enough to last, at least potentially, for many generations before being split by crossing-over; small enough to have a 'frequency' that can be changed by natural selection. Is it possible to choose it *too* small? I shall return to this question below, after approaching it from another direction.

I shall make no attempt to specify *exactly* how long a portion of chromosome can be permitted to be before it ceases to be usefully regarded as a replicator. There is no hard and fast rule, and we don't need one. It depends on the strength of the selection pressure of interest. We are not seeking an absolutely rigid definition, but 'a kind of fading-out definition, like the definition of "big" or "old"'. If the selection pressure we are discussing is very strong, that is if one replicator makes its possessors very much more likely to survive and reproduce than its alleles do, the replicator can be quite large and still be usefully regarded as a unit that is naturally selected. If, on the other hand, the difference in survival consequences between a putative replicator and its alleles is almost negligible, the replicators under discussion would have to be quite small if the difference in their survival values is to make itself felt. This is the rationale behind Williams's (1966, p. 25) definition: 'In evolutionary theory, a gene could be defined as any hereditary information for which there is a favorable or unfavorable selection bias equal to several or many times its rate of endogenous change.'

The possibility of strong linkage disequilibrium (Clegg 1978) does not weaken the case. It simply increases the size of the chunk of genome that we can usefully treat as a replicator. If, which seems doubtful, linkage disequilibrium is so strong that populations contain 'only a few gametic types' (Lewontin 1974, p. 312), the effective replicator will be a very large chunk of DNA. When what Lewontin calls lc, the 'characteristic length' (the 'distance over which coupling is effective'), is only 'a fraction of the chromosome length, each gene is out of linkage equilibrium only with its neighbors but is assorted essentially independently of other genes farther away. The characteristic length is, in some sense, the unit of evolution since genes within it are highly correlated. The concept is a subtle one, however. It does not mean that the genome is broken up into discrete adjacent chunks of length lc. Every locus is the center of such a correlated segment and evolves in linkage with the genes near it' (Lewontin 1974).

Similarly, Slatkin (1972) wrote that 'It is clear that when permanent linkage disequilibrium is maintained in a population, the higher order interactions are important and the chromosome tends to act as a unit. The degree to which this is true in any given system is a measure of whether the gene or the chromosome is the unit of selection, or, more accurately, what parts of the genome can be said to be acting in unison'. And Templeton *et al.* (1976) wrote that 'the unit of selection is a function in part of the intensity of selection: the more intense the selection, the more the whole genome tends to hold together as a unit'. It was in this spirit that I playfully contemplated titling an earlier work *The slightly selfish big bit of chromosome and the even more selfish little bit of chromosome* (Dawkins 1976a, p. 35).

It has often been suggested to me that a fatal objection to replicator selectionism is the existence of within-cistron crossing-over. If chromosomes were like bead necklaces, the argument runs, with crossing-over always breaking the necklace between beads and not within them, you might hope to define discrete replicators in the population, containing an integral number of cistrons. But since crossover can occur anywhere (Watson 1976), not just between beads, all hope of defining discrete units disappears.

This criticism underestimates the elasticity that the replicator concept is permitted by the purpose for which it was coined. As I have just shown, we are not looking for discrete units, but for pieces of chromosome of indeterminate length which become more or less numerous than alternatives of exactly the same length. Moreover, as Mark Ridley reminds me, most within-cistron crossovers are, in any case, indistinguishable in their effects from between-cistron crossovers. Obviously, if the cistron concerned happens to be homozygous, paired at meiosis with an identical allele, the two sets of genetic material exchanged in a crossover will be identical, and the crossover might just as well never have happened. If the cistrons concerned are heterozygous, differing at one nucleotide locus, any within-cistron crossover that occurs 'north' of the heterozygous nucleotide will be indistinguishable from one at the northern boundary of the cistron; any within-cistron crossover 'south' of the heterozygous nucleotide will be indistinguishable from one at the southern boundary of the cistron. Only if the cistrons differ at two loci, and the crossover occurs between them, will it be identifiable as a within-cistron crossover. The general point is that it does not particularly matter where crossovers occur in relation to cistron boundaries. What matters is where crossovers occur in relation to heterozygous nucleotides. If, for instance, a sequence of six adjacent cistrons happens to be homozygous throughout an entire breeding population, a crossover anywhere within the six will be exactly equivalent in effect to a crossover at either end of the six.

Natural selection can cause changes in frequency only at nucleotide loci that are heterozygous in the population. If there are large intervening chunks of nucleotide sequences that never differ among individuals, these cannot be subject to natural selection, for there is nothing to choose between them. Natural selection must focus its attention on heterozygous nucleotides. It is changes at the single nucleotide level that are responsible for evolutionarily significant phenotypic changes, although of course the unvarying remainder of the genome is necessary to produce a phenotype at all. Have we, then, arrived at an absurdly reductionistic *reductio ad absurdum?* Shall we write a book called *The Selfish Nucleotide?* Is adenine engaged in a remorseless struggle against cytosine for possession of locus number 30004?

At the very least, this is not a helpful way to express what is going on. It becomes downright misleading if it suggests to the student that adenine at one locus is, in some sense, allied with adenine at other loci, pulling together for an adenine team. If there is any sense in which purines and pyrimidines compete with each other for heterozygous loci, the struggle at each locus is insulated from the struggle at other loci. The molecular biologist may, for his own important purposes (Chargaff cited in Judson 1979), count adenines and cytosines in the genome as a whole, but to do so is an idle exercise for the student of natural selection. If they are competitors at all, they are competitors for each locus separately. They are indifferent to the fate of their exact replicas at other loci (see also Chapter 8).

But there is a more interesting reason for rejecting the concept of the selfish nucleotide, in favour of some larger replicating entity. The whole purpose of our search for a 'unit of selection' is to discover a suitable actor to play the leading role in our metaphors of purpose. We look at an adaptation and want to say, 'It is for the good of...'. Our quest in this chapter is for the right way to complete that sentence. It is widely admitted that serious error follows from the uncritical assumption that adaptations are for the good of the species. I hope I shall be able to show, in this book, that yet other theoretical dangers, albeit lesser ones, attend the assumption that adaptations are for the good of the individual organism. I am suggesting here that, since we must speak of adaptations as being for the good of something, the correct something is the active, germ-line replicator. And while it may not be strictly wrong to say that an adaptation is for the good of the nucleotide, i.e. the smallest replicator responsible for the phenotypic differences concerned in the evolutionary change, it is not helpful to do so.

We are going to use the metaphor of power. An active replicator is a chunk of genome that, when compared to its alleles, exerts phenotypic power over its world, such that its frequency increases or decreases relative to that of its alleles. While it is undoubtedly meaningful to speak of a single nucleotide as exerting power in this sense, it is much more useful, since the nucleotide only exerts a given type of power when embedded in a larger unit, to treat the larger unit as exerting power and hence altering the frequency of its copies. It might be thought that the same argument could be used to justify treating an even larger unit, such as the whole genome, as the unit that exerts the power. This is not so, at least for sexual genomes.

We reject the whole sexual genome as a candidate replicator, because of its high risk of being fragmented at meiosis. The single nucleotide does not suffer from this problem but, as we have just seen, it raises another problem. It cannot be said to have a phenotypic effect except in the context of the other nucleotides that surround it in its cistron. It is meaningless to speak of the phenotypic effect of adenine. But it is entirely sensible to speak of the phenotypic effect of substituting adenine for cytosine at a named locus within a named cistron. The case of a cistron within a genome is not analogous. Unlike a nucleotide, a cistron is large enough to have a consistent phenotypic effect, relatively, though not completely, independently of where it lies on the chromosome (but not regardless of what other genes share its genome). For a cistron, its sequential context *vis-à-vis* other cistrons is not overwhelmingly important in determining its phenotypic effect in comparison with its alleles. For a *nucleotide's* phenotypic effect, on the other hand, its sequential context is everything.

Bateson (1982) expresses the following misgiving about 'replicator selection'.

A winning character is defined in *relation* to another one while genetic replicators are thought about in *absolute* and atomistic terms. The difficulty is brought home if you ask yourself, what exactly is Dawkins' replicator? You might answer: 'That bit of genetic material making the difference between the winning and losing characters.' You would have stated that a replicator must be defined in relation to something else. Alternatively, your reply might be: 'A replicator consists of all the genes required for the expression of the surviving character'. In that case you are saddled with a complex and unwieldy concept. Either way your answer would show how misleading it is to think of replicators as the atoms of evolution.

I certainly would join Bateson in rejecting the second of his two alternative answers, the unwieldly one. The first of his alternatives, on the other hand, expresses exactly my position, and I do not share Bateson's misgivings about it. For my purposes a genetic replicator is defined by reference to its alleles, but this is not a weakness of the concept. Or, if it is deemed to be a weakness, it is a weakness that afflicts the whole science of population genetics, not just the particular idea of genetic units of selection. It is a fundamental truth, though it is not always realized, that *whenever* a geneticist studies a gene 'for' any phenotypic character, he is always referring to a *difference* between two alleles. This is a recurring theme throughout this book.

To prove the pudding, let me show how easy it is to use the gene as a conceptual unit of selection, while admitting that it is only defined by comparison with its alleles. It is now accepted that a particular major gene for dark coloration in the peppered moth *Biston betularia* has increased in frequency in industrial areas because it produces phenotypes that are superior in industrial areas (Kettlewell 1973). At the same time, we have to admit that this gene is only one of thousands that are necessary in order for the dark coloration to show itself. A moth cannot have dark wings unless it has wings, and it cannot have wings unless it has hundreds of genes and hundreds of equally necessary environmental factors. But this is all irrelevant. The *difference* between the *carbonaria* and the *typica* phenotype can still be due to a difference at one locus, even though the phenotypes themselves could not exist without the participation of thousands of genes. And it is the same difference that is the basis of the natural selection. Both geneticists *and* natural selection are concerned with differences! However complex the genetic basis of features that all members of a species have *in common*, natural selection is concerned with differences. Evolutionary change is a limited set of substitutions at identifiable loci.

Further difficulties will be dealt with in the next chapter. Meanwhile, I end this chapter with a small diversion which may be helpful in illustrating the replicator or 'gene's eye' view of evolution. An appealing aspect of the view appears if we look backwards in time. The replicators in frequent existence today constitute a relatively successful subset of those that have existed in the past. A given replicator in me could, in theory, be traced backwards through a straight line of ancestors. These ancestors, and the environments that they provided for the replicator, can be regarded as the replicator's 'past experience'.

The past experience of autosomal genetic fragments in a species is, statistically speaking, similar. It consists of an ensemble of typical species bodies, approximately 50 per cent male and 50 per cent female bodies, bodies which grew through a wide spectrum of ages at least up to reproductive age; and it includes a good random shuffling of 'companion' genes at other loci. The genes that exist today tend to be the ones that are good at surviving in that statistical ensemble of bodies, and in company with that statistical ensemble of companion genes. As we shall see, it is selection in favour of the qualities needed for survival *in company* with other, similarly selected, genes, that gives rise to the appearance of 'coadapted genomes'. I shall show in Chapter 13 that this is a much more illuminating interpretation of the phenomenon of coadaptation than its alternative, that 'the coadapted genome is the true unit of selection'.

Probably no two genes in an organism have identical past experiences, though a linked pair may come close, and, mutants aside, all the genes on a Y chromosome have travelled together through the same set of bodies for a large number of generations. But the exact nature of a gene's past experience is of less interest than the generalizations one can make about the past experiences of all genes that exist today. For instance, however variable the set of my ancestors may be, they all had in common that they survived at least to reproductive age, they copulated heterosexually and were fertile. The same generalization cannot be made about the historical set of bodies that were not my ancestors. The bodies that provided the past experience of existing genes are a non-random subset of all the bodies that have ever existed.

The genes that exist today reflect the set of environments that they have experienced in the past. This includes the internal environments provided by the bodies the genes have inhabited, and also external environments, desert, forest, seashore, predators, parasites, social companions, etc. This is, of course, not because the environments have imprinted their qualities on the genes—that would be Lamarckism (see Chapter 9)—but because the genes that exist today are a selected set, and the qualities that made them survive reflect the qualities of the environments in which they survived.

I said that a gene's experience consists of time spent in approximately 50 per cent male and 50 per cent female bodies, but this is not, of course, true of genes on sex chromosomes. In mammals, assuming no Y-chromosome crossing-over, a gene on a Y chromosome has experienced only male bodies, and a gene on an X chromosome has spent two-thirds of its history in female bodies and one-third in male bodies. In birds, Y-chromosome genes have experienced only female bodies, and in particular cases such as cuckoos we can say something further. Female *Cuculus canorus* are divided into 'gentes', each gens parasitizing a different species of host (Lack 1968). Apparently each female learns the qualities of her own foster parents and their nest, and returns when adult to parasitize the same species. Males do not seem to discriminate as to gens in their choice of mate, and therefore they act as vehicles for gene flow between gentes. Of the genes in a female cuckoo, therefore, those on autosomes and the X chromosome have probably had recent experience of all gentes in the population, and been 'reared' by foster parents belonging to all the species parasitized by the cuckoo population. But the Y chromosome, uniquely, is confined for long sequences of generations to one gens and one foster species. Of all the genes sitting in a robin's nest, one subset—robin genes and cuckoo Y-chromosome genes (and robin flea genes)—has sat in robins' nests for many generations back. Another subset—cuckoo autosomal and X-chromosome genes—has experienced a mixture of nests. Of course the first subset share only part of their experience, a long series of robin nests. In other aspects of their experience, cuckoo Y-chromosome genes will have more in common with other cuckoo genes than with robin genes. But as far as certain particular selection pressures found in nests are concerned, cuckoo Y-chromosome genes have more in common with robin genes than with cuckoo autosome genes. It is natural, then, that cuckoo Y chromosomes should have evolved to reflect their peculiar experience, while other cuckoo genes simultaneously evolved to reflect their more general experience—a kind of incipient intragenomic 'speciation' at the chromosomal level. It is, indeed, widely assumed for this reason that genes for foster species-specific egg mimicry must be carried on the Y chromosome, while genes for general parasitic adaptations might be carried on any chromosome.

I am not sure whether the fact has significance, but this backwards way of looking shows X chromosomes, too, to have a peculiar history. A gene on an autosome in a female cuckoo is as likely to have come from the father as from the mother, in which latter case it experiences the same host species for two generations running. A gene on an X chromosome in a female cuckoo is bound to have come from the father, and therefore is not especially likely to experience the same host species two generations running. A statistical 'runs test' on the sequence of foster species experienced by an autosomal gene would therefore reveal a slight runs effect, greater than that for a gene on an X chromosome, and much less than that for a gene on a Y chromosome.

In any animal, an inverted portion of chromosome may resemble a Y chromosome in being unable to cross over. The 'experience' of any part of the 'inversion supergene' therefore repeatedly includes the other parts of the supergene and their phenotypic consequences. A habitat selection gene anywhere in such a supergene, say a gene that makes individuals choose dry microclimates, would then provide a consistent habitat 'experience' for successive generations of the whole supergene. Thus a given gene may 'experience' consistently dry habitats for the same kind of reason as a gene on a cuckoo's Y chromosome consistently experiences meadow-pipit nests. This will provide a consistent selection pressure bearing upon that locus, favouring alleles that are adapted to a dry habitat in the same way as alleles for mimicking meadow-pipit eggs are favoured in Y chromosomes of female cuckoos of the meadow-pipit gens. This particular inversion supergene will tend to be found for generations in dry habitats, even though the rest of the genome may be randomly shuffled over the whole range of habitats available to the species. Many different loci on the inverted portion of chromosome may therefore come to be adapted to a dry climate, and again something akin to intragenomic incipient speciation may go on. I find this backwards way of looking at the past 'experience' of genetic replicators helpful.

Germ-line replicators, then, are units that actually survive or fail to survive, the difference constituting natural selection. Active replicators have some effect on the world, which influences their chances of surviving. It is the effects on the world of successful active germ-line replicators that we see as adaptations. Fragments of DNA qualify as active germ-line replicators. Where there is sexual reproduction, these fragments must not be defined too large if they are to retain the property of self-duplication. And they must not be defined too small if they are to be usefully regarded as active.

If there were sex but no crossing-over, each chromosome would be a replicator, and we should speak of adaptations as being for the good of the chromosome. If there is no sex we can, by the same token, treat the entire genome of an asexual organism as a replicator. But the organism itself is *not* a replicator. This is for two quite distinct reasons which should not be confused with each other. The first reason follows from the arguments developed in this chapter, and applies only where there is sexual reproduction and meiosis: meiosis and sexual fusion see to it that not even our genomes are replicators, so we ourselves are not replicators either. The second reason applies to asexual as well as sexual reproduction. It will be explained in the next chapter, which goes on to discuss what organisms, and also groups of organisms, *are*, given that they are not replicators.

6

ORGANISMS, GROUPS, AND MEMES

Replicators or Vehicles?

I have made so much of the fragmenting effects of meiosis as a reason for not regarding sexually reproduced organisms as replicators, that it is tempting to see this as the only reason. If this were true, it should follow that asexually reproduced organisms are true replicators, and that where reproduction is asexual we could legitimately speak of adaptations as 'for the good of the organism'. But the fragmenting effect of meiosis is not the only reason for denying that organisms are true replicators. There is a more fundamental reason, and it applies to asexual organisms as much as to sexual ones.

To regard an organism as a replicator, even an asexual organism like a female stick insect, is tantamount to a violation of the 'central dogma' of the non-inheritance of acquired characteristics. A stick insect looks like a replicator, in that we may lay out a sequence consisting of daughter, granddaughter, great-granddaughter, etc., in which each appears to be a replica of the preceding one in the series. But suppose a flaw or blemish appears somewhere in the chain, say a stick insect is unfortunate enough to lose a leg. The blemish may last for the whole of her lifetime, but it is not passed on to the next link in the chain. Errors that affect stick insects but not their genes are not perpetuated. Now lay out a parallel series consisting of daughter's genome, granddaughter's genome, great-granddaughter's genome, etc. If a blemish appears somewhere along *this* series it will be passed on to all subsequent links in the chain. It may also be reflected in the bodies of all subsequent links in the chain, because in each generation there are causal arrows leading from genes to body. But there is no causal arrow leading from body to genes. No part of the stick insect's phenotype is a replicator. Nor is her body as a whole. It is wrong to say that 'just as genes can pass on their structure in gene lineages, organisms can pass on their structure in organism lineages'.

I am sorry if I am about to labour this argument, but I fear it was my failure to be clear about it before that led to an unnecessary disagreement with Bateson, a disagreement which it is worth going to some trouble to sort out. Bateson (1978) made the point that genetic determinants of development are necessary but not sufficient. A gene may 'program' a particular bit of behaviour 'without it being the only thing to do so'. He goes on:

> Dawkins accepts all this but then reveals his uncertainty about which language he is using by immediately giving special status back to the gene as the programmer. Consider a case in which the ambient environmental temperature during development is crucial for the expression of a particular phenotype. If the temperature changes by a few degrees the survival machine is beaten by another one. Would not that give as much status to a necessary temperature value as to a necessary gene? The temperature value is also required for the expression of a particular phenotype. It is also stable (within limits) from one generation to the next. It may even be transmitted from one generation to the next if the survival machine makes a nest for its offspring. Indeed, using Dawkins' own style of teleological argument one could claim that the bird is the nest's way of making another nest [Bateson 1978].

I replied to Bateson, but did so too briefly, picking on the last remark about birds' nests and saying, 'A nest is not a true replicator because a [non-genetic] "mutation" which occurs in the construction of a nest, for example the accidental incorporation of a pine needle instead of the usual grass, is not perpetuated in future "generations of nests". Similarly, protein molecules are not replicators, nor is messenger RNA' (Dawkins 1978a). Bateson had taken the catchphrase about a bird being a gene's way of making another gene, and inverted it, substituting 'nest' for 'gene'. But the parallel is not a valid one. There is a causal arrow going from gene to bird, but none in the reverse direction. A changed gene may perpetuate itself better than its unmutated allele. A changed nest will do no such thing unless, of course, the change is due to a changed gene, in which case it is the gene that is perpetuated, not the nest. A nest, like a bird, is a gene's way of making another gene.

Bateson is worried that I seem to give 'special status' to genetic determinants of behaviour. He fears that an emphasis on the gene as the entity for whose benefit organisms labour, rather than the other way around, leads to an undue emphasis on the importance of genetic as opposed to environmental determinants of development. The answer to this is that when we are talking about *development* it is appropriate to emphasize non-genetic as well as genetic factors. But when we are talking about units of selection a different emphasis is called for, an emphasis on the properties of replicators. The special status of genetic factors rather than non-genetic factors is deserved for one reason only: genetic factors replicate themselves, blemishes and all, but non-genetic factors do not.

Let us grant with both hands that the temperature in the nest housing a developing bird is important both for its immediate survival and for the way it develops and therefore for its long-term success as an adult. The immediate effects of gene products on the biochemical springs of development may, indeed, closely resemble the effects of temperature changes (Waddington 1957). We could even imagine the enzyme products of genes as little Bunsen burners, selectively applied at crucial nodes of the branching biochemical tree of embryonic causation, controlling development by selective control of biochemical reaction rates. An embryologist rightly sees no fundamental distinction between genetic and environmental causal factors, and he correctly regards each as necessary but not sufficient. Bateson was putting the embryologist's point of view, and no ethologist is better qualified to do so. But I was not talking embryology. I was not concerned with the rival claims of determinants of development. I was talking about replicators surviving in evolutionary time, and Bateson certainly agrees that neither a nest, nor the temperature inside it, nor the bird that built it, are replicators. We can quickly see that they are not replicators by experimentally altering one of them. The change may wreak havoc on the animal, on its development and its chances of survival, *but the change will not be passed on to the next generation.* Now make a similar mutilation (mutation) to a gene in the germ-line: the change may or may not affect the bird's development and its survival, but it *can* be passed on to the next generation; it can be replicated.

As is so often the case, an apparent disagreement turns out to be due to mutual misunderstanding. I thought that Bateson was denying proper respect to the Immortal Replicator. Bateson thought that I was denying proper respect to the Great Nexus of complex causal factors interacting in development. In fact, each of us was laying legitimate stress on considerations which are important for two different major fields of biology, the study of development and the study of natural selection.

An organism, then, is not a replicator, not even (despite Lewontin 1970a—see Dawkins 1982) a crude replicator with poor copying fidelity. It is therefore better not to speak of adaptations as being for the good of the organism. What about larger units, groups of organisms, species, communities of species, etc.? Some of these larger groupings are clearly subject to a version of the 'internal fragmentation destroys copying fidelity' argument. The fragmenting agent in this case is not the recombining effects of meiosis, but immigration and emigration, the destruction of the integrity of groups by the movement of individuals into and out of them. As I have put it before, they are like clouds in the sky or dust-storms in the desert. They are temporary aggregations or federations. They are not stable through evolutionary time. Populations may last a long while, but they are constantly blending with other populations and so losing their identity. They are also subject to evolutionary change from within. A population is not a discrete enough entity to be a unit of natural selection, not stable and unitary enough to be 'selected' in preference to another population. But, just as the 'fragmentation' argument applied only to a subset of organisms, sexual ones, so it also applies only to a subset at the group level. It applies to groups capable of interbreeding, but it does not apply to reproductively isolated species.

Let us, then, examine whether species behave sufficiently like coherent entities, multiplying and giving rise to other species, to deserve to be called replicators. Note that that is not the same as Ghiselin's (1974b) logical claim that species are 'individuals' (see also Hull 1976). Organisms, too, are individuals in Ghiselin's sense, and I hope I have established that organisms are not replicators. Do species, or, to be more precise, do reproductively isolated gene-pools, really answer to the definition of replicators?

It is important to remember that mere immortality is not a sufficient qualification. A lineage, such as a sequence of parents and offspring from the long-unchanged brachiopod genus *Lingula*, is unending in the same sense, and to the same extent, as a lineage of genes. Indeed, for this example we perhaps need not have chosen a 'living fossil' like *Lingula*. Even a rapidly evolving lineage can, in a sense, be treated as an entity which is either extinct or extant at any moment in geological time. Now, certain kinds of lineage may be more likely to go extinct than others, and we may be able to discern statistical laws of extinction. For example, lineages whose females reproduce asexually may be systematically more or less likely to go extinct than lineages whose females stick to sex (Williams 1975; Maynard Smith 1978a). It has been suggested that ammonite and bivalve lineages with a high rate of evolving larger size (i.e. with a high rate of obeying Cope's Rule) are more likely to go extinct than more slowly evolving lineages (Hallam 1975). Leigh (1977) makes some excellent points about differential lineage extinction, and its relationship to lower levels of selection: 'those species are favored where selection within populations works more nearly for the good of the species'. Selection 'favors species that have, for whatever reason, evolved genetic systems where a gene's selective advantage more nearly matches its contribution to fitness'. Hull (1980a,b) is particularly clear about the logical status of the lineage, and about its distinction from the replicator and the interactor (Hull's name for what I am calling the 'vehicle').

Differential lineage extinction, though technically a form of selection, is not enough in itself to generate progressive evolutionary change. Lineages may be 'survivors', but this does not make them replicators. Grains of sand are survivors. Hard grains, made of quartz or diamond, will last longer than soft grains made of chalk. But nobody has ever invoked hardness selection among sand grains as the basis for an evolutionary progression. The reason, fundamentally, is that grains of sand do not multiply.

One grain may survive a long time, but it does not multiply and make copies of itself. Do species, or other groups of organisms, multiply? Do they replicate?

Alexander and Borgia (1978) assert that they do, and that they are therefore true replicators: 'Species give rise to species; species multiply.' The best case I can make for regarding species, or rather their gene-pools, as multiplying replicators arises from the theory of 'species selection' associated with the palaeontological idea of 'punctuated equilibria' (Eldredge & Gould 1972; Stanley 1975, 1979; Gould & Eldredge 1977; Gould 1977c, 1980a,b; Levinton & Simon 1980). I will take some time to discuss this body of theory, since 'species selection' is very relevant to this chapter. Another reason for taking the time is that I regard the suggestions of Eldredge and Gould as of great interest to biology generally, but I am anxious that they should not be oversold as more revolutionary than they actually are. Gould and Eldredge (1977, p. 117) are themselves conscious of this danger, though for different reasons.

My fear stems from the growing influence of a vigilant corps of lay critics of Darwinism, either religious fundamentalists or Shavian/Koestlerian Lamarckists who, for reasons that have nothing to do with science, eagerly seize upon anything that, with imperfect understanding, can be made to sound anti-Darwinian. Journalists are often only too ready to pander to the unpopularity of Darwinism in some lay circles. One of Britain's least disreputable daily newspapers (*The Guardian*, 21 November 1978) served up a journalistically garbled but still just recognizable version of the Eldredge/Gould theory in a leading article, as evidence that all is not well with Darwinism. Predictably, this elicited some uncomprehending fundamentalist glee in the letter columns of the paper, some of it from disquietingly influential sources, and the public could well have been left with the impression that even 'the scientists' themselves now have doubts about Darwinism. Dr Gould informs me that *The Guardian* did not favour him with a reply to his letter of protest. Another British newspaper, *The Sunday Times* (8 March 1981), in a much longer article called 'The new clues that challenge Darwin', sensationally exaggerated the difference between the Eldredge/Gould theory and other versions of Darwinism. The British Broadcasting Corporation also got in on the act around the same time, in two separate programmes made by rival production teams. They were called *The Trouble with Evolution* and *Did Darwin get it Wrong?*, and they differed hardly at all except that one had Eldredge and the other had Gould! The second programme actually went to the lengths of digging up some fundamentalists to comment on the Eldredge/Gould theory: not surprisingly, the ill-understood appearance of dissension within the ranks of Darwinists was meat and drink to them.

Journalistic standards are not unknown in learned periodicals too. *Science* (Vol. 210, pp. 883–887, 1980) reported on a recent conference on macroevolution under the dramatic heading, 'Evolutionary theory under fire', and with the equally sensational subtitle, 'An historic conference in Chicago challenges the four-decade long dominance of the Modern Synthesis' (see criticisms by Futuyma *et al.* 1981). As Maynard Smith is quoted as saying at the same conference, 'You are in danger of preventing understanding by suggesting that there is intellectual antagonism where none exists' (see also Maynard Smith 1981). In the face of all the ballyhoo, I am anxious to be very clear about, to quote one of their own section headings, exactly 'What Eldredge and Gould Did Not (And Did) Say'.

The theory of punctuated equilibria suggests that evolution does not consist in continuous smooth change, 'stately unfolding', but goes in jerks, punctuating long periods of stasis. Lack of evolutionary change in fossil lineages should not be written off as 'no data', but should be recognized as the norm, and as what we should really expect to see if we take our modern synthesis, particularly its embedded ideas of allopatric speciation, seriously: 'As a consequence of the allopatric theory, new fossil species do not originate in the place where their ancestors lived. It is extemely improbable that we shall be able to trace the gradual splitting of a lineage merely by following a certain species up through a local rock column' (Eldredge & Gould 1972, p. 94). Of course microevolution by ordinary natural selection, what I would call genetic replicator selection, goes on, but it is largely confined to brief bursts of activity around the crisis times known as speciation events. These bursts of microevolution are usually completed too fast for palaeontologists to track them. All we can see is the state of the lineage before and after the new species is formed. It follows that 'gaps' in the fossil record between species, far from being the embarrassment Darwinians have sometimes taken them to be, are exactly what we should expect.

The palaeontological evidence can be argued about (Gingerich 1976; Gould & Eldredge 1977; Hallam 1978), and I am not qualified to judge it. Approaching from a non-palaeontological direction, indeed in lamentable ignorance of the whole Eldredge/Gould theory, I once found the Wrightian/Mayrian idea of buffered gene-pools resisting change, but occasionally succumbing to genetic revolutions, satisfyingly compatible with one of my own enthusiasms, Maynard Smith's (1974) 'evolutionarily stable strategy' concept:

> The gene pool will become an *evolutionarily stable set* of genes, defined as a gene pool which cannot be invaded by any new gene. Most new genes which arise, either by mutation or reassortment or immigration, are quickly penalized by natural selection:

> the evolutionarily stable set is restored. Occasionally...there is a transitional period of instability, terminating in a new evolutionarily stable set...a population might have more than one alternative stable point, and it might occasionally flip from one to another. Progressive evolution may be not so much a steady upward climb as a series of discrete steps from stable plateau to stable plateau [Dawkins 1976a, p. 93].

I am also quite impressed with Eldredge and Gould's reiterated point about time-scales: 'How can we view a steady progression yielding a 10% increase in a million years as anything but a meaningless abstraction? Can this varied world of ours possibly impose such minute selection pressures so uninterruptedly for so long?' (Gould & Eldredge 1977). '[T]o see gradualism at all in the fossil record implies such an excruciatingly slow rate of per-generation change that we must seriously consider its invisibility to natural selection in the conventional mode—changes that confer momentary adaptive advantages' (Gould 1980a). I suppose the following analogy might be made. If a cork floats from one side of the Atlantic to the other, travelling steadily without deviating or going backwards, we might invoke the Gulf Stream or Trade Winds in explanation. This will seem plausible if the time the cork takes to cross the ocean is of the right order of magnitude, say a few weeks or months. But if the cork should take a million years to cross the ocean, again not deviating or going backwards but steadily inching its way across, we should not be satisfied with any explanation in terms of currents and winds. Currents and winds just don't move that slowly, or if they do they will be so weak that the cork will be overwhelmingly buffeted by other forces, backwards as much as forwards. If we found a cork steadily moving at such an extremely slow rate, we would have to seek a wholly different kind of explanation, an explanation commensurate with the time-scale of the phenomenon observed.

Incidentally, there is a mildly interesting historical irony here. One of the early arguments used against Darwin was that there wasn't enough time for the proposed amount of evolution to have happened. It seemed hard to imagine that selection pressures were strong enough to achieve all that evolutionary change in the short time then thought to be available. The argument of Eldredge and Gould just given is almost the exact opposite: it is hard to imagine a selection pressure *weak* enough to sustain such a slow rate of unidirectional evolution over such a long period! Perhaps we should take warning from this historical twist. Both arguments resort to the 'hard to imagine' style of reasoning that Darwin so wisely cautioned us against.

Although I find Eldredge and Gould's time-scale point somewhat plausible, I am less confident about it than they are, since I do fear the limitations of my own imagination. After all, the gradualist theory does not really need to assume *unidirectional* evolution for long periods. In terms of my cork analogy, what if the wind *is* so weak that the cork takes a million years to cross the Atlantic? Waves and local currents may indeed send it backwards almost as much as forwards. But when all is added up, the net statistical direction of the cork may still be determined by a slow and relentless wind.

I also wonder whether Eldredge and Gould give sufficient attention to the possibilities opened up by 'arms races' (Chapter 4). In characterizing the gradualist theory that they attack, they write: 'The postulated mechanism for gradual uni-directional change is "orthoselection", usually viewed as a constant adjustment to a uni-directional change in one or more features of the physical environment' (Eldredge & Gould 1972). If the winds and currents of the physical environment pressed steadily in the same direction over geological time-scales, it might indeed seem likely that animal lineages would reach the other side of their evolutionary ocean so fast that palaeontologists could not track their passage.

But change 'physical' to 'biological' and things may look different. If each small adaptive step in one lineage calls forth a counteradaptation in another lineage—say its predators—slow, directional orthoselection looks rather more plausible. The same is true of intraspecific competition, where the optimum size, say, for an individual can be slightly larger than the present population mode, *whatever the present population mode may be.* '[I]n the population as a whole there is a constant tendency to favor a size slightly above the mean. The slightly larger animals have a very small but in the long run, in large populations, decisive advantage in competition.... Thus, populations that are regularly evolving in this way are always well adapted as regards size in the sense that the optimum is continuously included in their normal range of variation, but a constant asymmetry in the centripetal selection favors a slow upward shift in the mean' (Simpson 1953, p. 151). Alternatively (and shamelessly having it both ways!), if evolutionary trends really are punctuated and stepped, perhaps this in itself could be explained by the arms race concept, given time-lags between adaptive advances on one side of the arms race and responses by the other.

But let us provisionally accept the theory of punctuated equilibria as an excitingly different way of looking at familiar phenomena, and turn to the other side of Gould and Eldredge's (1977) equation, 'punctuated equilibria + Wright's rule = species selection'. Wright's Rule (not his own coining) is 'the proposition that a set of morphologies produced by speciation events is essentially random with respect to the direction of evolutionary trends within a clade' (Gould & Eldredge 1977). For example, even if there is an overall trend towards larger size in a set of related lineages, Wright's Rule suggests that there is no systematic tendency for newly branched-off species to be larger than their parent species. The analogy with the 'randomness' of mutation is clear, and this leads directly to the right-hand side of the equation. If new species differ from their predecessors randomly with respect to major trends, major trends themselves must be due to differential extinction among those new species—'species selection', to use Stanley's (1975) term.

Gould (1980a) regards 'the testing of Wright's rule as a major task for macroevolutionary theory and paleobiology. For the theory of species selection, in its pure form, depends upon it. Consider, for example, a lineage displaying Cope's rule of increasing body size—horses, for example. If Wright's rule be valid, and new species of horses arise equally often at sizes smaller and larger than their ancestors, then the trend is powered by species selection. But if new species arise preferentially at sizes larger than their ancestors, then we don't require species selection at all, since random extinction would still yield the trend.' Gould here simultaneously sticks his neck out and hands Occam's Razor to his opponents! He could easily have claimed that even if his mutation-analogue (speciation) *was* directed, the trend might still be reinforced by species selection (Levinton & Simon 1980). Williams (1966, p. 99) in an interesting discussion which I have not seen cited in the literature on punctuated equilibria, considers a form of species selection acting *in opposition* to, and perhaps outweighing, the overall trend of evolution within species. He again uses the horse example, and the fact that earlier fossils tend to be smaller than later ones:

> From this observation, it is tempting to conclude that, at least most of the time and on the average, a larger than mean size was an advantage to an individual horse in its reproductive competition with the rest of its population. So the component

> populations of the Tertiary horse-fauna are presumed to have been evolving larger size most of the time and on the average. It is conceivable, however, that precisely the opposite is true. It may be that at any given moment during the Tertiary, most of the horse populations were evolving a smaller size. To account for the trend towards larger size it is merely necessary to make the additional assumption that group selection favoured such a tendency. Thus, while only a minority of the populations may have been evolving a larger size, it could have been this minority that gave rise to most of the populations of a million years later.

I do not find it hard to believe that some of the major macroevolutionary trends, of the Cope's Rule type (but see Hallam 1978), observed by palaeontologists are due to species selection, in the sense of this passage from Williams, which I think is the same as Eldredge and Gould's sense. As I believe all three authors would agree, this is a very different matter from accepting group selection as an explanation for individual self-sacrifice: adaptations for the good of the species. There we are talking about another kind of group selection model, where the group is really seen not as a replicator but as a vehicle for replicators. I shall come to this second kind of group selection later. Meanwhile I shall argue that a belief in the power of species selection to shape simple major trends is not the same as a belief in its power to put together complex adaptations such as eyes and brains.

The major trends of palaeontology, simple increases in absolute size, or of relative sizes of different parts of the body, are important and interesting, but they are, above all, simple. Accepting Eldredge and Gould's belief that natural selection is a general theory that can be phrased on many levels, the putting together of a certain quantity of evolutionary change demands a certain minimum number of selective replicator-eliminations. Whether the replicators that are selectively eliminated are genes or species, a simple evolutionary change requires only a few replicator substitutions. A large number of replicator substitutions, however, are needed for the evolution of a complex adaptation. The minimum replacement cycle time when we consider the gene as replicator is one individual generation, from zygote to zygote. It is measured in years or months, or smaller time units. Even in the largest organisms it is measured in only tens of years. When we consider the species as replicator, on the other hand, the replacement cycle time is the interval from speciation event to speciation event, and may be measured in thousands of years, tens of thousands, hundreds of thousands. In any given period of geological time, the number of selective species extinctions that can have taken place is many orders of magnitude less than the number of selective allele replacements that can have taken place.

Again it may be just a limitation of the imagination, but while I can easily see species selection shaping a simple size trend such as the Tertiary elongation of horse legs, I cannot see such slow replicator-elimination putting together a suite of adaptations such as those of whales to aquatic life. Now, it may be said, surely that is unfair. However complex the aquatic adaptations of whales may be when taken as a whole, can they not be broken down into a set of simple size trends with allometric constants of varying magnitude and sign in different parts of the body? If you can stomach one-dimensional elongation in horses' legs as due to species selection, why not a whole set of equally simple size trends advancing in parallel, and each one driven by species selection? The weakness in this argument is a statistical one. Suppose there are ten such parallel trends, which is surely a highly conservative estimate for the evolution of aquatic adaptations in whales. If Wright's Rule is to apply to all ten, in any one speciation event each of the ten trends is as likely to be reversed as to be advanced. The chance that all ten will be advanced in any one speciation event is one half to the power ten, which is less than one in a thousand. If there are twenty parallel trends, the chance of any one speciation event advancing all twenty simultaneously is less than one in a million.

Admittedly, some advancement towards complex multidimensional adaptation might be achieved through species selection even if not all ten (or twenty) trends were advanced together in any one selective event. After all, much the same critical point can be made about the selective deaths of individual organisms: it is rare to find one individual animal that is optimal in all the different dimensions of measurement. The argument finally returns to the difference in cycle times. We shall have to make a quantitative judgement taking into account the vastly greater cycle time between replicator deaths in the species selection case than in the gene selection case, and also taking into account the combinatorial problem raised above. I have neither the data nor the mathematical skills to undertake this quantitative judgement, though I have a dim feeling for the kind of methodology that would be involved in setting up an appropriate null hypothesis: it would fall under the general heading which I like to think of as 'What D'Arcy Thompson might have done with a computer', and programs of the appropriate type have already been written (Raup *et al.* 1973). My provisional guess is that species selection will not be found to be a generally satisfactory explanation of complex adaptation.

Consider another line of argument that the species selectionist might adopt. He might protest that it is unreasonable of me to base my combinatorial calculation on the assumption that the ten major trends in the evolution of whales are independent. Surely the number of combinations we have to consider will be drastically reduced by correlations among the different trends? It is important here to distinguish two different sources of correlation, which may be called incidental correlation and adaptive correlation. An incidental correlation is an intrinsic consequence of the facts of embryology. For example, elongation of the left foreleg is unlikely to be independent of elongation of the right foreleg. Any mutation that achieves one is intrinsically likely to achieve the other simultaneously. Slightly less obviously, the same might be partially true of elongation of the forelegs and the hindlegs. There are probably many similar cases that are less obvious still.

An adaptive correlation, on the other hand, does not follow directly from the mechanics of embryology. A lineage that moves from a terrestrial to an aquatic existence is likely to need changes to its locomotory system and its respiratory system, and there is no obvious reason to expect any intrinsic connection between them. Why *should* trends that convert walking limbs to flippers be intrinsically correlated with trends to boost the efficiency of the lungs in extracting oxygen? Of course it is *possible* that two such coadapted trends might be correlated as an incidental consequence of embryological mechanisms, but the correlation is no more likely to be positive than negative. We are back to our combinatorial calculation, albeit we must exercise care in counting our separate dimensions of change.

Finally, the species selectionist may retreat and invoke ordinary low-level natural selection to weed out ill-coadapted combinations of change, so that speciation events only serve up already tried and proved combinations to the sieve of species selection. But this 'species selectionist' is, by Gould's lights, no species selectionist at all! He has conceded that all the interesting evolutionary change results from inter-allele selection and not from inter-species selection, albeit it may be concentrated in brief bursts, punctuating stasis. He has conceded the violation of

Wright's Rule. And if 'Wright's Rule' now seems unfairly easy to violate, that is what I meant when I said that Gould had stuck his neck out. I should repeat that Wright himself is not responsible for naming the rule.

The theory of species selection, growing out of that of punctuated equilibria, is a stimulating idea which may well explain some single dimensions of quantitative change in macroevolution. I would be very surprised if it could be used to explain the sort of complex multidimensional adaptation that I find interesting, the 'Paley's watch', or 'Organs of extreme perfection and complication', kind of adaptation that seems to demand a shaping agent at least as powerful as a deity. Replicator selection, where the replicators are alternative alleles, may well be powerful enough. If the replicators are alternative *species*, however, I doubt if it is powerful enough, because it is too slow. Eldredge and Cracraft (1980, p. 269) appear to agree: 'The concept of natural selection (fitness differences, or differential reproduction of individuals within populations) appears to be a corroborated, within-population phenomenon, and constitutes the best available explanation for the origin, maintenance, and possible modification of adaptations.' If this is indeed the opinion of 'punctuationists' and 'species selectionists' generally, it is hard to see what all the fuss is about.

For simplicity I have discussed the theory of species selection as one in which the species is treated as a replicator. The reader will have noticed, however, that this is rather like speaking of an asexually reproducing organism as a replicator. Earlier in this chapter, we saw that the test of mutilation forces us strictly to limit the title of replicator to the *genome* of, say, a stick insect, not the stick insect itself. Similarly, in the species selection model, it is not the species that is the replicator but the *gene-pool*. It is tempting, now, to say, 'In that case, why not go the whole hog and regard the gene as replicator rather than some larger unit, even in the Eldredge/Gould model?' The answer is that if they are right about a gene-pool being a coadapted unit, homeostatically buffered against change, it might have the same kind of right to be treated as a single replicator as has the genome of a stick insect. The gene-pool has this right, however, only if it is reproductively isolated, just as the genome has that right only if it is reproduced asexually. Even then the right is a tenuous one.

Earlier in this chapter, we established that an organism is definitely not a replicator, although its genome may be if it is asexually reproduced. We have now seen that there may be a case for regarding the gene-pool of a reproductively isolated group, such as a species, as a replicator. If we provisonally accept the logic of this case, we can visualize evolution directed by selection among such replicators, but I have just concluded that this kind of selection is unlikely to explain complex adaptation. Apart from the small genetic fragment, which we discussed in the previous chapter, are there any other plausible candidates for the title of replicator?

I have previously supported the case for a completely non-genetic kind of replicator, which flourishes only in the environment provided by complex, communicating brains. I called it the 'meme' (Dawkins 1976a). Unfortunately, unlike Cloak (1975) but, if I understand them aright, like Lumsden and Wilson (1980), I was insufficiently clear about the distinction between the meme itself, as replicator, on the one hand, and its 'phenotypic effects' or 'meme products' on the other. A meme should be regarded as a unit of information residing in a brain (Cloak's 'i-culture'). It has a definite structure, realized in whatever physical medium the brain uses for storing information. If the brain stores information as a pattern of synaptic connections, a meme should in principle be visible under a microscope as a definite pattern of synaptic structure. If the brain stores information in 'distributed' form (Pribram 1974), the meme would not be localizable on a microscope slide, but still I would want to regard it as physically residing in the brain. This is to distinguish it from its phenotypic effects, which are its consequences in the outside world (Cloak's 'm-culture').

The phenotypic effects of a meme may be in the form of words, music, visual images, styles of clothes, facial or hand gestures, skills such as opening milk bottles in tits, or panning wheat in Japanese macaques. They are the outward and visible (audible, etc.) manifestations of the memes within the brain. They may be perceived by the sense organs of other individuals, and they may so imprint themselves on the brains of the receiving individuals that a copy (not necessarily exact) of the original meme is graven in the receiving brain. The new copy of the meme is then in a position to broadcast its phenotypic effects, with the result that further copies of itself may be made in yet other brains.

Returning, for clarification, to DNA as our archetypal replicator, its consequences on the world are of two important types. Firstly, it makes copies of itself, making use of the cellular apparatus of replicases, etc. Secondly, it has effects on the outside world, which influence the chances of its copies' surviving. The first of these two effects corresponds to the meme's use of the apparatus of inter-individual communication and imitation to make copies of itself. If individuals live in a social climate in which imitation is common, this corresponds to a cellular climate rich in enzymes for copying DNA.

But what about the second kind of effect of DNA, the kind conventionally called 'phenotypic'? How do a meme's phenotypic effects contribute to its success or failure in being replicated? The answer is the same as for the genetic replicator. Any effect that a meme has on the behaviour of a body bearing it may influence that meme's chance of surviving. A meme that made its bodies run over cliffs would have a fate like that of a gene for making its bodies run over cliffs. It would tend to be eliminated from the meme-pool. But just as promoting bodily survival is only part of what constitutes success in genetic replicators, so there are many other ways in which memes may work phenotypically for their own preservation. If the phenotypic effect of a meme is a tune, the catchier it is the more likely it is to be copied. If it is a scientific idea, its chances of spreading through the world's scientific brains will be influenced by its compatibility with the already established corpus of ideas. If it is a political or religious idea, it may assist its own survival if one of its phenotypic effects is to make its bodies violently intolerant of new and unfamiliar ideas. A meme has its own opportunities for replication, and its own phenotypic effects, and there is no reason why success in a meme should have any connection whatever with genetic success.

This is regarded by many of my biological correspondents as the weakest point in the whole meme theory (Greene 1978; Alexander 1980, p. 78; Staddon 1981). I don't see the problem; or, rather, I do see the problem but I don't think it is any greater for memes as replicators than it is for genes. Time and again, my sociobiological colleagues have upbraided me as a turncoat, because I will not agree with them that the *ultimate* criterion for the success of a meme must be its contribution to Darwinian 'fitness'. At bottom, they insist, a 'good meme' spreads because brains are receptive to it, and the receptiveness of brains is ultimately shaped by (genetic) natural selection. The very fact that animals imitate other animals at all must ultimately be explicable in terms of their Darwinian fitness.

But there is nothing magic about Darwinian fitness in the genetic sense. There is no law giving it priority as the fundamental quantity that is maximized. Fitness is just a way of talking about the survival of replicators, in this case genetic replicators. If another kind of entity arises, which answers to the definition of an active germ-line replicator, variants of the new replicator that work for their own survival will tend to become more numerous. To be consistent, we could invent a new kind of 'individual fitness', which measured the success of an individual in propagating his memes.

It is, of course, true that 'Memes are utterly dependent upon genes, but genes can exist and change quite independently of memes' (Bonner 1980). But this does not mean that the ultimate criterion for success in meme selection is gene survival. It does not mean that success goes to those memes that favour the genes of the individuals bearing them. To be sure, this will sometimes be so. Obviously a meme that causes individuals bearing it to kill themselves has a grave disadvantage, but not necessarily a fatal one. Just as a gene for suicide sometimes spreads itself by a roundabout route (e.g. in social insect workers, or parental sacrifice), so a suicidal meme can spread, as when a dramatic and well-publicized martyrdom inspires others to die for a deeply loved cause, and this in turn inspires others to die, and so on (Vidal 1955).

It is true that the relative survival success of a meme will depend critically on the social and biological climate in which it finds itself, and this climate will certainly be influenced by the genetic make-up of the population. But it will also depend on the memes that are already numerous in the meme-pool. Genetic evolutionists are already happy with the idea that the relative success of two alleles can depend upon which genes at other loci dominate the gene-pool, and I have already mentioned this in connection with the evolution of 'coadapted genomes'. The statistical structure of the gene-pool sets up a climate or environment which affects the success of any one gene relative to its alleles. Against one genetic background one allele may be favoured; against another genetic background its allele may be favoured. For example, if the gene-pool is dominated by genes that make animals seek dry places, this will set up selection pressures in favour of genes for an impermeable skin. But alleles for a more permeable skin will be favoured if the gene-pool happens to be dominated by genes for seeking damp places. The point is the obvious one that selection at any one locus is not independent of selection at other loci. Once a lineage begins evolving in a particular direction, many loci will fall into step, and the resulting positive feedbacks will tend to propel the lineage in the same direction, in spite of pressures from the outside world. An important aspect of the environment which selects between alleles at any one locus will be the genes that already dominate the gene-pool at other loci.

Similarly, an important aspect of selection on any one meme will be the other memes that already happen to dominate the meme-pool (Wilson 1975). If the society is already dominated by Marxist, or Nazi memes, any new meme's replication success will be influenced by its compatibility with this existing background. Positive feedbacks will provide a momentum which can carry meme-based evolution in directions unconnected with, or even contradictory to, the directions that would be favoured by gene-based evolution. I agree with Pulliam and Dunford (1980) that cultural evolution 'owes its origin and its rules to genetic evolution, but it has a momentum all its own'.

There are, of course, significant differences between meme-based and gene-based selection processes (Cavalli-Sforza & Feldman 1973, 1981). Memes are not strung out along linear chromosomes, and it is not clear that they occupy and compete for discrete 'loci', or that they have identifiable 'alleles'. Presumably, as in the case of genes, we can strictly only talk about phenotypic effects in terms of differences, even if we just mean the difference between the behaviour produced by a brain containing the meme and that of a brain not containing it. The copying process is probably much less precise than in the case of genes: there may be a certain 'mutational' element in every copying event, and this, by the way, is also true of the 'species selection' discussed earlier in the chapter. Memes may partially blend with each other in a way that genes do not. New 'mutations' may be 'directed' rather than random with respect to evolutionary trends. The equivalent of Weismannism is less rigid for memes than for genes: there may be 'Lamarckian' causal arrows leading from phenotype to replicator, as well as the other way around. These differences may prove sufficient to render the analogy with genetic natural selection worthless or even positively misleading. My own feeling is that its main value may lie not so much in helping us to understand human culture as in sharpening our perception of genetic natural selection. This is the only reason I am presumptuous enough to discuss it, for I do not known enough about the existing literature on human culture to make an authoritative contribution to it.

Whatever the claims of memes to be regarded as replicators in the same sense as genes, the first part of this chapter established that individual organisms are not replicators. Nevertheless, they are obviously functional units of great importance, and it is now necessary to establish exactly what their role is. If the organism is not a replicator, what is it? The answer is that it is a communal *vehicle* for replicators. A vehicle is an entity in which replicators (genes and memes) travel about, an entity whose attributes are affected by the replicators inside it, an entity which may be seen as a compound tool of replicator propagation. But individual organisms are not the only entities that might be regarded as vehicles in this sense. There is a hierarchy of entities embedded in larger entities, and in theory the concept of vehicle might be applied to any level of the hierarchy.

The concept of hierarchy is a generally important one. Chemists believe that matter is made of about a hundred different kinds of atoms, interacting with each other by means of their electrons. Atoms are gregarious, forming huge assemblages which are governed by laws at their own level. Without contradicting the laws of chemistry, therefore, we find it convenient to ignore atoms when we are thinking about large lumps of matter. When explaining the workings of a motor car we forget atoms and van der Waal's forces as units of explanation, and prefer to talk of cylinders and sparking plugs. This lesson applies not just to the two levels of atoms and cylinder heads. There is a hierarchy, ranging from fundamental particles below the atomic level up through molecules and crystals to the macroscopic chunks which our unaided sense organs are built to appreciate.

Living matter introduces a whole new set of rungs to the ladder of complexity: macromolecules folding themselves into their tertiary forms, intracellular membranes and organelles, cells, tissues, organs, organisms, populations, communities and ecosystems. A similar hierarchy of units embedded in larger units epitomizes the complex artificial products of living things—semiconductor crystals, transistors, integrated circuits, computers, and embedded units that can only be understood in terms of 'software'. At every level the units interact with each other following laws appropriate to that level, laws which are not conveniently reducible to laws at lower levels.

This has all been said many times before, and is so obvious as to be almost platitudinous. But one sometimes has to repeat platitudes in order to prove that one's heart is in the right place! Especially if one wishes to emphasize a slightly unconventional sort of hierarchy, for this may be mistaken for a 'reductionist' attack on the idea of hierarchy itself. Reductionism is a dirty word, and a kind of 'holistier than thou' self-righteousness has become fashionable. I enthusiastically follow this fashion when talking about mechanisms within individual bodies, and have advocated 'neuro-economic' and 'software' explanations of behaviour in preference to conventional neurophysiological ones (Dawkins 1976b). I would favour an analogous approach to individual development. But there are times when holistic preaching becomes an easy substitute for thought, and I believe the dispute about units of selection provides examples of this.

The neo-Weismannist view of life which this book advocates lays stress on the genetic replicator as a fundamental unit of explanation. I believe it has an atom-like role to play in functional, teleonomic explanation. If we wish to speak of adaptations as being 'for the good of' something, that something is the active, germ-line replicator. This is a small chunk of DNA, a single 'gene' according to some definitions of the word. But I am of course not suggesting that small genetic units work in isolation from each other, any more than a chemist thinks that atoms do. Like atoms, genes are highly gregarious. They are often strung together along chromosomes, chromosomes are wrapped up in groups in nuclear membranes, enveloped in cytoplasm and enclosed in cell membranes. Cells too are normally not isolated, but cloned to form the huge conglomerates we know as organisms. We are now plugged into the familiar embedded hierarchy, and need go no further. Functionally speaking, too, genes are gregarious. They have phenotypic effects on bodies, but they do not do so in isolation. I stress this over and over again in this book.

The reason I may sound reductionistic is that I insist on an atomistic view of units of selection, in the sense of the units that actually survive or fail to survive, while being whole-heartedly interactionist when it comes to the development of the phenotypic *means* by which they survive:

> Of course it is true that the phenotypic effect of a gene is a meaningless concept outside the context of many, or even all, of the other genes in the genome. Yet, however complex and intricate the organism may be, however much we may agree that the organism is a unit of *function*, I still think it is misleading to call it a unit of *selection*. Genes may interact, even 'blend', in their effects on embryonic development, as much as you please. But they do not blend when it comes to being passed on to future generations. I am not trying to belittle the importance of the individual phenotype in evolution. I am merely trying to sort out exactly what its role is. It is the all important instrument of replicator preservation: it is *not* that which is preserved [Dawkins 1978a, p. 69].

In this book I am using the word 'vehicle' for an integrated and coherent 'instrument of replicator preservation'.

A vehicle is any unit, discrete enough to seem worth naming, which houses a collection of replicators and which works as a unit for the preservation and propagation of those replicators. I repeat, a vehicle is not a replicator. A replicator's success is measured by its capacity to survive in the form of copies. A vehicle's success is measured by its capacity to propagate the replicators that ride inside it. The obvious and archetypal vehicle is the individual organism, but this may not be the only level in the hierarchy of life at which the title is applicable. We can examine as candidate vehicles chromosomes and cells below the organism level, groups and communities above it. At any level, if a vehicle is destroyed, all the replicators inside it will be destroyed. Natural selection will therefore, at least to some extent, favour replicators that cause their vehicles to resist being destroyed. In principle this could apply to groups of organisms as well as to single organisms, for if a group is destroyed all the genes inside it are destroyed too.

Vehicle *survival* is only part of the story, however. Replicators that work for the 'reproduction' of vehicles at various levels might tend to do better than rival replicators that work merely for vehicle survival. Reproduction at the organism level is familiar enough to need no further discussion. Reproduction at the group level is more problematic. In principle a group may be said to have reproduced if it sends off a 'propagule', say a band of young organisms who go out and found a new group. The idea of a nested hierarchy of levels at which selection might take place—vehicle selection in my terms—is emphasized in Wilson's (1975) chapter on group selection (e.g. his figure 5-1).

I have previously given reasons for sharing in the general scepticism about 'group selection' and selection at other high levels, and nothing in the recent literature tempts me to change my mind. But that is not the point at issue here. The point here is that we must be clear about the difference between those two distinct kinds of conceptual units, replicators and vehicles. I have suggested that the best way of understanding Eldredge and Gould's theory of 'species selection' is in terms of species as *replicators*. But the majority of models ordinarily called 'group selection', including all those reviewed by Wilson (1975), and most of those reviewed by Wade (1978), are implicitly treating groups as vehicles. The end result of the selection discussed is a change in gene frequencies, for example an increase of 'altruistic genes' at the expense of 'selfish genes'. It is still genes that are regarded as the replicators which actually survive (or fail to survive) as a consequence of the (vehicle) selection process.

As for group selection itself, my prejudice is that it has soaked up more theoretical ingenuity than its biological interest warrants.

I am informed by the editor of a leading mathematics journal that he is continually plagued by ingenious papers purporting to have squared the circle. Something about the fact that this has been proved to be impossible is seen as an irresistible challenge by a certain type of intellectual dilettante. Perpetual motion machines have a similar fascination for some amateur inventors. The case of group selection is hardly analogous: it has never been proved to be impossible, and never could be. Nevertheless, I hope I may be forgiven for wondering whether part of group selection's enduring romantic appeal stems from the authoritative hammering the theory has received ever since Wynne-Edwards (1962) did us the valuable service of bringing it out into the open. Anti-group selectionism has been embraced by the establishment as orthodox, and, as Maynard Smith (1976a) notes, 'It is in the nature of science that once a position becomes orthodox it should be subjected to criticism'. This is, no doubt, healthy, but Maynard Smith drily goes on: 'It does not follow that, because a position is orthodox, it is wrong'. More generous recent treatments of group selection are given by Gilpin (1975), E. O. Wilson (1975), Wade (1978), Boorman and Levitt (1980), and D. S. Wilson (1980, but see criticism by Grafen 1980).

I am not going to go again into the debate over group selection versus individual selection. This is because the main purpose of this book is to draw attention to the weaknesses of the whole vehicle concept, whether the vehicle is an individual organism or a group. Since even the staunchest group selectionist would agree that the individual organism is a far more coherent and important 'unit of selection', I shall concentrate my attack on the individual organism as my representative vehicle, rather than on the group. The case against the group should be strengthened by default.

It may seem that I have invented my own concept, the vehicle, as an Aunt Sally to be easily knocked down. This is not so.

I simply use the name vehicle to give expression to a concept which is fundamental to the predominant orthodox approach to natural selection. It is admitted that, in some fundamental sense, natural selection consists in the differential survival of genes (or larger genetic replicators). But genes are not naked, they work through bodies (or groups, etc.). Although the ultimate unit of selection may indeed be the genetic replicator, the proximal unit of selection is usually regarded as something larger, usually an individual organism. Thus Mayr (1963) devotes a whole chapter to demonstrating the functional coherence of the whole genome of an individual organism. I shall discuss Mayr's points in detail in Chapter 13. Ford (1975, p. 185) disdainfully writes off the 'error' that 'the unit of selection is the gene, whereas it is the individual'. Gould (1977b) says:

> Selection simply cannot see genes and pick among them directly. It must use bodies as an intermediary. A gene is a bit of DNA hidden within a cell. Selection views bodies. It favors some bodies because they are stronger, better insulated, earlier in their sexual maturation, fiercer in combat, or more beautiful to behold.... If, in favoring a stronger body, selection acted directly upon a gene for strength, then Dawkins might be vindicated. If bodies were unambiguous maps of their genes, then battling bits of DNA would display their colors externally and selection might act upon them directly. But bodies are no such thing.... Bodies cannot be atomized into parts, each constructed by an individual gene. Hundreds of genes contribute to the building of most body parts and their action is channeled through a kaleidoscopic series of environmental influences: embryonic and postnatal, internal and external.

Now if this were really a good argument, it would be an argument against the whole of Mendelian genetics, just as much as against the idea of the gene as the unit of selection. The

Lamarckian fanatic H. G. Cannon, indeed, explicitly uses it as such: 'A living body is not something isolated, neither is it a collection of parts as Darwin envisaged it, like, as I have said before, so many marbles in a box. That is the tragedy of modern genetics. The devotees of the neo-Mendelian hypothesis regard the organism as so many characters controlled by so many genes. Say what they like about polygenes—that is the essence of their fantastic hypothesis' (Cannon 1959, p. 131).

Most people would agree that this is not a good argument against Mendelian genetics, and no more is it a good argument against treating the gene as the unit of selection. The mistake which both Gould and Cannon make is that they fail to distinguish genetics from embryology. Mendelism is a theory of particulate inheritance, not particulate embryology. The argument of Cannon and Gould is a valid argument against particulate embryology and in favour of blending embryology. I myself give similar arguments elsewhere in this book (e.g. the cake analogy in the section of Chapter 9 called 'The poverty of preformationism'). Genes do indeed blend, as far as their effects on developing phenotypes are concerned. But, as I have already emphasized sufficiently, they do *not* blend as they replicate and recombine down the generations. It is this that matters for the geneticist, and it is also this that matters for the student of units of selection.

Gould goes on:

> So parts are not translated genes, and selection doesn't even work directly on parts. It accepts or rejects entire organisms because suites of parts, interacting in complex ways, confer advantages. The image of individual genes, plotting the course of their own survival, bears little relationship to developmental genetics as we understand it. Dawkins will need another metaphor: genes caucusing, forming alliances, showing deference for a chance to join a pact, gauging probable environments. But

> when you amalgamate so many genes and tie them together in hierarchical chains of action mediated by environments, we call the resultant object a body.

Gould comes closer to the truth here, but the truth is subtler, as I hope to show in Chapter 13. The point was alluded to in the previous chapter. Briefly, the sense in which genes may be said to 'caucus' and form 'alliances' is the following. Selection favours those genes which succeed *in the presence of other genes, which in turn succeed in the presence of them.* Therefore mutually compatible sets of genes arise in gene-pools. This is more subtle and more useful than to say that 'we call the resultant object a body'.

Of course genes are not directly visible to selection. *Obviously* they are selected by virtue of their phenotypic effects, and certainly they can only be said to *have* phenotypic effects in concert with hundreds of other genes. But it is the thesis of this book that we should not be trapped into assuming that those phenotypic effects are best regarded as being neatly wrapped up in discrete bodies (or other discrete vehicles). The doctrine of the extended phenotype is that the phenotypic effect of a gene (genetic replicator) is best seen as an effect upon the world at large, and only incidentally upon the individual organism—or any other vehicle—in which it happens to sit.

SELFISH WASP OR SELFISH STRATEGY?

This is a chapter about practical research methodology. There will be those who accept the thesis of this book at a theoretical level but who will object that, in practice, field workers find it useful to focus their attention on *individual* advantage. In a theoretical sense, they will say, it is right to see the natural world as a battleground of replicators, but in real research we are obliged to measure and compare the Darwinian fitness of individual organisms. I want to discuss a particular piece of research in detail to show that this is not necessarily the case. Instead of comparing the success of individual organisms, it is often in practice more useful to compare the success of 'strategies' (Maynard Smith 1974) or 'programs' or 'subroutines', averaged across the individuals that use them. Of the many pieces of research that I could have discussed, for instance the work on 'optimal foraging' (Pyke, Pulliam & Charnov 1977; Krebs 1978), Parker's (1978a) dungflies, or any of the examples reviewed by Davies (1982), I choose Brockmann's study of digger wasps purely because I am very familiar with it (Brockmann, Grafen & Dawkins 1979; Brockmann & Dawkins 1979; Dawkins & Brockmann 1980).

I shall use the word 'program' in exactly the same sense as Maynard Smith uses 'strategy'. I prefer 'program' to 'strategy' because experience warns that 'strategy' is quite likely to be misunderstood, in at least two different ways (Dawkins 1980). And incidentally, following the *Oxford English Dictionary* and standard American usage, I prefer 'program' to 'programme' which appears to be a nineteenth-century affectation imported from the French. A program (or strategy) is a recipe for action, a set of notional instructions that an animal seems to be 'obeying', just as a computer obeys its program. A computer programmer writes out his program in a language such as Algol or Fortran, which may look rather like imperative English. The machinery of the computer is so set up that it behaves as if obeying these quasi-English instructions. Before it can run, the program is translated (by computer) into a set of more elementary 'machine language' instructions, closer to the hardware and further from easy human comprehension. In a sense it is these machine language instructions that are 'actually' obeyed rather than the quasi-English program, although in another sense both are obeyed, and in yet another sense neither is!

A person watching and analysing the behaviour of a computer whose program had been lost might, in principle, be able to reconstruct the program or its functional equivalent. The last four words are crucial. For his own convenience he will write the reconstructed program in some particular language—Algol, Fortran, a flow chart, some particular rigorous subset of English. But there is no way of knowing in which, if any, of these languages the program was originally written. It may have been directly written in machine language, or 'hard-wired' into the machinery of the computer at manufacture. The end result is the same in any case: the computer performs some useful task such as calculating square roots, and a human can usefully treat the computer *as if* it was 'obeying' a set of imperative instructions written out in a language that is convenient for humans to understand. I think that for many purposes such 'software explanations' of behaviour mechanisms are just as valid and useful as the more obvious 'hardware explanations' favoured by neurophysiologists.

A biologist looking at an animal is in somewhat the same position as an engineer looking at a computer running a lost program. The animal is behaving in what appears to be an organized, purposeful way, as if it was obeying a program, an orderly sequence of imperative instructions. The animal's program has not actually been lost, for it never was written out. Rather, natural selection cobbled together the equivalent of a hard-wired machine code program, by favouring mutations that altered successive generations of nervous systems to behave (and to learn to change their behaviour) in appropriate ways. Appropriate means, in this case, appropriate for the survival and propagation of the genes concerned. Nevertheless, although no program was ever written down, just as in the case of the computer running a program which has been lost, it is convenient for us to think of the animal as 'obeying' a program 'written' in some easily understood language such as English. One of the things we can then do is to imagine alternative programs or subroutines which might 'compete' with each other for 'computer time' in the nervous systems of the population. Though we must treat the analogy with circumspection, as I shall show, we can usefully imagine natural selection as acting directly on a pool of alternative programs or subroutines, and treat individual organisms as temporary executors and propagators of these alternative programs.

In a particular model of animal fighting, for example, Maynard Smith (1972, p. 19) postulated five alternative 'strategies' (programs):

1 Fight conventionally; retreat if opponent proves to be stronger or if opponent escalates.
2 Fight at escalated level. Retreat only if injured.

3 Start conventionally. Escalate only if opponent escalates.
4 Start conventionally. Escalate only if opponent continues to fight conventionally.
5 Fight at escalated level. Retreat before getting hurt if opponent does likewise.

For the purpose of computer simulation it was necessary to define these five 'strategies' more rigorously, but for human understanding the simple imperative English notation is preferable. The important point for this chapter is that the five strategies were thought of as if they (rather than individual animals) were competing entities in their own right. Rules were set up in the computer simulation for the 'reproduction' of successful strategies (presumably individuals adopting successful strategies reproduced and passed on genetic tendencies to adopt those same strategies, but the details of this were ignored). The question asked was about strategy success, not individual success.

A further important point is that Maynard Smith was seeking the 'best' strategy in only a special sense. In fact he was seeking an 'evolutionarily stable strategy' or ESS. The ESS has been rigorously defined (Maynard Smith 1974), but it can be crudely encapsulated as a strategy that is successful when competing with copies of itself. This may seem an odd property to single out, but the rationale is really very powerful. If a program or strategy is successful, this means that copies of it will tend to become more numerous in the population of programs and will ultimately become almost universal. It will therefore come to be surrounded by copies of itself. If it is to remain universal, therefore, it must be successful when competing against copies of itself, successful compared with rare different strategies that might arise by mutation or invasion. A program that was not evolutionarily stable in this sense would not last long in the world, and would therefore not present itself for our explanation.

In the case of the five strategies listed above, Maynard Smith wanted to know what would happen in a population containing copies of all five programs. Was there one of the five which, if it came to predominate, would retain its numerical superiority against all comers? He concluded that program number 3 is an ESS: when it happens to be very numerous in the population no other program from the list does better than program 3 itself (actually there is a problem over this particular example—Dawkins 1980, p. 7—but I shall ignore it here). When we talk of a program as 'doing better' or as being 'successful' we are notionally measuring success as capacity to propagate copies of the same program in the next generation: in reality this is likely to mean that a successful program is one which promotes the survival and reproduction of the animal adopting it.

What Maynard Smith, together with Price and Parker (Maynard Smith & Price 1973; Maynard Smith & Parker 1976), has done is to take the mathematical theory of games and work out the crucial respect in which that theory must be modified to suit the Darwinist's purpose. The concept of the ESS, a strategy that does relatively well against copies of itself, is the result. I have already made two attempts at advocating the importance of the ESS concept and explaining its broad applicability in ethology (Dawkins 1976a, 1980), and I shall not repeat myself unnecessarily here. Here my purpose is to develop the relevance of this way of thinking for the subject of the present book, the debate about the level at which natural selection acts. I shall begin by recounting a specific piece of research that used the ESS concept. All the facts I shall give are from the field observations of Dr Jane Brockmann, reported in detail elsewhere and briefly mentioned in Chapter 3. I shall have to give a brief account of the research itself before I can relate it to the message of this chapter.

Sphex ichneumoneus is a solitary wasp, solitary in the sense that there are no social groups and no sterile workers, although the females do tend to dig their nests in loose aggregations. Each female lays her own eggs, and all labour on behalf of the young is completed before the egg is laid—the wasps are not 'progressive provisioners'. The female lays one egg in an underground nest which she has previously provisioned with stung and paralysed katydids (long-horned grasshoppers). Then she closes up that nest, leaving the larva to feed on the katydids, while she herself starts work on a new nest. The life of an adult female is limited to about six summer weeks. If one wished to measure it, the success of a female could be approximated as the number of eggs that she successfully lays on adequate provisions during this time.

The thing that especially interested us was that the wasps seemed to have two alternative ways of acquiring a nest. Either a female would dig her own nest in the ground, or she would attempt to take over an existing nest which another wasp had dug. We called these two behaviour patterns *digging* and *entering*, respectively. Now, how can two alternative ways of achieving the same end, in this case two alternative ways of acquiring a nest, coexist in one population? Surely one or the other ought to be more successful, and the less successful one should be removed from the population by natural selection? There are two general reasons why this might not happen, which I will express in the jargon of ESS theory: firstly, digging and entering might be two outcomes of one 'conditional strategy'; secondly, they might be equally successful at some critical frequency maintained by frequency-dependent selection—part of a 'mixed ESS' (Maynard Smith 1974, 1979). If the first possibility were correct, all wasps would be programmed with the same conditional rule: 'If X is true, dig, otherwise enter'; for instance, 'If you happen to be a small wasp, dig, otherwise use your superior size to take over another wasp's burrow'. We failed to find any evidence for a conditional program of this or any other kind. Instead, we convinced ourselves that the second possibility, the 'mixed ESS', fitted the facts.

There are in theory two kinds of mixed ESS, or rather two extremes with a continuum between. The first extreme is a balanced polymorphism. In this case, if we want to use the initials 'ESS', the final S should be thought of as standing for *state* of the population rather than strategy of the individual. If this possibility obtained, there would be two distinct kinds of wasps, diggers and enterers, who would tend to be equally successful. If they were not equally successful, natural selection would tend to eliminate the less successful one from the population. It is too much to hope that, by sheer coincidence, the net costs and benefits of digging would exactly balance the net costs and benefits of entering. Rather, we are invoking frequency-dependent selection. We postulate a critical equilibrium proportion of diggers, p^*, at which the two kinds of wasps are equally successful. Then, if the proportion of diggers in the population falls below the critical frequency, diggers will be favoured by selection, while if it rises above the critical frequency enterers will be favoured. In this way the population would hover around the equilibrium frequency.

It is easy to think of plausible reasons why benefit might be frequency-dependent in this way. Clearly, since new burrows come into existence only when diggers dig them, the fewer diggers there are in the population the stronger will be the competition among enterers for burrows, and the lower the benefit to a typical enterer. Conversely, when diggers are very numerous, available burrows abound and enterers tend to prosper. But, as I said, frequency-dependent polymorphism is only one end of a continuum. We now turn to the other end.

At the other end of the continuum there is no polymorphism among individuals. In the stable state all wasps obey the same program, but that program is itself a mixture. Every wasp is obeying the instruction, 'Dig with probability *p*, enter with probability $1-p$'; for instance, 'Dig on 70 per cent of occasions, enter on 30 per cent of occasions'. If this is regarded as a 'program', we could perhaps refer to digging and entering themselves as 'subroutines'. Every wasp is equipped with both subroutines. She is programmed to choose one or the other on each occasion with a characteristic probability, *p*.

Although there is no polymorphism of diggers and enterers here, something mathematically equivalent to frequency-dependent selection can go on. Here is how it would work. As before, there is a critical population frequency of digging, p^*, at which entering would yield exactly the same 'pay-off' as digging, p^* is, then, the evolutionarily stable digging probability. If the stable probability is 0.7, programs instructing wasps to follow a different rule, say 'Dig with probability 0.75', or 'Dig with probability 0.65', would do less well. There is a whole family of 'mixed strategies' of the form 'Dig with probability *p*, enter with probability $1-p$', and only one of these is the ESS.

I said that the two extremes were joined by a continuum. I meant that the stable population frequency of digging, p^* (70 per cent or whatever it is), could be achieved by any of a large number of combinations of pure and mixed individual strategies. There might be a wide distribution of *p* values in individual nervous systems in the population, including some pure diggers and pure enterers. But, provided the total frequency of digging in the population is equal to the critical value p^*, it would still be true that digging and entering were equally successful, and natural selection would not act to change the relative frequency of the two subroutines in the next generation. The population would be in an evolutionarily stable state. The analogy with Fisher's (1930a) theory of sex-ratio equilibrium will be clear.

Proceeding from the conceivable to the actual, Brockmann's data show conclusively that these wasps are not in any simple sense polymorphic. Individuals sometimes dug and sometimes entered. We could not even detect any statistical tendency for individuals to specialize in digging or entering. Evidently, if the wasp population is in a mixed evolutionarily stable state, it lies away from the polymorphism end of the continuum. Whether it is at the other extreme—all individuals running the same stochastic program—or whether there is some more complex mixture of pure and mixed individual programs, we do not know. It is one of the central messages of this chapter that, for our research purpose, we did not *need* to know. Because we refrained from talking about individual success, but thought instead about subroutine success averaged over all individuals, we were able to develop and test a successful model of the mixed ESS which left open the question of where along the continuum our wasps lay. I shall return to this point after giving some pertinent facts and after outlining the model itself.

When a wasp digs a burrow, she may either stay with it and provision it, or she may abandon it. Reasons for abandoning nests were not always obvious, but they included invasion by ants and other undesirable aliens. A wasp who moves into a burrow that another has dug may find that the original owner is still in residence. In this case she is said to have *joined* the previous owner, and the two wasps usually work on the same nest for a while, both independently bringing katydids to it. Alternatively, the entering wasp may be fortunate enough to hit upon a nest that has been abandoned by its original owner, in which case she has it to herself. The evidence indicates that entering wasps cannot distinguish a nest that has been abandoned from one that is still occupied by its previous owner. This fact is not so surprising as it may seem, since both wasps spend most of their time out hunting, so two wasps 'sharing' the same nest seldom meet. When they do meet they fight, and in any case only one of them succeeds in laying an egg in the nest under dispute.

Whatever had precipitated the abandoning of a nest by the original owner seemed usually to be a temporary inconvenience, and an abandoned nest constituted a desirable resource which was soon used by another wasp. A wasp who enters an abandoned burrow saves herself the costs associated with digging one. On the other hand, she runs the risk that the burrow she enters has not been abandoned. It may still contain the original owner, or it may contain another entering wasp who got there first. In either of these cases the entering wasp lets herself in for a high risk of a costly fight, and a high risk that she will not be the one to lay the egg at the end of a costly period of provisioning the nest.

We developed and tested a mathematical model (Brockmann, Grafen & Dawkins 1979) which distinguished four different 'outcomes' or fates that could befall a wasp in any particular nesting episode.

1 She could be forced to abandon the nest, say by an ant raid.
2 She could end up alone, in sole charge of the nest.
3 She could be joined by a second wasp.
4 She could join an already incumbent wasp.

Outcomes 1 to 3 could result from an initial decision to dig a burrow. Outcomes 2 to 4 could result from an initial decision to enter. Brockmann's data enabled us to measure, in terms of probability of laying an egg per unit time, the relative 'payoffs' associated with each of these four outcomes. For instance, in one study population in Exeter, New Hampshire, Outcome 4, 'joining', had a payoff score of 0.35 eggs per 100 hours. This score was obtained by averaging over all occasions when wasps ended up in that outcome. To calculate it we simply added up the total number of eggs laid by wasps who, on the occasion concerned, had joined an already incumbent wasp, and divided by the total time spent by wasps on nests that they had joined. The corresponding score for wasps who began alone but were subsequently joined was 1.06 eggs per 100 hours, and that for wasps who remained alone was 1.93 eggs per 100 hours.

If a wasp could control which of the four outcomes she ended up in, she should 'prefer' to end up alone since this outcome carried the highest payoff rate, but how might she achieve this? It was a key assumption of our model that the four outcomes did not correspond to decisions that were available to a wasp. A wasp can 'decide' to dig or to enter. She cannot decide to be joined or to be alone any more than a man can decide not to get cancer. These are outcomes that depend on circumstances beyond the individual's control. In this case they depend on what the other wasps in the population do. But just as a man may statistically reduce his chances of getting cancer, by taking the decision to stop smoking, so a wasp's 'task' is to make the only decision open to her—dig or enter—in such a way as to maximize her chances of ending up in a desirable outcome. More strictly, we seek the stable value of p, p^*, such that when p^* of the decisions in the population are digging decisions, no mutant gene leading to the adoption of some other value of p will be favoured by natural selection.

The probability that an entering decision will lead to some particular outcome, such as the desirable 'alone' outcome, depends on the overall frequency of entering decisions in the population. If a large amount of entering is going on in the population, the number of available abandoned burrows goes down, and the chances go up that a wasp that decides to enter will find herself in the undesirable position of joining an incumbent. Our model enables us to take any given value of *p*, the overall frequency of digging in the population, and predict the probability that an individual that decides to dig, or an individual that decides to enter, will end that episode in each of the four outcomes. The average payoff to a wasp that decides to dig, therefore, can be predicted for any named frequency of digging versus entering in the population as a whole. It is simply the sum, over the four outcomes, of the expected payoff yielded by each outcome, multiplied by the probability that a wasp that digs will end up in that outcome. The equivalent sum can be worked out for a wasp that decides to enter, again for any named frequency of digging versus entering in the population. Finally, making certain plausible additional assumptions which are listed in the original paper, we solve an equation to find the population digging frequency at which the average expected benefit to a wasp that digs is exactly equal to the average expected benefit to a wasp that enters. That is our predicted equilibrium frequency which we can compare with the observed frequency in the wild population. Our expectation is that the real population should either be sitting at the equilibrium frequency or else in the process of evolving towards the equilibrium frequency. The model also predicts the proportion of wasps ending up in each of the four outcomes at equilibrium, and these figures too can be tested against the observed data. The model's equilibrium is theoretically stable in that it predicts that deviations from equilibrium will be corrected by natural selection.

Brockmann studied two populations of wasps, one in Michigan and one in New Hampshire. The results were different in the two populations. In Michigan the model failed to predict the observed results, and we concluded that it was quite inapplicable to the Michigan population, for unknown reasons as discussed in the original paper (the fact that the Michigan population has now gone extinct is probably fortuitous!). The New Hampshire population, on the other hand, gave a convincing fit to the predictions of the model. The predicted equilibrium frequency of entering was 0.44, and the observed frequency was 0.41. The model also successfully predicted the frequency of each of the four 'outcomes' in the New Hampshire population. Perhaps most importantly, the average payoff resulting from digging decisions did not differ significantly from the average payoff resulting from entering decisions.

The point of my telling this story in the present book has now finally arrived. I want to claim that we would have found it difficult to do this research if we had thought in terms of *individual* success, rather than in terms of strategy (program) success averaged over all individuals. If the mixed ESS had happened to lie at the balanced polymorphism end of the continuum, it would, indeed, have made sense to ask something like the following. Is the success of wasps that dig equal to the success of wasps that enter? We would have classified wasps as diggers or enterers, and compared the total lifetime's egg-laying success of individuals of the two types, predicting that the two success scores should be equal. But as we have seen these wasps are not polymorphic. Each individual sometimes digs and sometimes enters.

It might be thought that it would have been easy to do something like the following. Classify all individuals into those that entered with a probability less than 0.1, those that entered with a probability between 0.1 and 0.2, those with a probability between 0.2 and 0.3, between 0.3 and 0.4, 0.4 and 0.5, etc. Then compare the lifetime reproductive successes of wasps in the different classes. But supposing we did this, exactly what would the ESS theory predict? A hasty first thought is that those wasps with a p value close to the equilibrium p^* should enjoy a higher success score than wasps with some other value of p: the graph of success against p should peak at an 'optimum' at p^*. But p^* is not really an optimum value, it is an evolutionarily stable value. The theory expects that, when p^* is achieved in the population as a whole, digging and entering should be equally successful. At equilibrium, therefore, we expect no correlation between a wasp's digging probability and her success. If the population deviates from equilibrium in the direction of too much entering, the 'optimum' choice rule becomes 'always dig' (not 'dig with probability p^*'). If the population deviates from equilibrium in the other direction, the 'optimum' policy is 'always enter'. If the population fluctuates at random around the equilibrium value, analogy with sex-ratio theory suggests that in the long run genetic tendencies to adopt exactly the equilibrium value, p^*, will be favoured over tendencies to adopt any other consistent value of p (Williams 1979). But in any one year this advantage is not particularly likely to be evident. The sensible expectation of the theory is that there should be no significant difference in success rates among the classes of wasps.

In any case this method of dividing wasps up into classes presupposes that there is some consistent variation among wasps in digging tendency. The theory gives us no particular reason to expect that there should be any such variation. Indeed, the analogy with sex-ratio theory just mentioned gives positive grounds for expecting that wasps should not vary in digging probability. In accordance with this, a statistical test on the actual data revealed no evidence of inter-individual variation in digging tendency. Even if there were some individual variation, the method of comparing the success of individuals with different p values would have been a crude and insensitive one for comparing the success rates of digging and entering. This can be seen by an analogy.

An agriculturalist wishes to compare the efficacy of two fertilizers, A and B. He takes ten fields and divides each of them into a large number of small plots. Each plot is treated, at random, with either A or B, and wheat is sown in all the plots of all the fields. Now, how should he compare the two fertilizers? The sensitive way is to take the yields of all the plots treated with A and compare them with the yields of all the plots treated with B, across all ten fields. But there is another, much cruder way. It happens that in the random allocation of fertilizers to plots, some of the ten fields received a relatively large amount of fertilizer A, while others happened to be given a relatively large amount of B. The agriculturalist could, then, plot the overall yield of each of the ten fields against the proportion of the field that was treated with fertilizer A rather than B. If there is a very pronounced difference in quality between the two fertilizers, this method might just show it up, but far more probably the difference would be masked. The method of comparing the yields of the ten fields would be efficient only if there is very high between-field variance, and there is no particular reason to expect this.

In the analogy, the two fertilizers stand for digging and entering. The fields are the wasps. The plots are the episodes of time that individual wasps devote either to digging or to entering. The crude method of comparing digging and entering is to plot the lifetime success of individual wasps against their proportionate digging tendency. The sensitive way is the one we actually employed.

We made a detailed and exhaustive inventory of the time spent by each wasp on each burrow with which she was associated. We divided each individual female's adult lifetime into consecutive episodes of known duration, each episode being designated a digging episode if the wasp concerned began her association with the burrow by digging it. Otherwise, it was designated an entering episode. The end of each episode was signalled by the wasp's leaving the nest for the last time. This instant was also treated as the start of the next burrow episode, even though the next burrow site had not, at the time, been chosen. That is to say, in our time accounting, the time spent searching for a new burrow to enter, or searching for a place to dig a new burrow, was designated retroactively as time 'spent' on that new burrow. It was added to the time subsequently spent provisioning the burrow with katydids, fighting other wasps, feeding, sleeping, etc., until the wasp left the new burrow for the last time.

At the end of the season, therefore, we could add up the total number of wasp hours spent on dug-burrow episodes, and also the total number of wasp hours spent on entered-burrow episodes. For the New Hampshire study these two figures were 8518.7 hours and 6747.4 hours, respectively. This is regarded as time spent, or invested, for a return, and the return is measured in numbers of eggs. The total number of eggs laid at the end of the dug-burrow episodes (i.e. by wasps that had dug the burrow concerned) in the whole New Hampshire population during the year of study was 82. The corresponding number for entered-burrow episodes was 57 eggs. The success rate of the digging subroutine was, therefore, 82/8518.7 = 0.96 eggs per 100 hours. The success rate of the entering subroutine was 57/6747.4 = 0.84 eggs per 100 hours. These success scores are averaged across all the individuals who used the two subroutines. Instead of counting the number of eggs laid in her lifetime by an individual wasp—the equivalent of measuring the wheat yield of each of the ten fields in the analogy—we count the number of eggs laid 'by' the digging (or entering) subroutine per unit 'running time' of the subroutine.

There is another respect in which it would have been difficult for us to have done this analysis if we had insisted on thinking in terms of individual success. In order to solve the equation to predict the equilibrium entering frequency, we had to have empirical estimates of the expected payoffs of each of the four 'outcomes' (abandons, remains alone, is joined, joins). We obtained payoff scores for the four outcomes in the same way as we obtained success scores for each of the two strategies, dig and enter. We averaged over all individuals, dividing the total number of eggs laid in each outcome by the total time spent on episodes that ended up in that outcome. Since most individuals experienced all four outcomes at different times, it is not clear how we could have obtained the necessary estimates of outcome payoffs if we had thought in terms of individual success.

Notice the important role of *time* in the computation of the 'success' of the digging and entering subroutines (and of the payoff given by each outcome). The total number of eggs laid 'by' the digging subroutine is a poor measure of success until it has been divided by the time spent on the subroutine. The number of eggs laid by the two subroutines might be equal, but if digging episodes are on average twice as long as entering episodes, natural selection will presumably favour entering. In fact rather more eggs were laid 'by' the digging subroutine than by the entering one, but correspondingly more time was spent on the digging subroutine so the overall success rates of the two were approximately equal. Notice too that we do not specify whether the extra time spent on digging is accounted for by a greater number of wasps digging, or by each digging episode lasting longer. The distinction may be important for some purposes, but it doesn't matter for the kind of economic analysis we undertook.

It was clearly stated in the original paper (Brockmann, Grafen & Dawkins 1979), and must be repeated here, that the method we used depended upon some assumptions. We assumed, for instance, that a wasp's choice of subroutine on any particular occasion did not affect her survival or success rate after the end of the episode concerned. Thus the costs of digging were assumed to be reflected totally in the time spent on digging episodes, and the costs of entering reflected in the time spent on entering episodes. If the act of digging had imposed some extra cost, say a risk of wear and tear to the limbs, shortening life expectation, our simple time-cost accounting would need to be amended. The success rates of the digging and the entering subroutines would have to be expressed, not in eggs per hour, but in eggs per 'opportunity cost'. Opportunity cost might still be measured in units of time, but digging time would have to be scaled up in a costlier currency than entering time, because each hour spent in digging shortens the expectation of effective life of the individual. Under such circumstances it might be necessary, in spite of all the difficulties, to think in terms of individual success rather than subroutine success.

It is for this kind of reason that Clutton-Brock *et al.* (1982) are probably wise in their ambition to measure the total lifetime reproductive success rates of their individual red deer stags. In the case of Brockmann's wasps, we have reason to think that our assumptions were correct, and that we were justified in ignoring individual success and concentrating on subroutine success. Therefore what N. B. Davies, in a lecture, jocularly called the 'Oxford method' (measuring subroutine success) and the 'Cambridge method' (measuring individual success) may each be justified in different circumstances. I am not saying that the Oxford method should always be used. The very fact that it is *sometimes* preferable is sufficient to answer the claim that field workers interested in measuring costs and benefits always have to think in terms of *individual* costs and benefits.

When computer chess tournaments are held, a layman might imagine that one computer plays against another. It is more pertinent to describe the tournament as being between programs. A good program will consistently beat a poor program, and it doesn't make any difference which physical computer either program is running on. Indeed the two programs could swap physical computers every other game, each one running alternately in an IBM and an ICL computer, and the result at the end of the tournament will be the same as if one program consistently ran in the IBM and the other consistently ran in the ICL. Similarly, to return to the analogy at the beginning of this chapter, the digging subroutine 'runs' in a large number of different physical wasp nervous systems. Entering is the name of a rival subroutine which also runs in many different wasp nervous systems, including some of the same physical nervous systems as, at other times, run the digging subroutine. Just as a particular IBM or ICL computer functions as the physical medium through which any of a variety of chess programs can act out their skills, so one individual wasp is the physical medium through which sometimes the digging subroutine, at other times the entering subroutine, acts out its characteristic behaviour.

As already explained, I call digging and entering 'subroutines', rather than programs, because we have already used 'program' for the overall lifetime choosing rules of an individual. An individual is regarded as being programmed with a rule for choosing the digging or the entering subroutine with some probability *p*. In the special case of a polymorphism, where each individual is either a lifelong digger or a lifelong enterer, *p* becomes 1 or 0, and the categories program and subroutine become synonymous. The beauty of calculating the egg-laying success rates of subroutines rather than of individuals is that the procedure we adopt is the same regardless of where on the mixed strategy continuum our animals are. Anywhere along the continuum we still predict that the digging subroutine should, at equilibrium, enjoy a success rate equal to that of the entering subroutine.

It is tempting, though rather misleading, to push this line of thought to what appears to be its logical conclusion, and think in terms of selection acting directly on subroutines in a subroutine pool. The population's nervous tissue, its distributed computer hardware, is inhabited by many copies of the digging subroutine and many copies of the entering subroutine. At any given time the proportion of running copies of the digging subroutine is *p*. There is a critical value of *p*, called p^*, at which the success rate of the two subroutines is equal. If either of the two becomes too numerous in the subroutine pool, natural selection penalizes it and the equilibrium is restored.

The reason this is misleading is that selection really works on the differential survival of alleles in a gene-pool. Even with the most liberal imaginable interpretation of what we mean by gene control, there is no useful sense in which the digging subroutine and the entering subroutine could be thought of as being controlled by alternative alleles. If for no other reason, this is because the wasps, as we have seen, are not polymorphic, but are programmed with a stochastic rule for choosing to dig or enter on any given occasion. Natural selection must favour genes that act on the stochastic program of individuals, in particular controlling the value of *p*, the digging probability. Nevertheless, although it is misleading if taken too literally, the model of subroutines competing directly for running time in nervous systems provides some useful short cuts to getting the right answer.

The idea of selection in a notional pool of subroutines also leads us to think about yet another time-scale on which an analogue of frequency-dependent selection might occur. The present model allows that from day to day the observed number of running copies of the digging subroutine might change, as individual wasps obeying their stochastic programs switch their hardware from one subroutine to another. So far I have implied that a given wasp is born with a built-in predilection to dig with a certain characteristic probability. But it is also theoretically possible that wasps might be equipped to monitor the population around them with their sense organs, and choose to dig or enter accordingly. In ESS jargon focusing on the individual level, this would be regarded as a conditional strategy, each wasp obeying an 'if clause' of the following form: 'If you see a large amount of entering going on around you, dig, otherwise enter.' More practically, each wasp might be programmed to follow a rule of thumb such as: 'Search for a burrow to enter; if you have not found one after a time *t*, give up and dig your own.' As it happens our evidence goes against such a 'conditional strategy' (Brockmann & Dawkins 1979), but the theoretical possibility is interesting. From the present point of view what is particularly interesting is this. We could still analyse the data in terms of a notional selection between subroutines in a subroutine pool, even though the selection process leading to the restoration of the equilibrium when perturbed would not be natural selection on a generational time-scale. It would be a developmentally stable strategy or DSS (Dawkins 1980) rather than an ESS, but the mathematics could be much the same (Harley 1981).

I must warn that analogical reasoning of this sort is a luxury that we dare not indulge unless we are capable of clearly seeing the limitations of the analogy. There are real and important distinctions between Darwinian selection and behavioural assessment, just as there were real and important distinctions between a balanced polymorphism and a true mixed evolutionarily stable strategy. Just as the value of *p*, the individual's digging probability, was considered to be adjusted by natural selection, so, in the behavioural assessment model, *t*, the individual's criterion for responding to the frequency of digging in the population, is presumably influenced by natural selection. The concept of selection among subroutines in a subroutine pool blurs some important distinctions while pointing up some important similarities: the weaknesses of this way of thinking are linked to its strengths. What I do remember is that, when we were actually wrestling with the difficulties of the wasp analysis, one of our main leaps forward occurred when, under the influence of A. Grafen, we kicked the habit of worrying about individual reproductive success and switched to an imaginary world where 'digging' competed directly with 'entering'; competed for 'running time' in future nervous systems.

This chapter has been an interlude, a digression. I have not been trying to argue that 'subroutines', or 'strategies' are really true replicators, true units of natural selection. They are not. Genes and fragments of genomes are true replicators. Subroutines and strategies can be thought of for certain purposes as if they were replicators, but when those purposes have been served we must return to reality. Natural selection really has the effect of choosing between alleles in wasp gene-pools, alleles which influence the probability that individual wasps will enter or dig. We temporarily laid this knowledge aside and entered an imaginary world of 'inter-subroutine selection' for a specific methodological purpose. We were justified in doing this because we were able to make certain assumptions about the wasps, and because of the already demonstrated mathematical equivalence between the various ways in which a mixed evolutionarily stable strategy can be put together.

As in the case of Chapter 4, the purpose of this chapter has been to undermine our confidence in the individual-centred view of teleonomy, in this case by showing that it is not always useful, in practice, to measure individual success if we are to study natural selection in the field. The next two chapters discuss adaptations which, by their very nature, we cannot even begin to understand if we insist on thinking in terms of individual benefit.

8

OUTLAWS AND MODIFIERS

Natural selection is the process whereby replicators out-propagate each other. They do this by exerting phenotypic effects on the world, and it is often convenient to see those phenotypic effects as grouped together in discrete 'vehicles' such as individual organisms. This gives substance to the orthodox doctrine that each individual body can be thought of as a unitary agent maximizing one quantity—'fitness', various notions of which will be discussed in Chapter 10. But the idea of individual bodies maximizing one quantity relies on the assumption that replicators at different loci within a body can be expected to 'cooperate'. In other words we must assume that the allele that survives best at any given locus tends to be the one that is best for the genome as a whole. This is indeed often the case. A replicator that ensures its own survival and propagation down the generations by conferring on its successive bodies resistance to a dangerous disease, say, will thereby benefit all the other genes in the successive genomes of which it is a member. But it is also easy to imagine cases where a gene might promote its own survival while harming the survival chances of most of the rest of the genome. Following Alexander and Borgia (1978) I shall call such genes *outlaws*.

I distinguish two main classes of outlaws. An 'allelic outlaw' is defined as a replicator that has a positive selection coefficient at its own locus but for which, at most other loci, there is selection in favour of reducing its effect at its own locus. An example is a 'segregation distorter' or 'meiotic drive' gene. It is favoured at its own locus through getting itself into more than 50 per cent of the gametes produced. At the same time, genes at other loci whose effect is to reduce the segregation distortion will be favoured by selection at their respective loci. Hence the segregation distorter is an outlaw. The other main class of outlaw, the 'laterally spreading outlaw', is less familiar. It will be discussed in the next chapter.

From the viewpoint of this book there is a sense in which we expect all genes to be potential outlaws, so much so that the very term might seem superfluous. On the other hand it can be argued, firstly, that outlaws are unlikely to be found in nature because at any given locus the allele that is best at surviving will nearly always turn out to be the allele that is best at promoting the survival and reproduction of the organism as a whole. Secondly various authors, following Leigh (1971), have argued that, even if outlaws did arise and were temporarily favoured by selection, they are likely, in Alexander and Borgia's words, 'to have their effects nullified, at least to the extent that they are outnumbered by the other genes in the genome'. It does, indeed, follow from the definition of an outlaw that this should tend to happen. The suggestion being made is that, whenever an outlaw arises, selection will favour modifier genes at so many other loci that no trace will remain of the phenotypic effects of the outlaw. Outlaws will therefore, the suggestion goes, be transient phenomena. This would not, however, make them negligible: if genomes are riddled with outlaw-suppressing genes, this, in itself, is an important effect of outlaw genes, even if no trace remains of their original phenotypic effects. I shall discuss the relevance of modifier genes in a later section.

There is a sense in which a 'vehicle' is worthy of the name in inverse proportion to the number of outlaw replicators that it contains. The idea of a discrete vehicle maximizing a unitary quantity—fitness—depends on the assumption that the replicators that it serves all stand to gain from the same properties and behaviour of their shared vehicle. If some replicators would benefit from the vehicle's doing act X, while other replicators would benefit from its doing act Y, that vehicle is correspondingly less likely to behave as a coherent unit. It will have the attributes of a human organization that is governed by a quarrelsome committee—pulled this way and that, and unable to show decisiveness and consistency of purpose.

There is a superficial analogy here with group selection. One of the problems of the theory that groups of organisms function as effective gene vehicles is that outlaws (from the group's point of view) are very likely to arise and be favoured by selection. If we are hypothesizing the evolution of individual restraint through group selection, a gene that makes individuals behave selfishly in an otherwise altruistic group is analogous to an outlaw. It is the near inevitability of such 'outlaws' arising that has dashed the hopes of many a group selection modeller.

The individual body is a much more persuasive gene vehicle than the group because, among other reasons, outlaw replicators within the body are not very likely to be strongly favoured over their alleles. The fundamental reason for this is the drilled formality of the mechanisms of individual reproduction, the 'gavotte of chromosomes' as Hamilton (1975b) called it. If all replicators 'know' that their only hope of getting into the next generation is via the orthodox bottleneck of individual reproduction, all will have the same 'interests at heart'; survival of the shared body to reproductive age, successful courtship and reproduction of the shared body, and a successful outcome to the parental enterprise of the shared body. Enlightened self-interest discourages outlawry when all replicators have an equal stake in the normal reproduction of the same shared body.

Where reproduction is asexual the stake is equal and total, for all replicators have the same 100 per cent chance of finding themselves in every child produced by their joint efforts. Where reproduction is sexual the corresponding chance for each replicator is only half as great, but the ritualized courtliness of meiosis, Hamilton's 'gavotte', largely succeeds in guaranteeing to each allele an equal chance of reaping the benefits of the success of the joint reproductive enterprise. It is, of course, another question *why* the gavotte of the chromosomes is so courtly. It is an immensely important question which I shall evade on a simple plea of cowardice. It is one of a set of questions about the evolution of genetic systems with which better minds than mine have wrestled more or less unsuccessfully (Williams 1975, 1980; Maynard Smith 1978a), a set of questions which moved Williams to remark that 'there is a kind of crisis at hand in evolutionary biology'. I don't understand why meiosis is the way it is, but given that it is much follows. In particular the organized fair-dealing of meiosis helps to account for the coherence and harmony which unites the parts of an individual organism. If, at the level of the *group* of individuals as potential vehicle, the privilege of reproduction was granted with the same scrupulous probity in a similarly well-disciplined 'gavotte of the organisms', group selection might become a more plausible theory of evolution. But, with the possible exception of the very special case of the social insects, group 'reproduction' is anarchical and favourable to individual outlawry. Even social insect colonies will never seem fully harmonious again after Trivers and Hare's ingenious analysis of sex-ratio conflicts (see Chapter 4).

This consideration tells us where we should look first if we want to discover outlaws within the individual body as vehicle. Any replicator that managed to subvert the rules of meiosis so that it enjoyed more than the ordained 50 per cent chance of ending up in a gamete would, other things being equal, tend to be favoured over its alleles in natural selection. Such genes are known to geneticists under the name of meiotic drive genes or segregation distorters. I have already used them as an example to illustrate my definition of an outlaw.

'Genes that beat the system'

The account of segregation distorters that I shall mainly follow is that of Crow (1979), who uses language congenial to the spirit of this book. His paper is called 'Genes that violate Mendel's rules', and it ends as follows: 'The Mendelian system works with maximum efficiency only if it is scrupulously fair to all genes. It is in constant danger, however, of being upset by genes that subvert the meiotic process to their own advantage.... There are many refinements of meiosis and sperm formation whose purpose is apparently to render such cheating unlikely. And yet some genes have managed to beat the system.'

Crow suggests that segregation distorters may be much more common than we ordinarily realize, for the methods of geneticists are not well geared to detecting them, especially if they produce only slight, quantitative effects. The *SD* genes in *Drosophila* are particularly well studied, and here there is some indication as to the actual mechanism of distortion. 'While the homologous chromosomes are still paired up during meiosis, the SD chromosome might *do something* to its normal partner (and rival) that later causes a dysfunction of the sperm receiving the normal chromosome.... SD might actually break the other chromosome'

(Crow 1979, my macabre emphasis). There is evidence that, in SD-heterozygous individuals, sperms not containing the SD chromosome have abnormal, and presumably faulty, tails. It might be thought, then, that the faulty tail results from some sabotage to the non-SD chromosome in the sperms containing it. This cannot be the whole story, as Crow points out, because sperms have been shown to be capable of developing normal tails without any chromosomes at all. Indeed, the whole sperm phenotype seems usually to be under the control of the diploid genotype of the father, not its own haploid genotype (Beatty & Gluecksohn-Waelsch 1972; see below). 'The effect of the SD chromosome on its homologue cannot, then, be simply to inactivate some function, because no function is required. SD must somehow induce its partner to commit a positive act of sabotage.'

Segregation distorters prosper when rare, because the chances are then good that their victims are alleles, not copies. When common, the distorter tends to occur homozygously, and therefore sabotages copies of itself, making the organism virtually sterile. The story is more complicated than this, but computer simulations described by Crow suggest that a stable proportion of segregation distorter genes will be maintained at a frequency somewhat greater than would be accounted for by recurrent mutation alone. There is some evidence that this is so in real life.

In order to qualify as an outlaw, a segregation distorter must do harm to most of the rest of the genome, not just to its alleles. Segregation distorters could have this effect by reducing the total gamete count of the individual. Even if they did not do this, there is a more general reason for expecting there to be selection at other loci in favour of suppressing them (Crow 1979). The argument needs to be developed stage by stage. Firstly many genes, when compared with their alleles, have several pleiotropic effects. Lewontin (1974) goes so far as to speak of 'the undoubted truth that every gene affects every character'. While to call this an 'undoubted truth' may be, to put it mildly, an enthusiastic exaggeration, for my purposes I need only assume that most new mutations have several pleiotropic effects.

Now it is reasonable to expect that most such pleiotropic effects will be deleterious—mutational effects usually are. If a gene is favoured by selection resulting from one beneficial effect, this will be because the advantages of its beneficial effect quantitatively outweigh the disadvantages of its other effects. Normally, by 'beneficial' and 'deleterious', we mean beneficial and deleterious to the whole organism. In the case of a segregation distorter, however, the beneficial effect we are talking about is beneficial to the gene alone. Any pleiotropic effects that it may have on the body are pretty likely to be deleterious to the whole body's survival and reproduction. Segregation distorters are therefore, on the whole, likely to be outlaws: we expect that selection will favour genes at other loci whose phenotypic effect is to reduce the segregation distortion. This brings us to the topic of modifiers.

Modifiers

The classic proving ground for the theory of modifier genes was R. A. Fisher's explanation of the evolution of dominance. Fisher (1930a, but see Charlesworth 1979) suggested that the beneficial effects of a given gene tend to become dominant through the selection of modifiers, while its deleterious effects tend to become recessive. He noted that dominance and recessiveness are not properties of genes themselves, but properties of their phenotypic effects. Indeed a given gene can be dominant in one of its pleiotropic effects and recessive in another. A phenotypic effect of a gene is the joint product of itself and its environment, an environment which includes the rest of the genome. This interactive view of gene action, which Fisher had to argue for at length in 1930, had become so well accepted by 1958 that he was able to take it for granted in the second edition of his great book. It follows from it that dominance or recessiveness, like any other phenotypic effect, may itself evolve, through the selection of other genes elsewhere in the genome, and this was the basis of Fisher's theory of dominance. Although these other genes are known as modifiers, it is now realized that there is no separate category of modifier genes as distinct from major genes. Rather, any gene may serve as modifier of the phenotypic effects of any other gene. Indeed, any given gene's phenotypic effects are susceptible to modification by many other genes in the genome, genes which themselves may have many other major and minor effects (Mayr 1963). Modifiers have been invoked for various other theoretical purposes, for instance in the Medawar/Williams/Hamilton progression of theories of the evolution of senescence (Kirkwood & Holliday 1979).

The relevance of modifiers to the subject of outlaw genes has already been alluded to. Since any gene's phenotypic effects may be subject to modification by genes at other loci, and since outlaws, by definition, work to the detriment of the rest of the genome, we should expect selection to favour genes that happen to have the effect of neutralizing the outlaw's deleterious effects on the body as a whole. Such modifiers would be favoured over alleles that did not influence the outlaw's effects. Hickey and Craig (1966), studying a sex-ratio distorting gene (see below) in the yellow-fever mosquito *Aedes aegypti*, found evidence for evolutionary diminution of the distorting effect which could be interpreted as resulting from the selection of modifiers (though their own interpretation was slightly different). If outlaws do, in general, call forth the selection of suppressing modifiers, there will presumably be an arms race between each outlaw and its modifiers.

As in any other arms race (Chapter 4), we now ask whether there is any general reason to expect one side to prevail over the other. Leigh (1971, 1977), Alexander and Borgia (1978), Kurland (1979, 1980), Hartung (1981), and others suggest that there is just such a general reason. Since, for any single outlaw, suppressing modifiers may arise anywhere in the genome, the outlaw will be outnumbered. As Leigh (1971) puts it, 'It is as if we had to do with a parliament of genes: each acts in its own self-interest, but if its acts hurt the others, they will combine together to suppress it.... However, at loci so closely linked to a distorter that the benefits of "riding its coattails" outweigh the damage of its disease, selection tends to enhance the distortion effect. Thus a species must have many chromosomes if, when a distorter arises, selection at most loci is to favor its suppression. Just as too small a parliament may be perverted by the Cabals of a few, a species with only one, slightly linked chromosome is an easy prey to distorters' (Leigh 1971, p. 249). I am not sure what I think about Leigh's point about chromosome numbers, but his more general point that there is some sense in which outlaws may be 'outnumbered' (Alexander & Borgia 1978, p. 458) by their modifiers seems to me to have promise.

I suppose 'outnumbering' could, in practice, work in two main ways. Firstly, if different modifiers each cause a quantitative diminution of the outlaw's effect, several modifiers might combine additively. Secondly, if any one of several modifiers would suffice to neutralize the outlaw, the chance of effective neutralization goes up with the number of available modifier loci. Alexander and Borgia's metaphor of 'outnumbering', and Leigh's metaphor of the power of the collective in a 'parliament' of the many, could be given meaning in either or both of these two ways. It is important for the argument that segregation distorters at different loci could not, in any obvious sense, 'pool their efforts'. They are not working for some common end of 'general segregation distortion'. Rather, each one is working to distort segregation in favour of itself, and this will hurt other segregation distorters just as much as it hurts non-distorters. Suppressors of segregation distorters, on the other hand, could, in a sense, pool their efforts.

The parliament of genes is one of those metaphors which, if we are not careful, tricks us into thinking that it explains more than it does. Like all humans, but unlike genes, human Members of Parliament are highly sophisticated computers capable of using foresight and language to conspire and reach agreements. Outlaws may seem to be suppressed by agreement of a collective in the parliament of genes, but what is really going on is the selection of modifier genes in preference to non-modifying alleles at their respective loci. Needless to say, Leigh and the other advocates of the 'parliament of genes' hypothesis are well aware of this. I now want to extend the list of outlaws.

Sex-linked outlaws

If a segregation distorter occurs on a sex chromosome, it is not only an outlaw in conflict with the rest of the genome and therefore subject to suppression by modifiers: it also, incidentally, threatens the whole population with extinction. This is because, in addition to ordinary detrimental side effects, it also distorts the sex ratio, and may even eliminate one sex from the population altogether. In one of Hamilton's (1967) computer simulations, a single mutant male with a 'driving Y' chromosome causing males to have only sons and no daughters, was introduced into a population of 1000 males and 1000 females. It took only fifteen generations to drive the model population extinct for lack of females. Something like this effect has been demonstrated in the laboratory (Lyttle 1977). The possibility of using a driving Y gene in the control of serious pests like the yellow-fever mosquito did not escape Hickey and Craig (1966). It is a method with sinister elegance because it is so cheap: all the work of spreading the pest control agent is done by the pests themselves together with natural selection. It is like 'germ warfare' except that the lethal 'germ' is not an extraneous virus but a gene in the species's own gene-pool. Perhaps the distinction is not a fundamental one anyway (Chapter 9).

X-linked drive is likely to have the same sort of detrimental effect on populations as Y-linked drive, but tends to take more generations to extinguish the population (Hamilton 1967). The driving gene on an X chromosome causes males to have daughters rather than sons (except in birds, Lepidoptera, etc.). As we saw in Chapter 4, if haploid male Hymenoptera could influence the amount of care devoted to their spouses' offspring, they would favour daughters rather than sons, since males pass no genes on to sons. The mathematics of this situation are analogous to the case of X-linked drive, the whole genome of the male hymenopteran functioning like an X chromosome (Hamilton 1967, p. 481 and footnote 18).

It is often the case that X chromosomes cross over with each other but not with Y chromosomes. It follows that all genes on X chromosomes could stand to gain from the presence in the gene-pool of a driving X gene which distorts gametogenesis in the heterogametic sex in favour of X gametes and against Y gametes. Genes on X chromosomes are, in a sense, united against Y genes, in a kind of 'anti linkage group', simply because they have no chance of finding themselves on a Y chromosome. Modifiers to suppress X-linked meiotic drive in the heterogametic sex might well not be favoured if they arose at other loci on X chromosomes. They would be favoured if they arose on autosomes. This is different from the case of segregation distorters on autosomes: here there might well be selection in favour of suppressing them by modifiers at other loci even on the same chromosome. X-linked distorters affecting gamete production in the heterogametic sex are, then, outlaws from the point of view of the autosomal part of the gene-pool, but not from the point of view of the rest of the X-chromosomal part of the gene-pool. This potential 'solidarity' among the genes on sex chromosomes suggests that the concept of the outlaw gene may be too simple. It conveys the image of a single rebel standing out against the rest of the genome. At times we might, instead, do better to think in terms of wars between rival gangs of genes, for instance the X-chromosome genes against the rest. Cosmides and Tooby (1981) coin the useful term 'coreplicon' for such a gang of genes that replicate together and therefore tend to work for the same ends. In many cases neighbouring coreplicons will blur into each other.

Ganging up by Y-chromosome genes is even more to be expected. As long as Y chromosomes do not cross over, it is clear that all genes on a Y chromosome stand to gain from the presence of a Y-linked distorter every bit as much as the distorter gene itself. Hamilton (1967) made the interesting suggestion that the reason for the well-known inertness of Y chromosomes (hairy ears seems to be the only conspicuous Y-linked trait in man) is that Y-suppressing modifiers have been positively selected elsewhere in the genome. It is not obvious how a modifier might go about suppressing the phenotypic activity of an entire chromosome, since the various phenotypic effects of a single chromosome are usually so heterogeneous. (Why would selection not suppress only the effects of the driving genes, leaving other Y-linked effects intact?) I suppose it might do it by physically deleting large chunks of Y chromosome, or by contriving to quarantine the Y chromosome from the cell's transcription machinery.

A weird example of a driving replicator which is probably not a gene in the ordinary sense of the word is given by Werren, Skinner, and Charnov (1981). They studied the parasitoid wasp *Nasonia vitripennis*, in which there is a variety of males called Dl, or 'daughterless'. Wasps being haplodiploid, males pass their genes only to daughters: a male's mate may have sons, but those sons are haploid and fatherless. When Dl males mate with females, they cause them to produce all male broods. Most of the sons of females mated to Dl males are themselves Dl males. Although no nuclear genes pass from father to son, therefore, the Dl factor somehow does pass from father to 'son'. The Dl factor rapidly spreads, in exactly the same way as a driving Y chromosome would. It is not known what the Dl factor physically consists of. It is certainly not nuclear genetic material, and it is theoretically possible that it is not even composed of nucleic acid, although Werren *et al.* suspect that it probably is cytoplasmically borne nucleic acid. Theoretically, *any* kind of physical or chemical influence of a Dl male on his mate, which causes her to have Dl sons, would spread like a driving Y chromosome, and would qualify as an active germ-line replicator in the sense of Chapter 5. It is also an outlaw *par excellence*, for it spreads itself at the total expense of all the nuclear genes in the males that bear it.

Selfish sperm

With some exceptions, all the diploid cells of an organism are genetically identical, but the haploid gametes it produces are all different. Only one out of many sperms in an ejaculate can fertilize an egg, and there is therefore potential for competition among them. Any gene that found phenotypic expression when in the haploid state in a sperm cell could be favoured over its alleles if it improved the competitive ability of the sperm. Such a gene would not necessarily be sex-linked: it could be found on any chromosome. If it was sex-linked it would have the effect of biasing the sex ratio, and would be an outlaw. If it was on an autosome it would still qualify as an outlaw for the general kind of reason already given for any segregation distorter: 'if there were genes affecting sperm-cell function there would be competition among sperm cells, and a gene that improved the ability to fertilize would increase in the population. If such a gene happened to cause, say, malfunction of the liver, that would be just too bad; the gene would increase anyway, since selection for good health is much less effective than selection by competition among sperm cells' (Crow 1979). There is, of course, no particular reason why a sperm competition gene should happen to cause malfunction of the liver but, as already pointed out, most mutations are deleterious, so some undesirable side effect is pretty likely.

Why does Crow assert that selection for good health is much less effective than selection by competition among sperm cells? There must inevitably be a quantitative trade-off involving the magnitude of the effect on health. But, that aside, and even allowing for the controversial possibility that only a minority of sperms are viable (Cohen 1977), the argument appears to have force because the competition between sperm cells in an ejaculate would seem to be so fierce.

A million million spermatozoa,
All of them alive:
Out of their cataclysm but one poor Noah
Dare hope to survive.

And of that billion minus one
Might have chanced to be
Shakespeare, another Newton, a new Donne—
But the One was Me.

Shame to have ousted your betters thus.
Taking ark while the others remained outside!
Better for all of us, froward Homunculus,
If you'd quietly died!

Aldous Huxley

One might imagine that a mutant gene that expressed itself when in the haploid genotype of a spermatozoon, causing increased competitive ability, say an improved swimming tail or the secretion of a spermicide to which the sperm itself was immune, would be immediately favoured by a selection pressure gigantic enough to outweigh all but the most catastrophic of deleterious side effects on the diploid body. But although it may be true that only one in hundreds of millions of sperms 'dare hope to survive', the calculation looks very different from the point of view of a single gene. If we forget linkage groups and brand-new mutations for a moment, however rare a gene may be in the gene-pool, if a given male has it in his diploid genotype, at least 50 per cent of his sperms must have it. If one sperm has received a gene giving it competitive advantage, 50 per cent of its rivals in the same ejaculate will have received the same gene. Only if the mutation has arisen *de novo* during the genesis of a single sperm will the selection pressure be astronomical in magnitude. Usually it will be a more modest selection pressure, not of millions to one but only two to one. If we take linkage effects into consideration the calculation is more complicated, and the selection pressure in favour of competitive sperms will increase somewhat.

In any case, it is a strong enough pressure for us to expect that, if genes expressed themselves when in the haploid genotype of the sperm, outlaws would be favoured, to the detriment of the rest of the genes in the diploid father's genome. It seems, to say the least, fortunate that sperm phenotypes are, as a matter of fact, usually not under the control of their own haploid genotypes (Beatty & Gluecksohn-Waelsch 1972). Of course sperm phenotypes must be under some genetic control, and natural selection has doubtless worked on the genes controlling sperm phenotypes to perfect sperm adaptation. But those genes seem to express themselves when in the diploid genotype of the father, not when in the haploid genotype of the sperm. When in the sperm they are passively carried.

The passivity of their genotypes may be an immediate consequence of the lack of cytoplasm in spermatozoa: a gene cannot achieve phenotypic expression except via cytoplasm. This is a proximal explanation. But it is at least worth toying with reversing the proposition to obtain an ultimate functional explanation: sperms are made small, as an adaptation to prevent the phenotypic expression of the haploid genotype. On this hypothesis we are proposing an arms race between (haploid-expressed) genes for increased competitive ability among spermatozoa on the one hand, and genes expressing themselves when in the diploid genotype of the father on the other hand, causing sperms to become smaller and therefore unable to give phenotypic expression to their own haploid genotypes. This hypothesis does not explain why eggs are larger than sperms; it assumes the basic fact of anisogamy, and therefore does not aspire to be an alternative to theories of the origins of anisogamy (Parker 1978b; Maynard Smith 1978a; Alexander & Borgia 1979). Moreover not all sperms are small, as Sivinski (1980) reminds us in a most intriguing review. But the present explanation still deserves consideration as an ancillary to others. It is analogous to Hamilton's (1967) explanation for the inertness of Y chromosomes, to which I have already alluded.

Green beards and armpits

Some of the outlaws I have been considering have been realistic, and are actually known to geneticists. Some of the suggested outlaws that I shall now come to are, frankly, pretty improbable. I make no apology for this. I see them as thought experiments. They play the same role in helping me to think straight about reality as imaginary trains travelling at nearly the speed of light do for physicists.

So, in this spirit of thought experiment, imagine a gene on a Y chromosome which makes its possessor kill his daughters and feed them to his sons. This is clearly a behavioural version of a driving Y-chromosome effect. If it arose it would tend to spread for the same reason, and it would be an outlaw in the same sense that its phenotypic effect would be detrimental to the rest of the male's genes. Modifiers, on any chromosome other than the Y chromosome, which tended to reduce the phenotypic effect of the daughter-killing gene would be favoured over their alleles. In a sense the outlaw gene is using the sex of the male's children as a convenient *label* for the presence or absence of itself: all sons are labelled as definite possessors of the gene, all daughters as definite non-possessors of it.

A similar argument can be made for X chromosomes. Hamilton (1972, p. 201) pointed out that in normal diploid species a gene on an X chromosome in the homogametic sex has three-quarters of a chance of being identical by descent with a gene in a sibling of the homogametic sex. Thus the 'X-chromosome relatedness' of human sisters is as high as the overall relatedness of hymenopteran sisters, and higher than the overall relatedness of human sisters. Hamilton went so far as to wonder whether an X-chromosome effect might account for the fact that helpers at the nest in birds seem, usually, to be elder brothers, rather than sisters, of the nestlings (the male sex is homogametic in birds). He noted that the X chromosome in birds accounts for some 10 per cent of the whole genome, and that it is therefore not too improbable that the genetic basis for brotherly care might lie on the X chromosome. If so, brotherly care might be favoured by the same kind of selection pressure as Hamilton had earlier suggested for sisterly care in Hymenoptera. Perhaps significantly, Syren and Luyckx (1977) point out that in some termites, the only non-haplodiploid group to have achieved full eusociality, 'approximately half the genome is maintained as a linkage group with the sex chromosome' (Lacy 1980).

Wickler (1977), commenting on Whitney's (1976) rediscovery of Hamilton's X-chromosome idea, suggests that Y-chromosome effects are potentially even more powerful than X-chromosome effects, but Y chromosomes as a rule don't make up such a high proportion of the genome. In any case 'sex-linked altruism' must be *discriminating*: individuals acting under the influence of their sex chromosomes should tend to show favouritism towards close relatives of the same sex rather than of the opposite sex. Genes for sex-blind sibling care would not be outlaws.

The value of the outlawed sex-chromosome thought experiment does not lie in its plausibility—which, like Hamilton, I do not rate highly—but in the fact that it focuses our attention on the importance of this discrimination. The sex of another individual is used as a *label* to identify it as a member of a class of individuals about whose genetics something is known. In the ordinary theory of kin selection, relatedness (or rather some proximate correlate of relatedness such as presence in one's own nest) is used as a label indicating higher than average probability of sharing a gene. From the point of view of a gene on a Y chromosome, the sex of a sibling is a label which signifies the difference between certainty of sharing the gene and certainty of not sharing it.

Note, by the way, the ineptness of notions of individual fitness, or even inclusive fitness as ordinarily understood, at dealing with situations like this. The normal calculation of inclusive fitness makes use of a coefficient of relationship which is some measure of the probability that a pair of relatives will share a particular gene, identical by descent. This is a good approximation provided the genes concerned have no better way of 'recognizing' copies of themselves in other individuals. If a gene is on a sex chromosome and can use the sex of relatives as a label, however, its best 'estimate' of the probability that a relative shares a copy of itself will be a better estimate than that provided by the coefficient of relationship. In its most general form, the principle of genes appearing to 'recognize' copies of themselves in other individuals has been dubbed the 'green-beard effect' (Dawkins 1976a, p. 96, following Hamilton 1964b, p. 25). Green-beard or 'recognition alleles' have been described in the literature as outlaws (Alexander & Borgia 1978; Alexander 1980), and they should therefore be discussed in this chapter, even though, as we shall see, their status as outlaws needs careful examination (Ridley & Grafen 1981).

The green-beard effect reduces the principle of 'gene self-recognition' to its bare essentials in an unrealistically hypothetical, but nevertheless instructive, way. A gene is postulated which has two pleiotropic effects. One effect is to confer a conspicuous label, the 'green beard'. The other effect is to confer a tendency to behave altruistically towards bearers of the label. Such a gene, if it ever arose, could easily be favoured by natural selection, although it would be vulnerable to a mutant arising which conferred the label without the altruism.

Genes are not conscious little devils, able to recognize copies of themselves in other individuals and to act accordingly. The only way for the green-beard effect to arise is by incidental pleiotropy. A mutation must arise which just happens to have two complementary effects: the label or 'green beard', and the tendency to behave altruistically towards labelled individuals. I have always thought such a fortuitous conjunction of pleiotropic effects too good to be true. Hamilton also noted the idea's inherent implausibility, but he went on 'exactly the same *a priori* objections might be made to the evolution of assortative mating which manifestly has evolved, probably many times independently and despite its obscure advantages' (Hamilton 1964b, p. 25). It is worth briefly examining this comparison with assortative mating, which for present purposes I shall take to mean the tendency of individuals to prefer to mate with individuals that genetically resemble them.

Why is it that the green-beard effect seems so much more far-fetched than assortative mating? It is not just that assortative mating is positively known to occur. I suggest another reason. This is that when we think of assortative mating we implicitly assume *self-inspection* as a means of facilitating the effect. If black individuals prefer to mate with black individuals, and white with white, we do not find this hard to believe because we tacitly assume that individuals perceive their *own* colour. Each individual, whatever his colour, is assumed to be obeying the *same* rule: inspect yourself (or members of your own family) and choose a mate of the same colour. This principle does not stretch our credulity by demanding that two specific effects—colour and behavioural preference—are controlled pleiotropically by the same gene. If there is a general advantage in mating with similar partners, natural selection will favour the self-inspection rule regardless of the exact nature of the recognition character used. It does not have to be skin colour. Any conspicuous and variable character would work with the identical behavioural rule. No far-fetched pleiotropism need be postulated.

Well then, will the same kind of mechanism work for the green-beard effect? Might animals obey a behavioural rule of the form: 'Inspect yourself and behave altruistically towards other individuals that resemble you'? The answer is that they might, but this would not be a true example of the green-beard effect. Instead, I call it the 'armpit effect'. In the paradigm hypothetical example, the animal is supposed to smell its own armpits, and behave altruistically towards other individuals with a similar smell. (The reason for choosing an olfactory name is that police dog trials have shown that dogs presented with handkerchiefs held under human armpits can distinguish the sweats of any two individual humans except identical twins (Kalmus 1955). This suggests that there is an enormously variable richness of genetic labelling in molecules of sweat. In the light of the result with identical twins, one feels inclined to bet that police dogs could be trained to sniff out the coefficient of relationship between pairs of humans and that, for instance, they could be trained to track down a criminal if given a sniff of his brother. Be that as it may, 'armpit effect' is here being used as a general name for any case of an animal inspecting himself, or a known close relative, and discriminating in favour of other individuals with a similar smell or with some other perceived similarity.)

The essential difference between the green-beard effect and the armpit self-inspection effect is as follows. The armpit self-inspection behavioural rule will lead to the detection of other individuals that are similar in some respect, perhaps in many respects, but it will not specifically lead to the detection of individuals that possess copies of the gene mediating the behavioural rule itself. The armpit rule might provide an admirable means of detecting true kin from non-kin, or of detecting whether a brother was a full brother or only a half brother. This could be very important, and it might provide the basis for selection in favour of self-inspecting behaviour, but the selection would be conventional, familiar kin selection. The self-inspection rule would be functioning simply as a kin-recognition device, analogous to a rule like: 'Behave altruistically towards individuals that grew up in your own nest.'

The green-beard effect is quite different. Here the important thing is that a gene (or close linkage group) programs the recognition specifically of copies of *itself*. The green-beard effect is not a mechanism for the recognition of kin. Rather, kin recognition and 'green-beard' recognition are alternative ways in which genes could behave as if discriminating in favour of copies of themselves.

To return to Hamilton's comparison with assortative mating, we can see that it does not really provide good grounds for optimism over the plausibility of the green-beard effect. Assortative mating is much more likely to involve self-inspection. If, for whatever reason, it is an advantage in general for like to mate with like, selection would favour an armpit type of behavioural rule: Inspect yourself, and choose a mate that resembles you. This will achieve the desired result—an optimal balance between outbreeding and inbreeding (Bateson 1983) or whatever the advantage may be—regardless of the exact nature of the characteristics by which individuals differ.

Assortative mating is not the only analogy Hamilton might have chosen. Another one is the case of cryptic moths choosing to sit on a background that matches their own colour. Kettlewell (1955) gave dark-coloured *carbonaria* and light-coloured typical morphs of the peppered moth *Biston betularia* the opportunity to sit on dark- or light-coloured backgrounds. There was a statistically significant tendency for moths to choose a background matching their own body colour. This could be due to pleiotropism (or genes for colour being closely linked to genes for background choice). If this were so, as Sargent (1969a) believes, it might, by analogy, reduce our scepticism about the inherent plausibility of the green-beard effect. Kettlewell believed, however, that the moths achieved the matching by the simpler mechanism of 'contrast conflict'. He suggested that a moth could see a small portion of its own body, and that it moved around until the observed contrast between its own body and the background reached a minimum. It is easy to believe that natural selection might favour the genetic basis for such a contrast-minimizing behavioural rule, because it would work automatically in conjunction with any colour, including a newly mutated colour. It is of course analogous to the 'armpit self-inspection' effect, and is plausible for the same reason.

Sargent's (1969a) intuition differs from Kettlewell's. He doubts the self-inspection theory, and thinks that the two morphs of *B. betularia* differ genetically in their background preference. He has no evidence about *B. betularia* itself, but has done some ingenious experiments on other species. For example, he took members of a dark species and of a light species, and painted the circum-ocular hairs in an attempt to 'fool' the moths into choosing a background that matched the painted hairs. They obstinately persisted in choosing a background that matched their genetically determined colour (Sargent 1968). Unfortunately, however, this interesting result was obtained with two different species, not with dark and light morphs of one species.

In another experiment which was done with a dimorphic species, *Phigalia titea*, Sargent (1969a) simply failed to get the result Kettlewell had with *B. betularia*. The *P. titea* individuals, whether the dark or the light morph, chose to sit on light backgrounds, presumably the appropriate background for the light ancestral form of the species. What is needed is for somebody to repeat Sargent's key experiment of painting the parts of a moth's body that it can see, but using a dimorphic species like *B. betularia* which is known to show morph-specific background choice. Kettlewell's theory would predict that moths painted black will select black backgrounds, and moths painted light will select light backgrounds, regardless of whether they are genetically *carbonaria* or *typica*. A purely genetic theory would predict that *carbonaria* will choose dark backgrounds and *typica* light backgrounds, regardless of how they are painted.

If the latter theory turned out to be right, would this result give aid and comfort to the green-beard theory? A little, perhaps, since it would suggest that a morphological character and behavioural recognition of something resembling that morphological character can be, or can rapidly become, closely linked genetically. It must be remembered, here, that in the moth crypsis example there is no suggestion that we are dealing with an outlaw effect. If there are two genes, one controlling colour and the other background choice behaviour, both stand to gain from the presence of the appropriate other one, and neither of the two is in any sense an outlaw. If the two genes started out distantly linked, selection would favour closer and closer linkage. It is not clear that selection would similarly favour the close linkage of a 'green-beard gene' and a gene for green-beard recognition. It seems that the association between the effects would have to be there by luck right from the start.

The green-beard effect is all about one selfish gene looking after copies of itself in other individuals, regardless of, indeed in spite of, the probability that those individuals will share genes in general. The green-beard gene 'spots' copies of itself, and it thereby seems to work against the interests of the rest of the genome. It appears to be an outlaw in the sense that it makes individuals work and pay costs, for the benefit of other individuals who are not particularly likely to share genes other than the outlaw itself. This is the reason Alexander and Borgia (1978) call it an outlaw, and it is one of the reasons for their scepticism about the existence of green-beard genes.

But it is actually not obvious that green-beard genes would, if they ever arose, be outlaws. The following caution is urged by Ridley and Grafen (1981). Our definition of an outlaw makes reference to its provoking of modifiers at other loci which tend to suppress its phenotypic effects. At first sight it seems clear that green-beard genes would indeed provoke suppressing modifiers, because the modifiers would not tend to have copies in the bodies of the (unrelated) green-bearded individuals cared for. But we must not forget that a modifier, if it is to have any effect on the phenotypic expression of a green-beard gene, is likely itself to be in a green-bearded body, and is therefore in a position to benefit from *receiving* altruism from other green-bearded individuals. Moreover, since those other green-bearded donors are not particularly likely to be relatives, the costs of their altruism will not be felt by copies of the would-be modifiers. A case could therefore be made that would-be modifiers at other loci will gain rather than lose from sharing a body with a green-beard gene. To this it cannot be objected that the costs of paying out altruism to other green-bearded individuals may outweigh the benefits of receiving altruism from other green-bearded individuals: if this were true there would be no question of the green-beard gene spreading in the first place. The essence of Ridley and Grafen's point is that *if* (which is unlikely) the green-beard gene has what it takes to spread through the population *at all*, the costs and benefits of the situation will be such as to favour modifiers that enhance rather than reduce the effect.

In evaluating this point, everything depends on the exact nature of what we are calling the green-beard phenotype. If the entire pleiotropic dual phenotype—green beard plus altruism towards green-bearded individuals—is regarded as a package deal, which modifiers can suppress or enhance only as a unit, then Ridley and Grafen are surely right that green-beard genes are not outlaws. But, of course, as they themselves stress, a modifier that could detach the two phenotypic effects from each other, suppressing the altruistic phenotype of the green-beard gene while not suppressing the green beard itself would certainly be favoured. A third possibility is the special case of a green-beard gene that caused parents to discriminate in favour of those of their children that happened to share the recognition character. Such a gene would be analogous to a meiotic-drive gene, and would be a true outlaw.

Whatever we may feel about Ridley and Grafen's point about the green-beard effect, it is clear that genes mediating altruism towards close kin, and favoured by conventional kin selection pressure, are definitely not outlaws. All the genes in the genome have the same statistical odds of gaining from the kin altruism behaviour, for all have the same statistical odds of being possessed by the individual benefited. A 'kin selection gene' is, in a sense, working for itself alone, but it benefits the other genes in its genome as well. There will therefore not be selection in favour of modifiers that suppress it. Armpit self-inspection genes would be a special case of kin-recognition genes, and are likewise not outlaws.

I have been negative about the plausibility of the green-beard effect. The postulated favouritism based on sex chromosomes, which I have already mentioned, is a special case of the green-beard effect, and is perhaps the least implausible one. I discussed it in the context of within-family favouritism: elder siblings were supposed to discriminate among their younger siblings on the basis of probability of sharing sex chromosomes, sex itself being used as the label ('green beard'). This is not too wildly improbable because, if Y chromosomes do not cross over, instead of having to postulate a single pleiotropic green-beard 'gene' we can postulate a whole 'green-beard chromosome'. It is sufficient that the genetic basis for sexual favouritism should occur anywhere on the sex chromosome concerned. One might make a similar argument for any substantial portion of a chromosome which, say because of inversion, did not cross over. It is conceivable, therefore, that a true green-beard effect of some kind might one day be discovered.

I suspect that all present examples of what might appear to be green-beard effects are, in fact, versions of the armpit self-inspection effect. Thus Wu *et al.* (1980) placed individual monkeys, *Macaca nemestrina*, in a choice apparatus where they had to choose to sit next to one of two offered companions. In each case one of the two companions offered was a half sibling, related through the father but not the mother; the other was an unrelated control. The result was a statistically significant tendency for individuals to choose to sit next to half siblings rather than unrelated controls. Note that the half siblings concerned were not related in the maternal line: this means that there is no possibility of their recognizing an odour acquired from the mother, say. Whatever the monkeys are recognizing, it comes from the shared father, and this suggests, in some sense, recognition of shared genes. My bet is that the monkeys recognize resemblances of relatives to perceived features of themselves. Wu *et al.* are of the same opinion.

Greenberg (1979) studied the primitively social sweat bee *Lasioglossum zephyrum*. (Seger, 1980, refers to this work under the picturesque heading, 'Do bees have green setae?'.) Where Wu *et al.* used choice of sitting partner as a behavioural assay, Greenberg used the decision of a guard worker whether to admit or exclude another worker seeking entrance to the nest. He plotted the probability of a worker's being admitted against her coefficient of relationship with the sentinel. Not only was there an excellent positive correlation: the slope of the line was almost exactly one, so the probability of a sentinel admitting a stranger was approximately *equal* to the coefficient of relationship! Greenberg's evidence convinced him that 'The genetic component is therefore in odor production and apparently not in the perceptual system' (p. 1096). In my terminology, Greenberg's words amount to the statement that he is dealing with the armpit effect, not with the green-beard effect. Of course the bees may, as Greenberg believes, have inspected relatives with whom they were already familiar rather than their own 'armpits' (Hölldobler & Michener 1980). It is still essentially an example of the armpit rather than the green-beard effect, in which case there is no question of the genes responsible being outlaws. A particularly elegant study which comes to a similar conclusion is that of Linsenmair (1972) on the family-specific chemical 'badge' of the social desert woodlouse *Hemilepistus reaumuri*. Similarly, Bateson (1983) provides intriguing evidence that Japanese quail discriminate their first cousins from their siblings and from more distant relatives, using learned visual cues.

Waldman and Adler (1979) have investigated whether tadpoles preferentially associate with siblings. Colour-marked tadpoles taken from two clutches were allowed to swim freely around a tank; then a grid was lowered into the tank, trapping each tadpole in one of sixteen compartments. There was a statistically significant tendency for tadpoles to end up closer to siblings than to non-siblings. Perhaps unfortunately, the experimental design does not rule out a possible confounding effect of genetically determined 'habitat selection'. If there was a genetically determined tendency to, say, hug the edges of the tank rather than the middle, genetic relatives might as a consequence be expected to end up in the same parts of the tank. The experiment does not, therefore, unequivocally demonstrate recognition of relatives or a preference for associating with relatives as such, but for many theoretical purposes this does not matter. The authors introduced their paper with reference to Fisher's (1930a) kin selection theory of the evolution of aposematism, and for purposes of that theory relatives simply have to end up together. It doesn't matter whether they are together because of shared habitat preference or because of true relative-recognition. For our present purpose, however, it is worth noting that, if further experiments confirm the rule of 'incidental habitat selection' for the tadpoles, this would rule out the 'armpit' theory but would not rule out the green-beard theory.

Sherman (1979) invokes the idea of genetic favouritism in an ingenious theory about chromosome numbers in social insects. He presents evidence that eusocial insects tend to have higher chromosome numbers than their nearest non-social phylogenetic relatives. Seger (1980) independently discovered the effect, and has his own theory to account for it. The evidence for the effect is somewhat equivocal and would perhaps benefit from a demanding analysis using the statistical methods developed by modern students of the comparative method (e.g. Harvey & Mace 1982). But what I am concerned with here is not the truth of the effect itself but Sherman's theory to account for it. He correctly notes that high chromosome numbers tend to reduce the variance in fraction of shared genes among siblings. To take an extreme case, if a species has only one chromosome pair with no crossing-over, any pair of full siblings will share (identical by descent) either all, none, or half of their genes with a mean of 50 per cent. If there are hundreds of chromosomes, on the other hand, the number of genes shared (identical by descent) among siblings will be narrowly distributed about the same mean of 50 per cent. Crossing-over complicates the issue, but it remains true that a high chromosome number in a species tends to go with low genetic variance among siblings in that species.

It follows that if social insect workers wished to discriminate in favour of those of their siblings with whom they happened to share the most genes, it would be easier for them to do so if the species's chromosome number were low than if it were high. Such preferential discrimination by workers would be detrimental to the fitness of the queen, who would 'prefer' a more even-handed treatment of her offspring. Sherman therefore suggests, in effect, that the high chromosome count in eusocial insects is an adaptation to cause 'offsprings' reproductive interests to more closely coincide with those of their mother'. We should not forget, by the way, that workers will not be unanimous. Each worker might show favouritism towards younger siblings that resemble her, but other workers will tend to resist her favouritism, for the same reason as the queen will resist it. The workers cannot be treated as a monolithic party in opposition to the queen, in the same way as Trivers and Hare (1976) were able to treat them in their theory of conflict over the sex ratio.

Sherman very fairly lists three weaknesses of his hypothesis, but there are two more serious problems with it, Firstly, unless we are careful to qualify it further, the hypothesis appears to come dangerously close to a fallacy which I have dubbed 'Misunderstanding Number 11' (Dawkins 1979a) or the 'Ace of Spades Fallacy' (Chapter 10). Sherman assumes that the extent of cooperation among conspecifics is related to the 'mean *fraction* of alleles they share' (my emphasis) whereas he should think in terms of the *probability* that a gene 'for' cooperation is shared (see also Partridge & Nunney 1977). On the latter assumption his hypothesis, as at present expressed, would not work (Seger 1980). Sherman could rescue his hypothesis from this particular criticism by invoking the 'armpit self-recognition' effect. I shall not spell the argument out in detail, because I suspect that Sherman accepts it. (The essential point is that the armpit effect can use weak linkage within a family, while the green-beard effect requires pleiotropy or linkage disequilibrium. If workers inspect themselves, and show favouritism to those of their reproductive siblings that share features that they perceive themselves to possess, ordinary linkage effects become all-important and Sherman's hypothesis could escape the 'Ace of Spades Fallacy'. It would also, incidentally, escape the first of Sherman's own objections, namely that the hypothesis 'depends on the existence of alleles that enable their bearers to recognize their alleles'. 'Such recognition alleles have never been discovered', and, by implication, are pretty implausible. Sherman could make things easier for himself by tying his hypothesis to the armpit rather than the green-beard effect.)

My second difficulty with Sherman's hypothesis was brought to my attention by J. Maynard Smith (personal communication). Taking the 'armpit' version of the theory, it is indeed conceivable that workers might be selected to inspect themselves, and show favouritism towards those of their reproductive siblings that share their own individual characteristics. It is also true that queens would then be selected to suppress this favouritism if they could, for instance by pheromonal manipulation. But, in order to be selected, any such move on the part of a queen would have to have an effect as soon as it arose as a mutation. Would this be true of a mutation that increased a queen's chromosome number? No it would not. An increase in chromosome number would change the selection pressures bearing on worker favouritism, and many generations later it might produce an evolutionary change which would be to the advantage of queens in general. But this would not help the original mutant queen, whose workers would follow their own genetic programming and would be oblivious to changes in selection pressures. Changes in selection pressures exert their effects over the longer time-scale of generations. A queen cannot be expected to initiate a program of artificial selection for the long-term benefit of future queens! The hypothesis may be rescued from this objection by suggesting that high chromosome numbers are not an adaptation to facilitate queen manipulation of workers, but rather are a preadaptation. Those groups that happened, for other reasons, to have high chromosome numbers were the most likely to evolve eusociality. Sherman mentions this version of his hypothesis, but sees no reason to favour it over the more positive, maternal-manipulation version. In conclusion, Sherman's hypothesis can be made theoretically sound, if it is phrased in terms of preadaptation rather than adaptation, and in terms of the armpit effect rather than the green-beard effect.

The green-beard effect may be implausible, but it is instructive. The student of kin selection who first understands the hypothetical green-beard effect, and then approaches kin selection theory in terms of its similarities to and differences from 'green-beard theory', is unlikely to fall prey to the many tempting opportunities for error that kin selection theory offers (Dawkins 1979a). Mastery of the green-beard model will convince him that altruism towards kin is not an end in itself, something that animals are mysteriously expected to practise in accordance with some clever mathematics that field workers don't understand. Rather, kinship provides just one way in which genes can behave as if they recognized and favoured copies of themselves in other individuals. Hamilton himself is emphatic on this point: 'kinship should be considered just one way of getting positive regression of genotype in the recipient, and . . . it is this positive regression that is vitally necessary for altruism. Thus the inclusive fitness concept is more general than "kin selection"' (Hamilton 1975a, p. 140–1).

Hamilton is here using what he earlier described as 'the extended meaning of inclusive fitness' (Hamilton 1964b, p. 25). The conventional meaning of inclusive fitness, the meaning upon which Hamilton himself based his detailed mathematics, is incapable of handling the green-beard effect and, indeed, outlaws such as meiotic-drive genes. This is because it is firmly tied to the idea of the *individual organism* as 'vehicle' or 'maximizing entity'. Outlaw genes demand to be treated as selfish maximizing entities in their own right, and they constitute a strong weapon in the case against the 'selfish organism' paradigm. Nowhere is this better exemplified than in Hamilton's own ingenious extensions of Fisher's sex-ratio theory (Hamilton 1967).

The green-beard thought experiment is instructive in further ways. Anyone who thinks about genes literally as molecular entities is in danger of being misled by passages like 'What is the selfish gene? It is not just one single physical bit of DNA…It is *all replicas* of a particular bit of DNA, distributed throughout the world…a distributed agency, existing in many different individuals at once…a gene might be able to assist *replicas* of itself which are sitting in other bodies.' The whole of kin selection theory rests on this general premise, yet it would be mystical and wrong to think that genes assist copies of each other *because* those copies are identical molecules to themselves. A green-beard thought experiment helps to explain this. Chimpanzees and gorillas are so similar that a gene in one species might be physically identical in its molecular details to a gene in the other. Is this molecular identity a sufficient reason to expect selection to favour genes in one species 'recognizing' copies of themselves in the other species, and extending them a helping hand? The answer is no, although a naive application of 'selfish gene' reasoning at the molecular level might lead us to think otherwise.

Natural selection at the gene level is concerned with competition among alleles for a particular chromosomal slot in a shared gene-pool. A green-beard gene in the chimpanzee gene-pool is not a candidate for a slot on any gorilla chromosome, and nor is any of its alleles. It is therefore indifferent to the fate of its structurally identical counterpart in the gorilla gene-pool. (It might not be indifferent to the fate of its phenotypically identical counterpart in the gorilla gene-pool, but that has nothing to do with molecular identity.) As far as the present argument is concerned, chimp genes and gorilla genes are not copies of each other in an important sense. They are copies of each other only in the trivial sense that they happen to have the identical molecular structure. The unconscious, mechanical laws of natural selection give us no grounds for expecting them to assist molecular copies, just because they *are* molecular copies.

Conversely, we might be justified in expecting to see genes assisting molecularly different alleles at their own locus within a species gene-pool, provided they had the same phenotypic effects. A phenotypically neutral mutation at a locus changes molecular identity but does nothing to weaken any selection there may be in favour of mutual assistance. Green-beard altruism could still increase the incidence of green-beard phenotypes in the population, even though genes were assisting other genes that were not strict copies of themselves in the molecular sense. It is the incidence of phenotypes that we are interested in explaining, not the incidence of molecular configurations of DNA. And if any reader thinks that last remark contradicts my basic thesis, I must have failed to make my basic thesis clear!

Let me use the green beard in one more instructive thought experiment, to clarify the theory of reciprocal altruism. I have called the green-beard effect implausible, with the possible exception of the sex chromosome special cases. But there is another special case which could conceivably have some counterpart in reality. Imagine a gene programming the behavioural rule: 'If you see another individual performing an altruistic act, remember the incident, and if the opportunity arises behave altruistically towards that individual in future' (Dawkins 1976a, p. 96). This may be called the 'altruism-recognition effect'. Using the legendary example of Haldane's (1955) jumping into the river to save a drowning person, the gene that I am postulating might spread because it was, in effect, recognizing copies of itself. It is, in fact, a kind of green-beard gene. Instead of using an incidental pleiotropic recognition character such as a green beard, it uses a non-incidental one: the behaviour pattern of altruistic rescuing itself. Rescuers tend to save only others who have rescued somebody in their time, so the gene tends to save copies of itself (setting aside problems of how the system could get started, etc.). My point in bringing this hypothetical example up here is to emphasize its distinctness from two other, superficially similar cases. The first is the one Haldane himself was illustrating, the saving of close kin; thanks to Hamilton we now understand this well. The second is reciprocal altruism (Trivers 1971). Any resemblance between true reciprocal altruism and the hypothetical altruism-recognition case I am now discussing is purely accidental (Rothstein 1981). The resemblance, however, sometimes muddles students of the theory of reciprocal altruism, which is why I am making use of green-beard theory to dispel the muddle.

In true reciprocal altruism, the 'altruist' stands to gain in the future from the presence of the *individual* beneficiary of his altruism. The effect works even if the two share no genes, and even (*contra* Rothstein 1981) if they belong to different species, as in Trivers's example of the mutualism between cleaner fish and their clients. Genes mediating such reciprocal altruism benefit the rest of the genome no less than they benefit themselves, and are clearly not outlaws. They are favoured by ordinary, familiar natural selection, albeit some people (e.g. Sahlins 1977, pp. 85–7) seem to have difficulty in understanding the principle, apparently because they overlook the frequency-dependent nature of the selection and the consequent need to think in game-theoretic terms (Dawkins 1976a, pp. 197–201; Axelrod and Hamilton 1981). The altruism-recognition effect is fundamentally different, though superficially similar. There is no need for the altruism recognizer individual to repay a good turn done to *himself*. He simply recognizes good deeds done to anybody, and singles out the altruist for his own later favours.

It would be impossible to give a sensible account of outlaws in terms of individuals maximizing their fitness. That is the reason for giving them prominence in this book. At the beginning of the chapter I divided outlaws into 'allelic outlaws' and 'laterally spreading outlaws'. All the suggested outlaws we have so far considered have been allelic: they are favoured over their alleles at their own loci, while being opposed by modifiers at other loci. I now turn to laterally spreading outlaws. These are outlaws unruly enough to break away altogether from the discipline of allelic competition within the confines of a locus. They spread to other loci, even create new loci for themselves by increasing the size of the genome. They are conveniently discussed under the heading 'Selfish DNA', a catch phrase which has recently acquired currency in the pages of *Nature*. This will be the subject of the first part of the next chapter.

9

SELFISH DNA, JUMPING GENES, AND A LAMARCKIAN SCARE

This chapter will be a somewhat miscellaneous one, gathering together the results of my brief and foolhardy incursions into the hinterlands of fields far from my own, molecular and cell biology, immunology, and embryology. The brevity I justify on the grounds that greater length would be even more foolhardy. The foolhardiness is less defensible, but may perhaps be forgiven on the grounds that an equally rash earlier raid yielded the germ of an idea which some molecular biologists now take seriously under the name of selfish DNA.

Selfish DNA

> ... it appears that the amount of DNA in organisms is more than is strictly necessary for building them: a large fraction of the DNA is never translated into protein. From the point of view of the individual organism this seems paradoxical. If the 'purpose' of DNA is to supervise the building of bodies, it is surprising to find a large quantity of DNA which does no such thing. Biologists are racking their brains trying to think what this surplus DNA is doing. But from the point of view of the selfish genes themselves, there is no paradox. The true 'purpose'

> of DNA is to survive, no more and no less. The simplest way to explain the surplus DNA is to suppose that it is a parasite, or at best a harmless but useless passenger, hitching a ride in the survival machines created by the other DNA [Dawkins 1976a, p. 47].

This idea was developed further and worked out more fully by molecular biologists in two stimulating papers published simultaneously in *Nature* (Doolittle & Sapienza 1980; Orgel & Crick 1980). These papers provoked considerable discussion, in later issues of *Nature* (symposia in Vol. 285, pp. 617–20 and Vol. 288, pp. 645–8) and elsewhere (e.g. BBC radio discussion). The idea is, of course, highly congenial to the whole thesis being advanced in this book.

The facts are as follows. The total amount of DNA in different organisms is very variable, and the variation does not make obvious sense in terms of phylogeny. This is the so-called 'C-value paradox'. 'It seems totally implausible that the number of radically different genes needed in a salamander is 20 times that found in a man' (Orgel & Crick 1980). It is equally implausible that salamanders on the West side of North America should need many times more DNA than congeneric salamanders on the East side. A large percentage of the DNA in eukaryotic genomes is never translated. This 'junk DNA' may lie between cistrons, in which case it is known as spacer DNA, or it may consist of unexpressed 'introns' within cistrons, interspersed with the expressed parts of the cistron, the 'exons' (Crick 1979). The apparently surplus DNA may be to varying extents repetitive and meaningless in terms of the genetic code. Some is probably never transcribed into RNA. Other portions may be transcribed into RNA, but then 'spliced out' before the RNA is translated into amino acid sequences. Either way, it is never expressed phenotypically, if by phenotypic expression we mean expression via the orthodox route of controlling protein synthesis (Doolittle & Sapienza 1980).

This does not mean, however, that the so-called junk DNA is not subject to natural selection. Various 'functions' for it have been proposed, where 'function' means adaptive benefit to the organism. The 'function' of extra DNA may 'simply be to separate the genes' (Cohen 1977, p. 172). Even if a stretch of DNA is not itself transcribed, it can increase the frequency of crossovers between genes simply by occupying space between them, and this is a kind of phenotypic expression. Spacer DNA might, therefore, in some sense be favoured by natural selection because of its effects on crossover frequencies. It would not, however, be compatible with conventional usage to describe a length of spacer DNA as equivalent to a 'gene for' a given recombination rate. To qualify for this title, a gene must have an effect on recombination rates *in comparison with its alleles*. It is meaningful to speak of a given length of spacer DNA as having alleles—differing sequences occupying the same space on other chromosomes in the population. But since the phenotypic effect of spacing out genes is a result solely of the length of the stretch of spacer DNA, all alleles at a given 'locus' must have the same 'phenotypic expression' if they all have the same length. If the 'function' of surplus DNA is 'to' space out genes, therefore, the word function is being used in an unusual way. The natural selection process involved is not the ordinary natural selection among alleles at a locus. Rather, it is the perpetuation of a feature of the genetic system—distance between genes.

Another possible 'function' for the non-expressed DNA is that suggested by Cavalier-Smith (1978). His theory is encapsulated in his title: 'Nuclear volume control by nucleoskeletal DNA, selection for cell volume and cell growth rate, and the solution of the DNA C-value paradox.' He thinks that '*K*-selected' organisms need larger cells than '*r*-selected' ones, and that varying the total amount of DNA per cell is a good way of controlling cell size. He asserts that 'there is a good correlation between strong *r*-selection, small cells, and low C-values on the one hand and between *K*-selection, large cells, and high C-values on the other'. It would be interesting to test for this correlation statistically, taking account of the difficulties inherent in quantitative comparative surveys (Harvey & Mace 1982). The *r*/*K* distinction itself, too, seems to arouse widespread doubts among ecologists, for reasons that have never been quite clear to me, and sometimes seem unclear to them too. It is one of those concepts that is often used, but almost always accompanied by a ritualistic apology, the intellectual equivalent of touching wood. Some objective index of a species's position on the *r*/*K* continuum would be needed before a rigorous test of the correlation could be undertaken.

While awaiting further evidence for and against hypotheses of the Cavalier-Smith variety, the thing to notice in the present context is that they are hypotheses made in the traditional mould; they are based on the idea that DNA, like any other aspect of an organism, is selected because it does the organism some good. The selfish DNA hypothesis is based on an inversion of this assumption: phenotypic characters are there because they help DNA to replicate itself, and if DNA can find quicker and easier ways to replicate itself, perhaps bypassing conventional phenotypic expression, it will be selected to do so. Even if the Editor of *Nature* (Vol. 285, p. 604, 1980) goes a bit far in describing it as 'mildly shocking', the theory of selfish DNA is in a way revolutionary. But once we deeply imbibe the fundamental truth that an organism is a tool of DNA, rather than the other way around, the idea of 'selfish DNA' becomes compelling, even obvious.

The living cell, especially the nucleus in eukaryotes, is packed with the active machinery of nucleic acid replication and recombination. DNA polymerase readily catalyses the replication of any DNA, regardless of whether or not that DNA is meaningful in terms of the genetic code. 'Snipping out' of DNA, and 'splicing in' of other bits of DNA, are also parts of the normal stock in trade of the cellular apparatus, for they occur every time there is a crossover or other type of recombination event. The fact that inversions and other translocations so readily occur further testifies to the casual ease with which chunks of DNA may be cut out of one part of the genome, and spliced into another part. Replicability and 'spliceability' seem to be among the most salient features of DNA in its natural environment (Richmond 1979) of cellular machinery.

Given the possibilities of such an environment, given the existence of cellular factories set up for the replication and splicing of DNA, it is only to be expected that natural selection would favour DNA variants that are able to exploit the conditions to their own advantage. Advantage, in this case, simply means multiple replication in germ-lines. Any variety of DNA whose properties happen to make it readily replicated will automatically become commoner in the world.

What might such properties be? Paradoxically, we are most familiar with the more indirect, elaborate, and roundabout methods by which DNA molecules secure their future. These are their phenotypic effects on bodies, achieved by the proximal route of controlling protein synthesis, and hence by the more distal routes of controlling embryonic development of morphology, physiology, and behaviour. But there are also much more direct and simple ways in which varieties of DNA can spread at the expense of rival varieties. It is becoming increasingly evident that, in addition to the large, orderly chromosomes with their well-regimented gavotte, cells are home to a motley riff-raff of DNA and RNA fragments, cashing in on the perfect environment provided by the cellular apparatus.

These replicating fellow-travellers go by various names depending on size and properties: plasmids, episomes, insertion sequences, plasmons, virions, transposons, replicons, viruses. Whether they should be regarded as rebels who have broken away from the chromosomal gavotte, or as invading parasites from outside, seems to matter less and less. To take a parallel, we may regard a pond, or a forest, as a community with a certain structure, and even a certain stability. But the structure and stability are maintained in the face of a constant turnover of participants. Individuals immigrate and emigrate, new ones are born and old ones die. There is a fluidity, a jumping in and out of component parts, so that it becomes meaningless to try to distinguish 'true' community members from foreign invaders. So it is with the genome. It is not a static structure, but a fluid community. 'Jumping genes' immigrate and emigrate (Cohen 1976).

> Since the range of possible hosts in nature at least for transforming DNA and for plasmids such as RP4 is so large, one feels that at least in Gram-negative bacteria all populations may indeed be connected. It is known that bacterial DNA can be expressed in widely different host species.... It may indeed be impossible to view bacterial evolution in terms of simple family trees; rather, a network, with converging as well as diverging junctions, may be a more appropriate metaphor [Broda 1979, p. 140].

Some authors speculate that the network is not confined to bacterial evolution (e.g. Margulis 1976).

> There is substantial evidence that organisms are not limited for their evolution to genes that belong to the gene pool of their species. Rather it seems more plausible that in the time-scale of evolution the whole of the gene pool of the biosphere is available to all organisms and that the more dramatic steps and apparent discontinuities in evolution are in fact attributable to very rare events involving the adoption of part or all of a foreign

> genome. Organisms and genomes may thus be regarded as compartments of the biosphere through which genes in general circulate at various rates and in which individual genes and operons may be incorporated if of sufficient advantage [Jeon & Danielli 1971].

That eukaryotes, including ourselves, may not be insulated from this hypothetical genetic traffic is suggested by the fast-growing success of the technology of 'genetic engineering', or gene manipulation. The legal definition of gene manipulation in Britain is 'the formation of new combinations of heritable material by the insertion of nucleic acid molecules, produced by whatever means outside the cell, into any virus, bacterial plasmid or other vector system so as to allow their incorporation into a host organism in which they do not naturally occur but in which they are capable of continued propagation' (Old & Primrose 1980, p. 1). But of course human genetic engineers are beginners in the game. They are just learning to tap the expertise of the natural genetic engineers, the viruses and plasmids that have been selected to make their living at the trade.

Perhaps the greatest feat of natural genetic engineering on the grand scale is the complex of manipulations associated with sexual reproduction in eukaryotes: meiosis, crossing-over, and fertilization. Two of our foremost modern evolutionists have failed to explain to their own satisfaction the advantage of this extraordinary procedure for the individual organism (Williams 1975; Maynard Smith 1978a). As Maynard Smith (1978a, p. 113) and Williams (1979) both note, this may be one area where, if nowhere else, we shall have to switch our attention away from individual organisms to true replicators. When we try to solve the paradox of the cost of meiosis, perhaps instead of worrying about how sex helps the *organism* we should search for replicating 'engineers' of meiosis, intracellular agents which actually cause meiosis to happen. These hypothetical engineers, fragments of nucleic acid which might lie inside or outside the chromosomes, would have to achieve their own replication success as a byproduct of forcing meiosis upon the organism. In bacteria, recombination is achieved by a separate fragment of DNA or 'sex factor' which, in older textbooks, was treated as a part of the bacterium's own adaptive machinery, but which is better regarded as a replicating genetic engineer working for its own good. In animals, centrioles are thought to be self-replicating entities with their own DNA, like mitochondria, although unlike mitochondria they often pass through the male as well as the female line. Although at present it is just a joke to picture chromosomes being dragged kicking and screaming into the second anaphase by ruthlessly selfish centrioles or other miniature genetic engineers, stranger ideas have become commonplace in the past. And, after all, orthodox theorizing has so far failed to dent the paradox of the cost of meiosis.

Orgel and Crick (1980) say much the same about the lesser paradox of variable C values and about the selfish DNA theory to account for it: 'The main facts are, at first sight, so odd that only a somewhat unconventional idea is likely to explain them.' By a combination of fact and daydreamed extrapolation, I have tried to set a stage on to which selfish DNA itself can enter almost unnoticed; to paint a backdrop against which selfish DNA will appear, not unconventional but almost inescapable. DNA that is not translated into protein, whose codons would spell meaningless gibberish if they ever were translated, may yet vary in its replicability, its spliceability, and its resistance to detection and deletion by the debugging routines of the cellular machinery. 'Intragenomic selection' can therefore lead to an increase in the amount of certain types of meaningless, or untranscribed, DNA, littered around and cluttering up the chromosomes. Translated DNA, too, may be subject to this kind of selection, although here the intragenomic selection pressures will probably be swamped by more powerful pressures, positive and negative, resulting from conventional phenotypic effects.

Conventional selection results in changes in the frequency of replicators relative to their alleles at defined loci on the chromosomes of populations. Intragenomic selection of selfish DNA is a different kind of selection. Here we are not dealing with the relative success of alleles at one locus in a gene-pool, but with the spreadability of certain kinds of DNA to *different* loci or the creation of new loci. Moreover, the selection of selfish DNA is not limited to the time-scale of individual generations; it can selectively increase in any mitotic cell division in the germ-lines of developing bodies.

In conventional selection, the variation on which selection acts is produced, ultimately, by mutation, but we usually think of it as mutation within the constraints of an orderly system of loci: mutation produces a variant gene at a named locus. It is therefore possible to think of selection as choosing among alleles at such a discrete locus. Mutation in the larger sense, however, includes more radical changes in the genetic system, minor ones such as inversions, and major ones such as changes in chromosome number or ploidy, and changes from sexuality to asexuality and vice versa. These larger mutations 'change the rules of the game' but they are still, in various senses, subject to natural selection. Intragenomic selection for selfish DNA belongs in the list of unconventional types of selection, not involving choice among alleles at a discrete locus.

Selfish DNA is selected for its power to spread 'laterally', to get itself duplicated into new loci elsewhere in the genome. It does not spread at the expense of a particular set of alleles, in the way that, say, a gene for melanism in moths spreads in industrial areas at the expense of its alleles at the same locus. It is this that distinguishes it, as a 'laterally spreading outlaw', from the 'allelic outlaws' which were the subject of the previous chapter. Lateral spreading to new loci is like the spread of a virus through a population, or like the spread of cancer cells through a body. Orgel and Crick indeed refer to the spreading of functionless replicators as a 'cancer of the genome'.

As for the qualities which are actually likely to be favoured in the selection of selfish DNA, I would have to be a molecular biologist in order to predict them in detail. One does not have to be a molecular biologist, however, to surmise that they might be classified into two main classes: qualities that make for ease of duplication and insertion, and qualities that make it difficult for defence mechanisms of the cell to seek them out and destroy them. Just as a cuckoo egg seeks protection by mimicking the host eggs which legitimately inhabit the nest, so selfish DNA might evolve mimetic qualities that 'make it more like ordinary DNA and so, perhaps, less easy to remove' (Orgel & Crick 1980). Just as a full understanding of cuckoo adaptations is likely to require a knowledge of the perceptual systems of hosts, so a full appreciation of the details of selfish DNA adaptations will require detailed knowledge of exactly how DNA polymerase works, exactly how snipping and splicing occur, exactly what goes on in molecular 'proof-reading'. While full knowledge of these matters can only come from detailed research work of the kind that molecular biologists have so brilliantly succeeded in before, it is, perhaps, not too much to hope that molecular biologists might be aided in their research by the realization that DNA is not working for the good of the cell but for the good of itself. The machinery of replication, splicing, and proof-reading may be better understood if it is seen as in part the product of a ruthless arms race. The point may be emphasized with the aid of an analogy.

Imagine that Mars is a Utopia in which there is complete trust, total harmony, no selfishness and no deceit. Now imagine a scientist from Mars trying to make sense of human life and technology. Suppose he studied one of our large data processing centres—an electronic computer with its associated machinery of duplication, editing, and error-correction. If he made the assumption—natural to his own society—that the machinery had been designed for the common good, he would go a long way towards understanding it. Error-correcting devices, for instance, would clearly be designed to combat the inevitable and non-malevolent Second Law of Thermodynamics. But certain aspects would remain puzzling. He would make no sense of the elaborate and costly systems of security and protection: secret passwords and code numbers that have to be typed in by computer users. If our Martian examined a military electronic communication system he might diagnose its purpose as the rapid and efficient transmission of useful information, and he might therefore be baffled by the trouble and expense to which the system seems to go in order to encode its messages in a way which is obscure and hard to decode. Is this not wanton and absurd inefficiency? Brought up as he is in a trusting Utopia, it might require a major flash of revolutionary insight for our Martian to see that much of human technology only makes sense when you realize that humans *distrust* each other, that some humans work against the best interests of other humans. There is a struggle between those who wish to obtain illicit information from a communication system and those who wish to withhold that information from them. Much of human technology is the product of arms races and can only be understood in those terms.

Spectacular as their achievements have been, is it possible that molecular biologists have hitherto, like biologists at other levels, been in something like the position of our Martian? By assuming that the cell is a place where molecular machinery whirrs for the good of the organism, they have come a long way. They may go further if they now cultivate a more cynical view and countenance the possibility that some molecules may be up to no good from the point of view of the rest. Obviously they already do this when they contemplate viruses and other invading parasites. All that is needed is to turn the same cynical eye on the cell's 'own' DNA. It is because they are starting to do just this that I find the papers of Doolittle and Sapienza and Orgel and Crick so exciting, in comparison with the objections of Cavalier-Smith (1980), Dover (1980), and others, although of course the objectors may be right in the particular detailed points they make. Orgel and Crick summarize the message well:

> In short, we may expect a kind of molecular struggle for existence within the DNA of the chromosomes, using the process of natural selection. There is no reason to believe that this is likely to be much simpler or more easy to predict than evolution at any other level. At bottom, the existence of selfish DNA is possible because DNA is a molecule which is replicated very easily and because selfish DNA occurs in an environment in which DNA replication is a necessity. It thus has the opportunity of subverting these essential mechanisms to its own purpose.

In what sense is selfish DNA an outlaw? It is an outlaw to the extent that organisms would be better off without it. Perhaps it wastes space and molecular raw materials, perhaps it wastefully consumes valuable running time on the duplicating and proofreading machinery. In any event we may expect that selection will tend to eliminate selfish DNA from the genome. We may distinguish two kinds of 'anti-selfish DNA' selection. Firstly, selection might favour positive adaptations to rid organisms of selfish

DNA. For example, the already discovered proof-reading principle might be extended. Long sequences might be examined for 'sense' and excised if found wanting. In particular, highly repetitive DNA might be recognized by its statistical uniformity. These positive adaptations are what I had in mind in my above discussion of arms races, 'mimicry', etc. We are talking, here, about the evolution of anti-selfish DNA machinery which could be as elaborate and as specialized as the antipredator adaptations of insects.

There is, however, a second kind of selection that could act against selfish DNA, which is much simpler and cruder. Any organism that happened to experience a random deletion of part of its selfish DNA would, by definition, be a mutant organism. The deletion itself would be a mutation, and it would be favoured by natural selection to the extent that organisms possessing it benefited from it, presumably because they did not suffer the economic wastage of space, materials, and time that selfish DNA brings. Mutant organisms would, other things being equal, reproduce at a higher rate than the loaded down 'wild-type' individuals, and the deletion would consequently become more common in the gene-pool. Note that I am not now talking about selection in favour of the *capacity* to delete selfish DNA: that was the subject of the previous paragraph. Here we are recognizing that the deletion itself, the *absence* of the selfish DNA, is itself a replicating entity (a replicating absence!), which can be favoured by selection.

It is tempting to include under the heading of outlaws somatic mutations that cause cells to out-reproduce the non-mutant cells of a body, to the ultimate detriment of the body itself. But although there is a kind of quasi-Darwinian selection that can go on in cancer tumours, and Cairns (1975) has ingeniously drawn attention to what appear to be bodily adaptations to forestall such within-body selection, I think that to apply the outlaw concept here would not be helpful. Not, that is to say, unless the mutant genes concerned somehow manage to propagate themselves indefinitely. They could do this either by getting themselves transported in virus-like vectors, say through the air, or by somehow burrowing into the nuclear germ-line. In either of these two cases they would qualify as 'germ-line replicators' as defined in Chapter 5 and the title of outlaw would be appropriate.

There has been one startling recent suggestion that genes which are beneficiaries of somatic selection might indeed burrow into the germ-line, though in this case they are not cancerous and not necessarily outlaws. I want to mention this work because it has been given publicity as a would-be resuscitator of the so-called 'Lamarckian' theory of evolution. Since the theoretical position adopted in this book is fairly-describable as 'extreme Weismannism', I am bound to see any serious revival of Lamarckism as undermining my position. It is therefore necessary to discuss it.

A Lamarckian scare

I use the word 'scare' because, to be painfully honest, I can think of few things that would more devastate my world view than a demonstrated need to return to the theory of evolution that is traditionally attributed to Lamarck. It is one of the few contingencies for which I might offer to eat my hat. It is therefore all the more important to give a full and fair hearing to some claims made on behalf of Steele (1979) and Gorczynski and Steele (1980, 1981). Before Steele's (1979) book was available in Britain, *The Sunday Times* of London (13 July 1980) printed a full-page article about his ideas and 'astonishing experiment which seems to challenge Darwinism and resurrect Lamarck'. The BBC have given the results similar publicity, in at least two television programmes and several radio programmes: as we have already seen, 'scientific' journalists are constantly on the alert for anything that sounds like a challenge to Darwin. No less a scientist than Sir Peter Medawar forced us to take Steele's work seriously by doing so himself. He was quoted as being properly cautious about the need to repeat the work, and as concluding: 'I have no idea what the outcome will be, but I hope Steele is right' (*The Sunday Times*).

Naturally any scientist hopes that the truth, whatever it is, will out. But a scientist is also entitled to his innermost hopes as to what that truth will turn out to be—a revolution in one's head is bound to be a painful experience—and I confess that my own hopes did not initially coincide with Sir Peter's! I had my doubts about whether his could really coincide with those attributed to him, until I recalled his, to me always slightly puzzling, remark that 'The main weakness of modern evolutionary theory is its lack of a fully worked out theory of variation, that is, of *candidature* for evolution, of the form in which genetic variants are proffered for selection. We have therefore no convincing account of evolutionary progress—of the otherwise inexplicable tendency of organisms to adopt ever more complicated solutions of the problems of remaining alive' (Medawar 1967). Medawar is one of those who have, more recently, tried very hard, and yet failed, to replicate Steele's findings (Brent *et al.* 1981).

To anticipate the conclusion I shall come to, I now view with equanimity, if with dwindling expectation (Brent *et al.* 1981; McLaren *et al.* 1981), the prospect of Steele's theory being upheld, because I now realize that, in the deepest and fullest sense, it is a Darwinian theory; a variety of Darwinian theory moreover which, like the theory of jumping genes, is particularly congenial to the thesis of this book, since it stresses selection at a level other than that of the individual organism. Though pardonable, the claim that it challenges Darwinism turns out to be just a journalistic gloat, provided Darwinism is understood in the way that I think it ought to be understood. As for Steele's theory itself, even if the facts do not uphold it, it will have done us the valuable service of forcing us to sharpen our perception of Darwinism. I am not qualified to evaluate the technical details of Steele's experiments and those of his critics (a good evaluation is given by Howard 1981), and will concentrate on discussing the impact of his theory, should the facts eventually prove to support it.

Steele forges a threefold union of Burnet's (1969) theory of clonal selection, Temin's (1974) provirus theory, and his own attack on the sanctity of Weismann's germ-line. From Burnet he gets the idea of somatic mutation leading to genetic diversity among the cells of the body. Natural selection within the body then sees to it that the body becomes populated by successful varieties of cells at the expense of unsuccessful varieties. Burnet limits the idea to a special class of cells within the immune system, and 'success' means success in neutralizing invading antigens, but Steele would generalize it to other cells. From Temin he gets the idea of RNA viruses serving as intercellular messengers, transcribing genes in one cell, carrying the information to another cell, and reverse-transcribing it back into DNA in the second cell using reverse transcriptase.

Steele uses the Temin theory, but with an important additional emphasis on *germ-line* cells as recipients of reverse-transcribed genetic information. He wisely limits most of his discussion to the immune system, although his ambitions for his theory are larger. He cites four studies on 'idiotypy' in the rabbit. If injected with a foreign substance, different individual rabbits combat it by making different antibodies. Even if members of a genetically identical clone are subjected to the same antigen, each individual responds with its own unique 'idiotype'. Now, if the rabbits really are genetically identical, the difference in their idiotypes must be due to environmental or chance differences, and should not, according to orthodoxy, be inherited. Of the four studies cited, one gave a surprising result. The idiotype of a rabbit turned out to be inherited by its children, although not shared by its clone-mates. Steele stresses the fact that in this study the parent rabbits were exposed to the antigen *before* mating to produce the off-spring. In the other three studies the parents were injected with antigen *after* mating, and the offspring did not inherit their idio-types. If idiotype were inherited as a part of an inviolate germ-plasm, it should not have made any difference whether the rabbits mated before or after injection.

Steele's interpretation begins with the Burnet theory. Somatic mutation generates genetic diversity in the population of immune cells. Clonal selection favours those genetic varieties of cell that satisfactorily destroy the antigen, and they become very numer-ous. There is more than one solution to any antigenic problem, and the end result of the selection process is different in every rab-bit. Now Temin's proviruses step in. They transcribe a random sample of the genes in the immune cells. Because cells carrying successful antibody genes outnumber the others, these successful genes are statistically most likely to be transcribed. The provi-ruses cart these genes off to the germ cells, burrow into the germ-line chromosomes, and leave them there, presumably snipping out the incumbent occupants of the locus as they do so. The next generation of rabbits is thereby able to benefit directly from the immunological experience of its parents, without having to expe-rience the relevant antigens themselves, and without the painfully slow and wasteful intervention of selective organism death.

The really impressive evidence only became available after Steele's theory was cut and dried and published, a striking and rather surprising instance of science proceeding in the way philosophers think it proceeds. Gorczynski and Steele (1980) investigated the inheritance, via the father, of immune tolerance in mice. Using an extremely high-dosage version of the classic Medawar method, they exposed baby mice to cells from another strain, thereby rendering them tolerant as adults to subsequent grafts from the same donor strain. They then bred from these tolerant males, and concluded that their tolerance was inherited by about half their children, who were not exposed as infants to the foreign antigens. Furthermore, the effect seemed to carry over to the grandchild generation.

Subject to confirmation, we have here a prima-facie case for the inheritance of acquired characteristics. Gorczynski and Steele's brief discussion of their experiment, and of extended experiments reported more recently (Gorczynski & Steele 1981), resembles Steele's interpretation of the rabbit work, paraphrased above. The main differences between the two cases are firstly that the rabbits could have inherited something in maternal cytoplasm while the mice could not; and secondly that the rabbits were alleged to have inherited an acquired immunity, while the mice are supposed to have inherited an acquired tolerance. These differences are probably important (Brent *et al.* 1981), but I shall not make much of them, since I am not attempting to evaluate the experimental results themselves. I shall concentrate on the question of whether, in any case, Steele is really offering 'a Lamarckian challenge to Darwinism'.

There are some historical points to get out of the way first. The inheritance of acquired characteristics is not the aspect of his theory that Lamarck himself emphasized and, *contra* Steele (1979, p. 6), it is not true that the idea originated with him: he simply took over the conventional wisdom of his time and grafted to it other principles like 'striving' and 'use and disuse'. Steele's viruses seem more reminiscent of Darwin's own pangenetic 'gemmules' than of anything postulated by Lamarck. But I mention history just to get it out of the way. We give the name Darwinism to the theory that undirected variation in an insulated germ-line is acted upon by selection of its phenotypic consequences. We give the name Lamarckism to the theory that the germ-line is not insulated, and that environmentally imprinted improvements may directly mould it. In this sense, is Steele's theory Lamarckian and anti-Darwinian?

By inheriting their parents' acquired idiotypes, rabbits would undoubtedly benefit. They would begin life with a head start in the immunological battle against plagues that their parents met, and that they themselves are likely to meet. This is a directed, adaptive change, then. But is it really imprinted by the environment? If antibody formation worked according to some kind of 'instructive' theory, the answer would be yes. The environment, in the shape of antigenic protein molecules, would then directly mould antibody molecules in parent rabbits. If the offspring of those rabbits turned out to inherit a predilection to make the same antibodies, we would have full-blooded Lamarckism. But the conformations of the antibody proteins would, on this theory, somehow have to be reverse-translated into nucleotide code. Steele (p. 36) is adamant that there is no suggestion of such reverse translation, only reverse *transcription* from RNA to DNA. He is *not* proposing any violation of Crick's central dogma, although of course others are at liberty to do so (I shall return to this point in a more general context later).

The very essence of Steele's hypothesis is that the adaptive improvement comes about through selection of initially random variation. It is about as Darwinian a theory as it is possible to be, provided we think of the replicator and not the organism as the unit of selection. Nor is it just vaguely analogous to Darwinism, in the way of, say, the 'meme' theory, or Pringle's (1951) theory that learning results from selection among a pool of oscillation frequencies in a population of neuronal coupled oscillators. Steele's replicators are DNA molecules in cell nuclei. They are not just *analogous* to the replicators of Darwinism. They are the very same replicators. The scheme for natural selection which I outlined in Chapter 5 needs no modification to plug straight into the Steele theory. Steele's kind of Lamarckism only seems like the imprinting of environmental features on the germ-line if we think at the level of the individual organism. It is true that he is claiming that characteristics acquired by the *organism* are inherited. But if we look at the lower level of genetic replicators, it is clear that the adaptation comes about through selection not 'instruction' (see below). It just happens to be selection within the organism. Steele (1979, p. 43) would not disagree: 'it depends very much on essential Darwinian principles of natural selection'.

Despite Steele's avowed indebtedness to Arthur Koestler, there is naught here for the comfort of those, usually non-biologists, whose antipathy to Darwinism is fundamentally provoked by the bogy of 'blind chance'. Or, for that matter, the twin bogy of a ruthlessly indifferent grim reaper, mocking us as the sole First Cause of our exalted persons, modifying 'all things by blindly starving and murdering everything that is not lucky enough to survive in the universal struggle for hogwash' (Shaw 1921). If Steele proves right, we should hear no triumphant chuckles from the shade of Bernard Shaw! Shaw's live spirit rebelled passionately against the Darwinian 'chapter of accidents': 'it seems simple, because you do not at first realize all that it involves. But when its whole significance dawns on you, your heart sinks into a heap of sand within you. There is a hideous fatalism about it, a ghastly and damnable reduction of beauty and intelligence, of strength and purpose, of honor and aspiration'. If we *must* place emotion before truth, I have always found natural selection to have an inspiring, if grim and austere, poetry of its own—a 'grandeur in this view of life' (Darwin 1859). All I am saying here is that if you are squeamish about 'blind chance', do not look to Steele's theory for an escape. But perhaps it is not too much to hope that a proper understanding of Steele's theory may help to show that 'blind chance' is not the adequate epitome of Darwinism that Shaw, Cannon (1959), Koestler (1967), and others think it is.

Steele's theory, then, is a version of Darwinism. The cells which are selected, according to the Burnet theory, are vehicles for active replicators, namely the somatically mutated genes within them. They are active, but are they *germ-line* replicators? The essence of what I am saying is that the answer is an emphatic yes, if Steele's addition to the Burnet theory is true. They do not belong to what we have conventionally thought of as the germ-line, but it is a logical implication of the theory that we have simply been mistaken as to what the germ-line truly is. Any gene in a 'somatic' cell which is a candidate for proviral conveyance into a germ cell is, *by definition*, a germ-line replicator. Steele's book might be retitled *The Extended Germ-line!* Far from being uncomfortable for neo-Weismannists, it turns out to be deeply congenial to us.

Perhaps, then, it is not really all that ironic that, apparently unknown to Steele, something bearing more than a passing resemblance to his theory was adopted by, of all people, Weismann himself, in 1894. The following account is taken from Ridley (1982; Maynard Smith 1980, also noted the precedent). Weismann developed an idea from Roux, which he called 'intra-selection'. I quote from Ridley: 'Roux had argued that there is a struggle for food between the parts of an organism just like the struggle for existence between organisms.... Roux's theory was that the struggle of the parts, together with the inheritance of acquired characters, was sufficient to explain adaptation.' Substitute 'clones' for 'parts', and you have Steele's theory. But, as might be expected, Weismann did not go all the way with Roux in postulating the literal inheritance of acquired characteristics. Instead, in his theory of 'germinal selection', he invoked the pseudo-Lamarckian principle which later became known as the 'Baldwin Effect' (Weismann was not the only one to discover the idea before Baldwin). Weismann's use of the theory of intra-selection in explaining coadaptation will be dealt with below, for it closely parallels one of Steele's own preoccupations.

Steele does not venture far from his own field of immunology, but he would like a version of his theory to apply elsewhere, and particularly to the nervous system and the adaptive improvement mechanism known as learning. 'If [the hypothesis] is to have any general applicability to the evolutionary adaptation process, it *must account* for the adaptive potential of the neuronal networks of the brain and central nervous system' (Steele 1979, p. 49, his rather surprising emphasis). He seems a little vague as to exactly what might be selected within the brain, and, in case he can do something with it, I offer him a free gift of my own theory of 'selective neurone death as a possible memory mechanism' (Dawkins 1971).

But is the theory of clonal selection really likely to apply outside the domain of the immune system? Is it limited by the very special circumstances of the immune system, or might it be linked up with the old Lamarckian principle of use and disuse? Could clonal selection be embraced by the blacksmith's arms? Could the adaptive changes brought about by muscular exercise be inherited? I doubt it very much: the conditions are not right for natural selection to work within the blacksmith's arms in favour of, say, cells that flourish in an aerobic environment over those preferring anaerobic biochemistry, the successful genes being reverse-transcribed into just the right chromosomal locus in the germ-line. But even if this kind of thing were conceivable for some example outside the immune system, there is a major theoretical difficulty.

The problem is this. The qualities that make for success in clonal selection would necessarily be those that give *cells* an advantage over rival cells in the same body. These qualities need have no connection with what is good for the body as a whole, and our discussion of outlaws suggests that they might well actively conflict with what is good for the body as a whole. Indeed, to me a slightly unsatisfactory aspect of the Burnet theory itself is that the selection process at its heart is contrived *ad hoc*. It is assumed that those cells whose antibodies neutralize invading antigens will propagate at the expense of other cells. But this propagation is not due to any intrinsic cellular advantage: on the contrary, cells that did not risk their lives smothering antigens but selfishly left the task to their colleagues should, on the face of it, have a built-in advantage. The theory has to introduce an arbitrary and unparsimonious selection rule, imposed, as it were, from above, so that those cells which benefit the body as a whole become more numerous. It is as though a human dog-breeder deliberately selected for altruistic heroism in the face of danger. He could probably achieve it, but *natural* selection could not. Unadulterated clonal selection should favour selfish cells whose behaviour conflicts with the best interests of the body as a whole.

In the terms of Chapter 6, what I am saying is that vehicle selection at the cellular level is likely, under the Burnet theory, to come into conflict with vehicle selection at the organism level. This, of course, does not worry me, since I carry no brief for the organism as pre-eminent vehicle; I simply add one more entry to the list of 'outlaws I have known'; one more ingenious byway of replicator propagation, along with jumping genes and selfish DNA. But it should worry anybody, such as Steele, who sees clonal selection as a supplementary means by which *bodily* adaptations come about.

The problem goes deeper than that. It is not just that clonally selected genes would tend to be outlaws as far as the rest of the body is concerned. Steele looks to clonal selection to speed up evolution. Conventional Darwinism proceeds by differential individual success, and its speed, other things being equal, will be limited by individual generation time. Clonal selection would be limited by cellular generation time, which is perhaps two orders of magnitude shorter. This is why it might be thought to speed up evolution but, to anticipate the argument of my final chapter, it raises a deep difficulty. The success of a complex, multicellular organ, like an eye, cannot be judged in advance of the eye's starting to work. Cellular selection could not improve the design of an eye, because the selective events all take place in the prefunctional eye of an embryo. The embryo's eye is closed, and never sees an image until after cellular selection, if it existed, would be complete. The general point is that cellular selection cannot achieve the speeding up of evolution attributed to it, if the adaptation of interest has to develop on the slow scale of multicellular cooperation.

Steele has a point to make about coadaptation. As Ridley (1982) comprehensively documents, multidimensional coadaptation was one of the bugbears of the early Darwinians. For instance, to take the eye again, J. J. Murphy said, 'It is probably no exaggeration that, in order to improve such an organ as the eye at all, it must be improved in ten different ways at once' (1866, quoted by Ridley). It may be remembered that, speaking of the evolution of whales, I used a similar premise for a different purpose in Chapter 6. Fundamentalist orators still find the eye one of their most telling standbys. Incidentally, *The Sunday Times* (13 July 1980) and *The Guardian* (21 November 1978) both raise the debating point of the eye as though it were a new one, the latter paper reassuring us that an eminent philosopher (!) was rumoured to be giving the problem his best attention. Steele seems to have been initially attracted by Lamarckism because of his unease about coadaptation, and he believes that his clonal selection theory could in principle alleviate the difficulty, if difficulty there be.

Let us make use of that other chestnut of the schoolroom, the giraffe's neck, discussing it in conventional Darwinian terms first. A mutation to elongate the ancestral neck might work on, say, the vertebrae, but it seems to a naive observer too much to hope that the same mutation will simultaneously elongate the arteries, veins, nerves, etc. Actually, whether it is too much to hope depends on details of embryology which we should learn to be more aware of: a mutation that acts sufficiently early in development could easily have all those parallel effects simultaneously. However, let us play along with the argument. The next step is to say that it is difficult to imagine a mutant giraffe with elongated vertebrae being able to exploit its treetop browsing advantage, because its nerves, blood vessels, etc. are too short for its neck. Conventional Darwinian selection, naively understood, has to wait for an individual that is fortunate enough to combine all the necessary coadapted mutations simultaneously. This is where clonal selection could come to the rescue. One major mutation, say the elongation of the vertebrae, sets up conditions in the neck which select for clones of cells that can flourish in that environment. Maybe the elongated vertebrae provide an overstretched, highly strung, tense environment in the neck, where only elongated cells flourish. If there is genetic variation among the cells, genes 'for' elongated cells survive and are passed on to the giraffe's children. I have put the point facetiously, but a more sophisticated version of it could, I suppose, follow from the clonal selection theory.

I said I would return to Weismann at this point, for he too saw the utility of within-body selection as a solution to the coadaptation problem. Weismann thought that 'intra-selection'—the selective struggle among parts within the body—'would ensure that all the parts inside the organism were of the best mutual proportions' (Ridley 1982). 'If I am not mistaken, the phenomenon which Darwin called *correlation*, and justly regarded as an important factor in evolution, is for the most part an effect of intra-selection' (Weismann, quoted in Ridley). As already stated, Weismann, unlike Roux, did not go on to invoke the direct inheritance of intra-selected varieties. Rather, 'in each separate individual the necessary adaptation will be temporarily accomplished by intra-selection.... Time would thus be gained till, in the course of generations, by constant selection of those germs the primary constituents of which are best suited to one another, the greatest possible degree of harmony may be reached.' I think I find Weismann's 'Baldwin Effect' version of the theory more plausible than Steele's Lamarckian version, and just as satisfying an explanation of coadaptation.

I used the word 'scare' in the heading of this section, and went so far as to say that a true revival of Lamarckism would devastate my world-view. Yet the reader may now feel that the statement was a hollow one, like that of a man who dramatically threatens to eat his hat, while knowing full well that his hat is made of pleasantly flavoured ricepaper. A Lamarckian zealot may complain that the last resort of a Darwinian, having failed to discredit awkward experimental results, is to claim them for his own; to make his own theory so resilient that there is no experimental result that could falsify it. I am sensitive to this criticism, and must reply to it. I must show that the hat that I threatened to eat really is tough and distasteful. So, if Steele's kind of Lamarckism is really Darwinism in disguise, what kind of Lamarckism would not be?

The key issue is the origin of adaptedness. Gould (1979) makes a related point when he says that the inheritance of acquired characters, *per se*, is not Lamarckian: 'Lamarckism is a theory of *directed* variation' (my emphasis). I distinguish two classes of theory of the origin of adaptedness. For fear of getting caught up in historical details of exactly what Lamarck and Darwin said, I shall not any more call them Lamarckism and Darwinism. Instead, I shall borrow from immunology and call them the instruction theory and the selection theory. As Young (1957), Lorenz (1966), and others have emphasized, we recognize adaptedness as an informational match between organism and environment. An animal that is well adapted to its environment can be regarded as embodying information about its environment, in the way that a key embodies information about the lock that it is built to undo. A camouflaged animal has been said to carry a picture of its environment on its back.

Lorenz distinguished two kinds of theory for the origin of this kind of fit between organism and environment, but both his theories (natural selection and reinforcement learning) are subdivisions of what I am calling the selection theory. An initial pool of variation (genetic mutation or spontaneous behaviour) is worked upon by some kind of selection process (natural selection or reward/punishment), with the end result that only the variants fitting the environmental lock remain. Thus adaptedness improves by selection. The instruction theory is quite different. Whereas the selective key-maker begins with a large random pool of keys, tries them all in the lock, and discards those that don't fit, the instructive key-maker simply takes a wax impression of the lock, and makes up a good key directly. The instructively camouflaged animal resembles its environment because the environment directly imprints its appearance on the animal, as elephants merge into the background because they are covered by its dust. It has been alleged that French mouths eventually become permanently deformed into a shape suitable for pronouncing French vowels. If so, this would be an instructive adaptation. So, perhaps, is background resemblance in chameleons, though of course the *capacity* to change colour adaptively is presumably a selective adaptation. The adaptive changes in physiology to which we give names like acclimatization and training, the effects of exercise, use and disuse, are probably all instructive. Complex and elaborate adaptive fits can be achieved by instruction, as in the learning of a particular human language. As already explained, it is clear that in Steele's theory the adaptedness comes not from instruction but from selection, and genetic replicator selection at that. My world-view will be overturned if somebody demonstrates the genetic inheritance, not just of an 'acquired characteristic' but of an instructively acquired adaptation. The reason is that the inheritance of an instructively acquired adaptation would violate the 'central dogma' of embryology.

The poverty of preformationism

Oddly, my belief in the inviolability of the central dogma is not a dogmatic one! It is based on reason. I must be cautious here, and distinguish two forms of central dogma, the central dogma of molecular genetics and the central dogma of embryology. The first is the one stated by Crick: genetic information may be translated from nucleic acid to protein, but not the other way. Steele's theory, as he himself is careful to point out, does not violate this dogma. He makes use of reverse transcription from RNA to

DNA, but not reverse translation from protein to RNA. I am not a molecular biologist, and so cannot judge the extent to which the theoretical boat would be rocked if such reverse translation were ever discovered. It does not seem to me obviously impossible in principle, because the translation from nucleic acid to protein, or from protein to nucleic acid, is a simple dictionary look-up procedure, only slightly more complex than the DNA/RNA transcription. In both cases there is a one-to-one mapping between the two codes. If a human, or a computer, equipped with the dictionary, can translate from protein to RNA, I don't see why nature should not. There may be a good theoretical reason, or it may just be an empirical law that has not yet been violated. There is no need for me to pursue the matter because, in any case, a very good theoretical case can be made against violating the other central dogma, the central dogma of embryology. This is the dogma that the macroscopic form and behaviour of an organism may be, in some sense, coded in the genes, but the code is irreversible. If Crick's central dogma states that protein may not be translated back into DNA, the central dogma of embryology states that bodily form and behaviour may not be translated back into protein.

If you sleep in the sun with your hand over your chest, a white image of your hand will be imprinted on your otherwise tanned body. This image is an acquired characteristic. In order for it to be inherited, gemmules or RNA viruses, or whatever agent of reverse translation is postulated, would have to scan the *macroscopic* outline of the hand image and translate it into the molecular structure of DNA necessary to program the development of a similar hand image. It is suggestions of this kind that constitute a violation of the central dogma of embryology.

The central dogma of embryology does not follow inevitably from common sense. Rather, it is a logical implication of rejecting the preformationist view of development. I suggest, indeed, that there is a close link between the epigenetic view of development and the Darwinian view of adaptation, and between preformationism and the Lamarckian view of adaptation. You may believe in inheritance of Lamarckian (i.e. 'instructive') adaptations, but only if you are prepared to embrace a preformationistic view of embryology. If development were preformationistic, if DNA really *were* a 'blueprint for a body', really were a codified homunculus, reverse development—looking-glass embryology—would be conceivable.

But the blueprint metaphor of the textbooks is dreadfully misleading, for it implies a one-to-one mapping between bits of body and bits of genome. By inspecting a house, we may reconstruct a blueprint from which somebody else could build an identical house, using the same building technique as was used for the original house. The informational arrows from blueprint to house are reversible. The relative positions of the ink lines in the blueprint and of the brick walls in the house are transformable, one into the other, by a few simple scaling rules. To go from blueprint to house, you multiply all measurements by, say twenty. To go from house to blueprint, you divide all measurements by twenty. If the house somehow acquires a new feature, say a west wing, a simple, automatic procedure could be written down for adding a scaled-down map of the west wing to the blueprint. If the genome were a blueprint with a one-to-one mapping from genotype to phenotype, it would not be inconceivable that the white imprint of a hand on an otherwise tanned chest could be mapped on to a sort of miniature genetic shadow of itself, and so inherited.

But this is utterly alien to everything we now understand about the way development works. The genome is not, in any sense whatsoever, a scale model of the body. It is a set of instructions which, if faithfully obeyed in the right order and under the right conditions, will result in a body. I have previously used the metaphor of a cake. When you make a cake you may, in some sense, say that you are 'translating' from recipe to cake. But it is an irreversible process. You cannot dissect a cake and thereby reconstruct the original recipe. There is no one-to-one, reversible mapping from words of recipe to crumbs of cake. This is not to say that a skilled cook could not achieve a passable reversal, by taking a cake presented to him and matching its taste and properties against his own past experience of cakes and recipes, and then reconstructing the recipe. But that would be a kind of mental selection procedure, and would in no sense be a *translation* from cake to recipe (a good discussion of the difference between reversible and irreversible codes, in the context of the nervous system, is given by Barlow (1961)).

A cake is the consequence of the obeying of a series of instructions, when to mix the various ingredients, when to apply heat, etc. It is not true that the cake *is* those instructions rendered into another coding medium. It is not like a translation of the recipe from French into English, which is in principle reversible (give or take a few *nuances*). A body, too, is the consequence of the obeying of a series of instructions; not so much when to apply heat as when to apply enzymes speeding up particular chemical reactions. If the process of embryonic development is correctly set in motion in the right environment, the end result will be a well-formed adult body, many of whose attributes will be interpretable as consequences of its genes. But you cannot reconstruct an individual's genome by inspecting his body, any more than you could reconstruct William Shakespeare by decoding his collected works. Cannon's and Gould's false argument of pp. 176–7 is validly adapted to embryology.

Let me put the matter in another way. If a man is particularly fat, there are many ways in which this might have come about.

He might have a genetic predisposition to metabolize his food particularly thoroughly. Or he might have been overfed. The end result of an excess of food may be identical to the end result of a particular gene. In both cases the man is fat. But the routes by which the two causal agents produced their common effects are totally different. For a man who is artificially stuffed with food to pass on his acquired fatness to his children genetically, some mechanism would have to exist that sensed his fatness, then located a 'fatness gene' and caused it to mutate. But how could such a fatness gene be located? There is nothing intrinsic in the nature of the gene that makes it recognizable as a fatness gene. It has its obese effect only as a result of the long and complex unfolding sequence that is epigenetic development. The only way, in principle, to recognize a 'fatness gene' for what it is, is to allow it to exert its effects on the normal processes of development, and this means development in the normal, forward direction.

This is why bodily adaptations can come about by selection. Genes are allowed to exert their normal effects on development. Their developmental consequences—phenotypic effects—feed back on those genes' chances of surviving, and as a result gene frequencies change in succeeding generations in adaptive directions. Selective theories of adaptation, but not instructive theories, can cope with the fact that the relationship between a gene and its phenotypic effect is not an intrinsic property of the gene, but a property of the forward developmental consequences of the gene when interacting with the consequences of many other genes and many external factors.

Complex adaptation to an environment may arise in individual organisms through instruction from that environment. In many cases this certainly happens. But, given an assumption of epigenetic, not preformationistic embryology, to expect such complex adaptations to be translated into the medium of the genetic code, by some means other than the selection of undirected variation, is a gross violation of all that I hold rational.

There are other examples of what looks like true Lamarckian 'instruction' from the environment being inherited. Non-genetic anomalies that appear, or even that are surgically induced, in the cortex of ciliates, may be directly inherited. This has been demonstrated by Sonneborn and others. By Bonner's account, they cut out a small portion of the cortex of *Paramecium* and reversed it. 'The result is a *Paramecium* with part of one row of basal bodies in which the fine structure and details are all pointed 180° away from the rest of the surface. This anomalous kinety is now inherited; it appears to be a permanent fixture of the progeny (which have been carried through 800 generations)' (Bonner 1974, p. 180). The inheritance appears to be non-genic and is obviously non-nuclear. '[T]he cortex is made up of macromolecules that assemble in a particular pattern, and... this pattern, even in a disturbed state, is directly inherited.... we have a large and exceedingly complex cortex whose pattern of fitting together is a property of the macromolecules at the cortex and is not directly under nuclear control. Over what must have been a long time and a vast number of cell cycles, a surface structure evolved. The structure itself had properties such that its immediate form is independent of the nucleus; at the same time, it is totally dependent on the nucleus, we presume, for the synthesis of its specifically shaped building blocks' (Bonner 1974).

As in the case of Steele's work, whether we regard this as the inheritance of acquired characteristics depends upon our definition of the germ-line. If we direct our attention to the individual body, a surgical mutilation of its cortex is clearly an acquired characteristic having nothing to do with the nuclear germ-line. If, on the other hand, we look at underlying replicators, in this case perhaps the basal bodies of cilia, the phenomenon falls under the general heading of replicator propagation. Given that macromolecular structures in the cortex are true replicators, surgically rotating a portion of cortex is analogous to cutting out a portion of chromosome, inverting it, and putting it back. Naturally the inversion is inherited, because it is part of the germ-line. It appears that elements of the cortex of *Paramecium* have a germ-line of their own, although a particularly remarkable one in that the information transmitted does not seem to be encoded in nucleic acids. We should definitely predict that natural selection might act directly on this non-genic germ-line, shaping surface structure for the adaptive benefit of the replicating units in the surface itself. If there is any conflict between the interests of these surface replicators and nuclear genes, the resolution of the conflict should make a fascinating study.

This is by no means the only example of non-nuclear inheritance. It is becoming increasingly clear that non-nuclear genes, either in organelles such as mitochondria or loose in cytoplasm, exert noticeable effects on phenotypes (Grun 1976). I had intended to include a section called *The Selfish Plasmagene*, discussing the expected consequences of selection acting on cytoplasmic replicators, and the likely results of conflicts with nuclear genes. However, I had got no further than some brief remarks on 'selfish mitochondria' (now in Chapter 12), when two papers arrived (Eberhard 1980; Cosmides & Tooby 1981) which, independently, say everything I might have said and much more. To give just one example, 'The migration of egg mitochondria to cluster around the egg nucleus, so as to favor their inclusion in the "neocytoplasm" of the proembryo in the gymnosperms *Larix* and *Pseudotsuga*... may result from competition for inclusion in the embryo' (Eberhard 1980, p. 238). Rather than extensively duplicate what is in them I would prefer simply to recommend readers to consult both these excellent papers. I will only add that both papers are good examples of the kind of discussion that I believe will become commonplace, once the replicator replaces the individual organism as the fundamental conceptual unit in our thinking about natural selection. One does not have to be clairvoyant to prophesy, for example, the rise of a flourishing new discipline of 'prokaryotic sociobiology'.

Neither Eberhard nor Cosmides and Tooby explicitly justify or document the gene's eye view of life: they simply *assume* it: 'The recent shift towards viewing the gene as the unit of selection, coupled with a recognition of the different modes of genetic inheritance makes the concept of parasitism, symbiosis, conflict, cooperation, and coevolution—which were developed with reference to whole organisms—relevant to genes within an organism' (Cosmides & Tooby). These papers have what I can only describe as the flavour of post-revolutionary normal science (Kuhn 1970).

10

AN AGONY IN FIVE FITS

The reader may remark that we have come thus far with scarcely a mention of 'fitness'. This was deliberate. I have misgivings about the term, but I have been holding my fire. Several of the earlier chapters have been aimed, in their different ways, at exposing the weaknesses in the individual organism's candidacy for the title of optimon, the unit for whose benefit adaptations may be said to work. 'Fitness', as it is normally used by ecologists and ethologists, is a verbal trick, a device contrived to make it possible to talk in terms of individuals, as opposed to true replicators, as beneficiaries of adaptation. The word is therefore a kind of verbal symbol of the position that I am trying to argue against. More than that, the word is actively confusing because it has been used in so many different ways. It is therefore fitting to end the critical section of the book with a discussion of fitness.

Herbert Spencer's (1864) term 'survival of the fittest' was adopted by Darwin (1866) at the urging of Wallace (1866). Wallace's argument makes fascinating reading today, and I cannot resist quoting him at some length:

> My dear Darwin,—I have been so repeatedly struck by the utter inability of numbers of intelligent persons to see clearly, or at

> all, the self-acting and necessary effects of Natural Selection, that I am led to conclude that the term itself, and your mode of illustrating it, however clear and beautiful to many of us, are yet not the best adapted to impress it on the general naturalist public...in Janet's recent work on the 'Materialism of the Present Day'...he considers your weak point to be that you do not see that 'thought and direction are essential to the action of Natural Selection'. The same objection has been made a score of times by your chief opponents, and I have heard it as often stated myself in conversation. Now, I think this arises almost entirely from your choice of the term Natural Selection, and so constantly comparing it in its effects to man's selection, and also to your so frequently personifying nature as 'selecting', as 'preferring', as 'seeking only the good of the species', etc., etc. To the few this is as clear as daylight, and beautifully suggestive, but to many it is evidently a stumbling block. I wish, therefore, to suggest to you the possibility of entirely avoiding this source of misconception in your great work (if now not too late), and also in any future editions of the 'Origin', and I think it may be done without difficulty and very effectually by adopting Spencer's term (which he generally uses in preference to Natural Selection), viz. 'Survival of the Fittest'. This term is the plain expression of the fact; 'Natural Selection' is a metaphorical expression of it, and to a certain degree *indirect* and *incorrect*, since, even personifying Nature, she does not so much select special variations as exterminate the most unfavourable ones...[The Wallace–Darwin Correspondence].

It may seem hard to believe that anyone could have been misled in the way that Wallace indicates, but Young (1971) provides ample evidence confirming that Darwin's contemporaries often were so misled. Even today the confusion is not unknown, and an analogous muddle arises over the catch phrase 'selfish gene': 'This is an ingenious theory but far-fetched. There is no reason for imputing the complex emotion of selfishness to molecules' (Bethell 1978); 'Genes cannot be selfish or unselfish, any more than atoms can be jealous, elephants abstract or biscuits teleological' (Midgley 1979; see reply in Dawkins 1981).

Darwin (1866) was impressed by Wallace's letter, which he found 'as clear as daylight', and he resolved to incorporate 'survival of the fittest' into his writings, although he cautioned that 'the term Natural Selection has now been so largely used abroad and at home that I doubt whether it could be given up, and with all its faults I should be sorry to see the attempt made. Whether it will be rejected must now depend on the "survival of the fittest"' (Darwin clearly understood the 'meme' principle). 'As in time the term must grow intelligible, the objections to its use will grow weaker and weaker. I doubt whether the use of any term would have made the subject intelligible to some minds.... As for M. Janet, he is a metaphysician, and such gentlemen are so acute that I think they often misunderstand common folk.'

What neither Wallace nor Darwin could have foreseen was that 'survival of the fittest' was destined to generate more serious confusion than 'natural selection' ever had. A familiar example is the attempt, rediscovered with almost pathetic eagerness by successive generations of amateur (and even professional) philosophers ('so acute that they misunderstand common folk'?), to demonstrate that the theory of natural selection is a worthless tautology (an amusing variant is that it is unfalsifiable and therefore false!). In fact the illusion of tautology stems entirely from the phrase 'survival of the fittest', and not from the theory itself at all. The argument is a remarkable example of the elevation of words above their station, in which respect it resembles St Anselm's ontological proof of the existence of God. Like God, natural selection is too big a theory to be proved or disproved by word-games. God and natural selection are, after all, the only two workable theories we have of why we exist.

Briefly, the tautology idea is this. Natural selection is defined as the survival of the fittest, and the fittest are defined as those that survive. Therefore the whole of Darwinism is an unfalsifiable tautology and we don't have to worry our heads about it any more. Fortunately, several authoritative replies to this whimsical little conceit are available (Maynard Smith 1969; Stebbins 1977; Alexander 1980), and I need not labour my own. I will, however, chalk up the tautology idea on my list of muddles attributable to the concept of fitness.

It is, as I have said, a purpose of this chapter to show that fitness is a very difficult concept, and that there might be something to be said for doing without it whenever we can. One way I shall do this is to show that the word has been used by biologists in at least five different senses. The first and oldest meaning is the one closest to everyday usage.

Fit the First

When Spencer, Wallace, and Darwin originally used the term 'fitness', the charge of tautology would not have occurred to anyone. I shall call this original usage fitness[1]. It did not have a precise technical meaning, and the fittest were not *defined* as those that survive. Fitness meant, roughly, the capacity to survive and reproduce, but it was not defined and measured as precisely synonymous with reproductive success. It had a range of specific meanings, depending upon the particular aspect of life that one was examining. If the subject of attention was efficiency in grinding vegetable food, the fittest individuals were those with the hardest teeth or the most powerful jaw muscles. In different contexts the fittest individuals would be taken to mean those with the keenest eyes, the strongest leg muscles, the sharpest ears, the swiftest reflexes. These capacities and abilities, along with countless others, were supposed to improve over the generations, and natural selection effected that improvement. 'Survival of the fittest' was a general characterization of these particular improvements. There is nothing tautological about that.

It was only later that fitness was adopted as a technical term. Biologists thought they needed a word for that hypothetical quantity that tends to be maximized as a result of natural selection. They could have chosen 'selective potential', or 'survivability', or 'W' but in fact they lit upon 'fitness'. They did the equivalent of recognizing that the definition they were seeking must be 'whatever it takes to make the survival of the fittest into a tautology'. They redefined fitness accordingly.

But the tautology is not a property of Darwinism itself, merely of the catchphrase we sometimes use to describe it. If I say that a train travelling at an average velocity of 120 m.p.h. will reach its destination in half the time it takes a train travelling at 60 m.p.h., the fact that I have uttered a tautology does not prevent the trains from running, nor does it stop us asking meaningful questions about what makes one train faster than the other: does it have a larger engine, superior fuel, a more streamlined shape, or what? The concept of velocity is *defined* in such a way as to make statements such as the one above tautologically true. It is this that makes the concept of velocity useful. As Maynard Smith (1969) witheringly put it: 'Of course Darwinism contains tautological features: any scientific theory containing two lines of algebra does so.' And when Hamilton (1975a), speaking of 'survival of the fittest', said that 'accusations of tautology seem hardly fair on this small phrase itself', he was putting it mildly. Given the purpose for which fitness was redefined, 'survival of the fittest' *had* to become a tautology.

To redefine fitness in a special technical sense might have done no harm, other than to give some earnest philosophers a field day, but unfortunately its exact technical meaning has varied widely, and this has had the more serious effect of confusing some biologists too. The most precise and unexceptionable of the various technical meanings is that adopted by population geneticists.

Fit the Second

For population geneticists, fitness is an operational measure, exactly defined in terms of a measurement procedure. The word is applied not really to a whole individual organism but to a genotype, usually at a single locus. The fitness W of a genotype, say *Aa*, may be defined as $1 - s$, where s is the coefficient of selection against the genotype (Falconer 1960). It may be regarded as a measure of the number of offspring that a typical individual of genotype *Aa* is expected to bring up to reproductive age, when all other variation is averaged out. It is usually expressed relative to the corresponding fitness of one particular genotype at the locus, which is arbitrarily defined as 1. Then there is said to be selection, at that locus, in favour of genotypes with higher fitness, relative to genotypes with lower fitness. I shall call this special population geneticists' meaning of the term fitness[2]. When we say that brown-eyed individuals are fitter than blue-eyed individuals we are talking about fitness[2]. We assume that all other variation among the individuals is averaged out, and we are, in effect, applying the word fitness to two genotypes at a single locus.

Fit the Third

But while population geneticists are interested directly in changes in genotype frequencies and gene frequencies, ethologists and ecologists look at whole organisms as integrated systems that appear to be maximizing something. Fitness[3], or 'classical fitness', is a property of an individual organism, often expressed as the product of survival and fecundity. It is a measure of the individual's reproductive success, or its success in passing its genes on to future generations. For instance, as mentioned in Chapter 7, Clutton-Brock *et al.* (1982) are conducting a long-term study of a red deer population on the island of Rhum, and part of their aim is to compare the lifetime reproductive successes or fitness[3] of identified individual stags and hinds.

Notice the difference between the fitness[3] of an individual and the fitness[2] of a genotype. The measured fitness[2] of the brown-eyed genotype will contribute to the fitness[3] of an individual who happens to have brown eyes, but so will the fitness[2] of his genotype at all other loci. Thus the fitness[2] of a genotype at a locus can be regarded as an average of the fitness[3]s of all individuals possessing that genotype. And the fitness[3] of an individual can be regarded as influenced by the fitness[2] of his genotype, averaged over all his loci (Falconer 1960).

It is easy to measure the fitness[2] of a genotype at a locus, because each genotype, *AA*, *Aa*, etc., occurs a countable number of times in successive generations in a population. But the same is not true of the fitness[3] of an organism. You can't count the number of times an organism occurs in successive generations, because he only occurs once, ever. The fitness[3] of an organism is often measured as the number of his offspring reared to adulthood, but there is some dispute over the usefulness of this. One problem is raised by Williams (1966) criticizing Medawar (1960) who had said: 'The genetical usage of "fitness" is an extreme attenuation of the ordinary usage: it is, in effect, a system of pricing the endowments of organisms in the currency of offspring, i.e., in terms of net reproductive performance. It is a genetic valuation of goods, not a statement about their nature or quality.' Williams is worried that this is a retrospective definition, suitable for particular individuals who have existed. It suggests a posthumous evaluation of particular animals as ancestors, not a way of evaluating the qualities that can be expected to make for success in general. 'My main criticism of Medawar's statement is that it focuses attention on the rather trivial problem of the degree to which an organism actually achieves reproductive survival. The central biological problem is not survival as such, but design for survival' (Williams 1966, p. 158). Williams is, in a sense, hankering after the pre-tautological virtues of fitness[1], and there is much to be said in his favour. But the fact is that fitness[3] has become widely used by biologists in the sense described by Medawar. Medawar's passage was addressed to laymen and was surely an attempt to enable them to follow standard biological terminology while avoiding the otherwise inevitable confusion with common-usage 'athletic' fitness.

The concept of fitness has the power to confuse even distinguished biologists. Consider the following misunderstanding of Waddington (1957) by Emerson (1960). Waddington had used the word 'survival' in the sense of reproductive survival or fitness[3]: 'survival does not, of course, mean the bodily endurance of a single individual... that individual "survives" best which leaves most offspring'. Emerson quotes this, then goes on: 'Critical data on this contention are difficult to find, and it is likely that much new investigation is needed before the point is either verified or refuted.' For once, the ritual lip-service to the need for more research is utterly inappropriate. When we are talking about matters of definition, empirical research cannot help us. Waddington was clearly *defining* survival in a special sense (the sense of fitness[3]), not making a proposition of fact subject to empirical verification or falsification. Yet Emerson apparently thought

Waddington was making the provocative statement that those individuals with the highest capacity to survive tend also to be the individuals with the largest number of offspring. His failure to grasp the technical concept of fitness[3] is indicated by another quotation from the same paper: 'It would be extremely difficult to explain the evolution of the uterus and mammary glands in mammals...as the result of natural selection of the fittest individual.' In accordance with the influential Chicago-based school of thought of which he was a leader (Allee, Emerson *et al.* 1949), Emerson used this as an argument in favour of group selection. Mammary glands and uteruses were, for him, adaptations for the continuation of the species.

Workers who correctly use the concept of fitness[3] admit that it can be measured only as a crude approximation. If it is measured as the number of children born it neglects juvenile mortality and fails to account for parental care. If it is measured as number of offspring reaching reproductive age it neglects variation in reproductive success of the grown offspring. If it is measured as number of grandchildren it neglects...and so on *ad infinitum*. Ideally we might count the relative number of descendants alive after some very large number of generations. But such an 'ideal' measure has the curious property that, if carried to its logical conclusion, it can take only two values; it is an all-or-none measure. If we look far enough into the future, either I shall have no descendants at all, or all persons alive will be my descendants (Fisher 1930a). If I am descended from a particular individual male who lived a million years ago, it is virtually certain that you are descended from him too. The fitness of any particular long-dead individual, as measured in present-day descendants, is either zero or total. Williams would presumably say that if this is a problem it is so only for people that wish to measure the *actual* reproductive success of particular individuals. If, on the other hand, we are interested in qualities that tend, on average, to make individuals *likely* to end up in the set of ancestors, the problem does not arise. In any case a more biologically interesting shortcoming of the concept of fitness[3] has led to the development of two newer usages of the technical term fitness.

Fit the Fourth

Hamilton (1964a,b), in a pair of papers which we can now see to have marked a turning point in the history of evolutionary theory, made us aware of an important deficiency in classical fitness[3], the measure based on the reproductive success of an organism. The reason reproductive success matters, as opposed to mere individual survival, is that reproductive success is a measure of success in passing on genes. The organisms we see around us are descended from ancestors, and they have inherited some of the attributes that made those individuals ancestors as opposed to non-ancestors. If an organism exists it contains the genes of a long line of successful ancestors. The fitness[3] of an organism is its success as an ancestor, or, according to taste, its capacity for success as an ancestor. But Hamilton grasped the central importance of what, previously, had been only glancingly referred to in stray sentences of Fisher (1930a) and Haldane (1955). This is that natural selection will favour organs and behaviour that cause the individual's genes to be passed on, whether or not the individual is, himself, an ancestor. An individual that assists his brother to be an ancestor may thereby ensure the survival in the gene-pool of the genes 'for' brotherly assistance. Hamilton saw that parental care is really only a special case of caring for close relatives with a high probability of containing the genes for caring. Classical fitness[3], reproductive success, was too narrow. It had to be broadened to *inclusive fitness*, which will here be called fitness[4].

It is sometimes supposed that the inclusive fitness of an individual is his own fitness[3] plus half the fitnesses[3] of each brother plus one-eighth of the fitness[3] of each cousin, etc. (e.g. Bygott *et al.* 1979). Barash (1980) explicitly defines it as 'the sum of individual fitness (reproductive success) and the reproductive success of an individual's relatives, with each relative devalued in proportion as it is more distantly related'. This would not be a sensible measure to try to use, and, as West-Eberhard (1975) emphasizes, it is not the measure Hamilton offered us. The reason it would not be sensible can be stated in various ways. One way of putting it is that it allows children to be counted many times, as though they had many existences (Grafen 1979). Then again, if a child is born to one of a set of brothers, the inclusive fitness of all the other brothers would, according to this view, immediately rise an equal notch, regardless whether any of them lifted a finger to feed the infant. Indeed the inclusive fitness of another brother yet unborn would theoretically be increased by the birth of his elder nephew. Further, this later brother could be aborted shortly after conception and still, according to this erroneous view, enjoy a massive 'inclusive fitness' through the descendants of his elder brothers. To push to the *reductio ad absurdum*, he needn't even bother to be conceived, yet still have a high 'inclusive fitness'!

Hamilton clearly saw this fallacy, and therefore his concept of inclusive fitness was more subtle. The inclusive fitness of an organism is not a property of himself, but a property of his actions or effects. Inclusive fitness is calculated from an individual's own reproductive success plus his *effects* on the reproductive success of his relatives, each one weighed by the appropriate coefficient of relatedness. Therefore, for instance, if my brother emigrates to Australia, so I can have no effect, one way or the other, on his reproductive success, my inclusive fitness does not go up each time he has a child!

Now 'effects' of putative causes can only be measured by comparison with other putative causes, or by comparison with their absence. We cannot, then, think of the effects of individual A on the survival and reproduction of his relatives in any absolute sense. We could compare the effects of his choosing to perform act X rather than act Y. Or we could take the effects of his lifetime's set of deeds and compare them with a hypothetical lifetime of total inaction—as though he had never been conceived. It is this latter usage that is normally meant by the inclusive fitness of an individual organism.

The point is that inclusive fitness is not an absolute property of an organism in the same way as classical fitness[3] could in theory be, if measured in certain ways. Inclusive fitness is a property of a triple consisting of an organism of interest, an act or set of acts of interest, and an alternative set of acts for comparison. We aspire to measure, then, not the absolute inclusive fitness of the organism I, but the effect on I's inclusive fitness of his doing act X in comparison with his doing act Y. If 'act' X is taken to be I's entire life story, Y may be taken as equivalent to a hypothetical world in which I did not exist. An organism's inclusive fitness, then, is defined so that it is not affected by the reproductive success of relatives on another continent whom he never meets and whom he has no way of affecting.

The mistaken view that the inclusive fitness of an organism is a weighted sum of the reproductive successes of all its relatives everywhere, who have ever lived, and who will ever live, is extremely common. Although Hamilton is not responsible for the errors of his followers, this may be one reason why many people have so much difficulty dealing with the concept of inclusive fitness, and it may provide a reason for abandoning the concept at some time in the future. There is yet a fifth meaning of fitness, which was designed to avoid this particular difficulty of inclusive fitness, but which has difficulties of its own.

Fit the Fifth

Fitness[5] is 'personal fitness' in the sense of Orlove (1975, 1979). It can be thought of as a kind of backwards way of looking at inclusive fitness. Where inclusive fitness[4] focuses on the effects the individual of interest has on the fitness[3] of his relatives, personal fitness[5] focuses on the effects that the individual's relatives have on his fitness[3]. The fitness[3] of an individual is some measure of the number of his offspring or descendants. But Hamilton's logic has shown us that we may expect an individual to end up with more offspring then he could rear himself, because his relatives contribute towards rearing some of his offspring. An animal's fitness[5] may be briefly characterized as 'the same as his fitness[3] but don't forget that this must include the extra offspring he gets as a result of help from his relatives'.

The advantage, in practice, of using personal fitness[5] over inclusive fitness[4] is that we end up simply counting offspring, and there is no risk of a given child's being mistakenly counted many times over. A given child is a part of the fitness[5] of his parents only. He potentially corresponds to a term in the inclusive fitness[4] of an indefinite number of uncles, aunts, cousins, etc., leading to a danger of his being counted several times (Grafen 1979; Hines & Maynard Smith 1979).

Inclusive fitness[4] when properly used, and personal fitness[5], give equivalent results. Both are major theoretical achievements and the inventor of either of them would deserve lasting honour. It is entirely characteristic that Hamilton himself quietly invented both in the same paper, switching from one to another with a swiftness that bewildered at least one later author (Cassidy 1978, p. 581). Hamilton's (1964a) name for fitness[5] was 'neighbour-modulated fitness'. He considered that its use, though correct, would be unwieldy, and for this reason he introduced inclusive fitness[4] as a more manageable alternative approach. Maynard Smith (1982) agrees that inclusive fitness[4] is often easier to use than neighbour-modulated fitness[5], and he illustrates the point by working through a particular hypothetical example using both methods in turn.

Notice that both these fitnesses, like 'classical' fitness[3], are firmly tied to the idea of the individual organism as 'maximizing agent'. It is only partly facetiously that I have characterized inclusive fitness as 'that property of an individual organism which will appear to be maximized when what is really being maximized is gene survival' (Dawkins 1978a). (One might generalize this principle to other 'vehicles'. A group selectionist might define his own version of inclusive fitness as 'that property of a *group* which will appear to be maximized when what is really being maximized is gene survival'!)

Historically, indeed, I see the concept of inclusive fitness as the instrument of a brilliant last-ditch rescue attempt, an attempt to save the individual organism as the level at which we think about natural selection. The underlying spirit of Hamilton's (1964a,b) papers on inclusive fitness is gene selectionist. The brief note of 1963 which preceded them is explicitly so: 'Despite the principle of "survival of the fittest" the ultimate criterion that determines whether *G* will spread is not whether the behavior is to the benefit of the behaver but whether it is to the benefit of the gene *G*'. Together with Williams (1966), Hamilton could fairly be regarded as one of the fathers of gene selectionism in modern behavioural and ecological studies:

> A gene is being favored in natural selection if the aggregate of its replicas forms an increasing fraction of the total gene pool. We are going to be concerned with genes supposed to affect the social behavior of their bearers, so let us try to make the

> argument more vivid by attributing to the genes, temporarily, intelligence and a certain freedom of choice. Imagine that a gene is considering the problem of increasing the number of its replicas and imagine that it can choose between causing purely self-interested behavior by its bearer A (leading to more reproduction by A) and causing 'disinterested' behavior that benefits in some way a relative, B [Hamilton 1972].

Having made use of his intelligent gene model, Hamilton later explicitly abandons it in favour of the inclusive fitness effect of an *individual* on the propagation of copies of his genes. It is part of the thesis of this book that he might have done better to have stuck by his 'intelligent gene' model. If individual organisms can be assumed to work for the aggregate benefit of all their genes, it doesn't matter whether we think in terms of genes working to ensure their survival, or of individuals working to maximize their inclusive fitness. I suspect that Hamilton felt more comfortable with the individual as the agent of biological striving, or perhaps he surmised that most of his colleagues were not yet ready to abandon the individual as agent. But, of all the brilliant theoretical achievements of Hamilton and his followers, which have been expressed in terms of inclusive fitness[4] (or personal fitness[5]), I cannot think of any that could not have been more simply derived in terms of Hamilton's 'intelligent gene', manipulating bodies for its own ends (Charnov 1977).

Individual-level thinking is superficially attractive because individuals, unlike genes, have nervous systems and limbs which render them capable of working in obvious ways to maximize something. It is therefore natural to ask what quantity, in theory, they might be expected to maximize, and inclusive fitness is the answer. But what makes this so dangerous is that it, too, is really a metaphor. Individuals do not consciously strive to maximize anything; they behave *as if* maximizing something. It is exactly the same 'as if' logic that we apply to 'intelligent genes'. Genes manipulate the world as if striving to maximize their own survival. They do not really 'strive', but my point is that in this respect they do not differ from individuals. Neither individuals nor genes really strive to maximize anything. Or, rather, individuals may strive for something, but it will be a morsel of food, an attractive female, or a desirable territory, not inclusive fitness. To the extent that it is useful for us to think of individuals working as if to maximize fitness, we may, with precisely the same licence, think of genes as if they were striving to maximize their survival. The difference is that the quantity the genes may be thought of as maximizing (survival of replicas) is a great deal simpler and easier to deal with in models than the quantity individuals may be thought of as maximizing (fitness). I repeat that if we think about individual animals maximizing something, there is a serious danger of our confusing ourselves, since we may forget whether we are using 'as if' language or whether we are talking about the animals consciously striving for some goal. Since no sane biologist could imagine DNA molecules consciously striving for anything, the danger of this confusion ought not to exist when we talk of genes as maximizing agents.

It is my belief that thinking in terms of individuals striving to maximize something has led to outright error, in a way that thinking in terms of genes striving to maximize something would not. By outright error, I mean conclusions that their perpetrators would admit are wrong after further reflection. I have documented these errors in the section labelled 'Confusion' of Dawkins (1978a), and in Dawkins (1979a, especially Misunderstandings 5, 6, 7, and 11). These papers give detailed examples, from the published literature, of errors which, I believe, result from 'individual-level' thinking. There is no need to harp on them again here, and I will just give one example of the kind of thing I mean, without mentioning names, under the title of the 'Ace of Spades Fallacy'.

The coefficient of relationship between two relatives, say grandfather and grandson, can be taken to be equivalent to two distinct quantities. It is often expressed as the mean *fraction* of a grandfather's genome that is expected to be identical by descent with that of the grandson. It is also the *probability* that a named gene of the grandfather will be identical by descent with a gene in the grandson. Since the two are numerically the same, it might seem not to matter which we think in terms of. Even though the probability measure is logically more appropriate, it might seem that either measure could be used for thinking about how much 'altruism' a grandfather 'ought' to dispense to his grandson. It does matter, however, when we start thinking about the variance as well as the mean.

Several people have pointed out that the fraction of genome overlap between parent and child is exactly equal to the coefficient of relationship, whereas for all other relatives the coefficient of relationship gives only the mean figure; the actual fraction shared might be more and it might be less. It has been said, therefore, that the coefficient of relationship is 'exact' for the parent/child relationship, but 'probabilistic' for all others. But this uniqueness of the parent/child relationship applies only if we think in terms of *fractions* of genomes shared. If, instead, we think in terms of *probabilities* of sharing particular genes, the parent/child relationship is just as 'probabilistic' as any other.

This still might be thought not to matter, and indeed it does not matter until we are tempted to draw false conclusions. One false conclusion that has been drawn in the literature is that a parent, faced with a choice between feeding its own child and feeding a full sibling exactly the same age as its own child (and with exactly the same mean coefficient of relationship), should favour its own child purely on the grounds that its genetic relatedness is a 'sure thing' rather than a 'gamble'. But it is only the *fraction* of genome shared that is a sure thing. The *probability* that a particular gene, in this case a gene for altruism, is identical by descent with one in the offspring is just as chancy as in the case of the full sibling.

It is next tempting to think that an animal might try to use cues to estimate whether a particular relative happens to share many genes with itself or not. The reasoning is conveniently expressed in the currently fashionable style of subjective metaphor: 'All my brothers share, on average, half my genome, but some of my brothers share more than half and others less than half. If I could work out which ones share more than half, I could show favouritism towards them, and thereby benefit my genes. Brother A resembles me in hair colour, eye colour, and several other features, whereas brother B hardly resembles me at all. Therefore A probably shares more genes with me. Therefore I shall feed A in preference to B.'

That soliloquy was supposed to be spoken by an individual animal. The fallacy is quickly seen when we compose a similar soliloquy, this time to be spoken by one of Hamilton's 'intelligent' *genes*, a gene 'for' feeding brothers: 'Brother A has clearly inherited my gene colleagues from the hair-colour department and from the eye-colour department, but what do I care about them? The great question is, has A or B inherited a copy of *me*? Hair colour and eye colour tell me nothing about that unless I happen to be linked to those other genes.' Linkage is, then, important here, but it is just as important for the 'deterministic' parent/offspring relationship as for any 'probabilistic' relationship.

The fallacy is called the Ace of Spades Fallacy because of the following analogy. Suppose it is important to me to know whether your hand of thirteen cards contains the ace of spades. If I am given no information, I know that the odds are thirteen in fifty-two, or one in four, that you have the ace. This is my first guess as to the probability. If somebody whispers to me that you have a very strong hand in spades, I would be justified in revising upwards my initial estimate of the probability that you have the ace. If I am told that you have the king, queen, jack, 10, 8, 6, 5, 4, 3 and 2, I would be correct in concluding that you have a very strong hand in spades. But, so long as the deal was honest, I would be a mug if I therefore placed a bet on your having the ace! (Actually the analogy is a bit unfair here, because the odds of your having the ace are now three in forty-two, substantially lower than the prior odds of one in four.) In the biological case we may assume that, linkage aside, knowledge of a brother's eye colour tells us nothing, one way or the other, about whether he shares a particular gene for brotherly altruism.

There is no reason to suppose that the theorists who have perpetrated the biological versions of the Ace of Spades Fallacy are bad gamblers. It wasn't their probability theory they got wrong, but their biological assumptions. In particular, they assumed that an individual organism, as a coherent entity, works on behalf of copies of all the genes inside it. It was as if an animal 'cared' about the survival of copies of its eye-colour genes, hair-colour genes, etc. It is better to assume that only genes 'for caring' care, and they only care about copies of themselves.

I must stress that I am not suggesting that errors of this kind follow inevitably from the inclusive fitness approach. What I do suggest is that they are traps for the unwary thinker about individual-level maximization, while they present no danger to the thinker about gene-level maximization, however unwary. Even Hamilton has made an error, afterwards pointed out by himself, which I attribute to individual-level thinking.

The problem arises in Hamilton's calculation of coefficients of relationship, *r*, in hymenopteran families. As is now well known, he made brilliant use of the odd *r* values resulting from the haplodiploid sex-determining system of Hymenoptera, notably the curious fact that *r* between sisters is $\frac{3}{4}$. But consider the relationship between a female and her father. One half of the female's genome is identical by descent with that of her father: the 'overlap' of her genome with his is $\frac{1}{2}$, and Hamilton correctly gave $\frac{1}{2}$ as the coefficient of relationship between a female and her father. The trouble comes when we look at the same relationship the other way round. What is the coefficient of relationship between a male and his daughter? One naturally expects it to be reflexive, $\frac{1}{2}$ again, but there is a difficulty. Since a male is haploid, he has half as many genes as his daughter in total. How, then, can we calculate the 'overlap', the fraction of genes shared? Do we say that the male's genome overlaps with half his daughter's genome, and therefore that *r* is $\frac{1}{2}$? Or do we say that every single one of the male's genes will be found in his daughter, and therefore *r* is 1?

Hamilton originally gave $\frac{1}{2}$ as the figure, then in 1971 changed his mind and gave 1. In 1964 he had tried to solve the difficulty of how to calculate an overlap between a haploid and a diploid genotype by arbitrarily treating the male as a kind of honorary diploid. 'The relationships concerning males are worked out by assuming each male to carry a "cipher" gene to make up his diploid pair, one "cipher" never being considered identical by descent with another' (Hamilton 1964b). At the time, he recognized that this procedure was 'arbitrary in the sense that some other value for the fundamental mother–son and father–daughter link would have given an equally coherent system'. He later pronounced this method of calculation positively erroneous and, in an appendix added to a reprinting of his classic paper, gave the correct rules for calculating *r* in haplodiploid systems

(Hamilton 1971b). His revised method of calculation gives r between a male and his daughter as 1 (not $\frac{1}{2}$), and r between a male and his brother as $\frac{1}{2}$ (not $\frac{1}{4}$). Crozier (1970) independently corrected the error.

The problem would never have arisen, and no arbitrary 'honorary diploid' method would have been called for, had we all along thought in terms of selfish genes maximizing their survival rather than in terms of selfish individuals maximizing their inclusive fitness. Consider an 'intelligent gene' sitting in the body of a male hymenopteran, 'contemplating' an act of altruism towards a daughter. It knows for certain that the daughter's body contains a copy of itself. It does not 'care' that her genome contains twice as many genes as its present, male, body. It ignores the other half of her genome, secure in the knowledge that when the daughter reproduces, making grandchildren for the present male, it, the intelligent gene itself, has a 50 per cent chance of getting into each grandchild. To the intelligent gene in a haploid male, a grandchild is as valuable as an ordinary offspring would be in a normal diploid system. By the same token, a daughter is twice as valuable as a daughter would be in a normal diploid system. From the intelligent gene's point of view, the coefficient of relationship between father and daughter is indeed 1, not $\frac{1}{2}$.

Now look at the relationship the other way round. The intelligent gene agrees with Hamilton's original figure of $\frac{1}{2}$ for the coefficient of relationship between a female hymenopteran and her father. A gene sits in a female and contemplates an act of altruism towards the father of that female. It knows that it has an equal chance of having come from the father or from the mother of the female in which it sits. From its point of view, then, the coefficient of relationship between its present body and either of the two parent bodies is $\frac{1}{2}$.

The same kind of reasoning leads to an analogous non-reflexiveness in the brother–sister relationship. A gene in a female sees a sister as having $\frac{3}{4}$ of a chance of containing itself, and a brother as having $\frac{1}{4}$ of a chance of containing itself. A gene in a male, however, looks at the sister of that male and sees that she has $\frac{1}{2}$ a chance of containing a copy of itself, not as Hamilton's original cipher gene ('honorary diploid') method gave.

I believe it will be admitted that had Hamilton used his own 'intelligent-gene' thought experiment when calculating these coefficients of relationship, instead of thinking in terms of *individuals* as agents maximizing something, he would have got the right answer the first time. If these errors had been simple miscalculations it would obviously be pedantic to discuss them, once their original author had pointed them out. But they were not miscalculations, they were based on a highly instructive conceptual error. The same is true of the numbered 'Misunderstandings of Kin Selection' that I quoted before.

I have tried to show in this chapter that the concept of fitness as a technical term is a confusing one. It is confusing because it can lead to admitted error, as in the case of Hamilton's original calculation of haplodiploid coefficients of relationship, and as in the case of several of my '12 Misunderstandings of Kin Selection'. It is confusing because it can lead philosophers to think the whole theory of natural selection is a tautology. And it is confusing even to biologists because it has been used in at least five different senses, many of which have been mistaken for at least one of the others.

Emerson, as we have seen, confused fitness[3] with fitness[1]. I now give an example of a confusion of fitness[3] with fitness[2]. Wilson (1975) provides a useful glossary of terms needed by sociobiologists. Under 'fitness' he refers us to 'genetic fitness'. We turn to 'genetic fitness' and find it defined as 'The contribution to the next generation of one genotype in a population relative to the contributions of other genotypes.' Evidently 'fitness' is being used in the sense of the population geneticist's fitness[2]. But then, if we look up 'inclusive fitness' in the glossary we find: 'The sum of an individual's own fitness plus all its influence on fitness in its relatives other than direct descendants'. Here, 'the individual's own fitness' must be 'classical' fitness[3] (since it is applied to individuals), not the genotypic fitness (fitness[2]) which is the only 'fitness' defined in the glossary. The glossary is, then, incomplete, apparently because of a confusion between fitness of a genotype at a locus (fitness[2]) and reproductive success of an individual (fitness[3]).

As if my fivefold list were not confusing enough already, it may need extending. For reasons concerned with an interest in biological 'progress', Thoday (1953) seeks the 'fitness' of a long-term lineage, defined as the probability that the lineage will continue for a very long time such as 10^8 generations, and contributed to by such 'biotic' factors (Williams 1966) as 'genetic flexibility'. Thoday's fitness does not correspond to any of my list of five. Then again, the fitness[2] of population geneticists is admirably clear and useful, but many population geneticists are, for reasons best known to themselves, very interested in another quantity which is called the mean fitness of a population. Within the general concept of 'individual fitness', Brown (1975; Brown & Brown 1981) wishes to make a distinction between 'direct fitness' and 'indirect fitness'. Direct fitness is the same as what I am calling fitness[3]. Indirect fitness can be characterized as something like fitness[4] minus fitness[3], i.e. the component of inclusive fitness that results from the reproduction of collateral relatives as opposed to direct descendants (I presume grandchildren count in the direct component, though the decision is arbitrary). Brown himself is clear about the meaning of the terms, but I believe they have considerable power to confuse. For instance, they appear to lend weight to the view (not held by Brown, but held by a distressing number of other authors, e.g. Grant 1978, and several writers on 'helpers at the nest' in birds) that there is something unparsimonious about 'kin selection' (the 'indirect component') as compared with 'individual selection' (the 'direct component'), a view which I have criticized sufficiently before (Dawkins 1976a, 1978a, 1979a).

The reader may have been bewildered and irritated at my drawn out list of five or more separate meanings of fitness. I have found this a painful chapter to write, and I am aware that it will not have been easy to read. It may be the last resort of a poor writer to blame his subject matter, but I really do believe it is the concept of fitness itself which is responsible for the agony in this case. The population geneticists' fitness[2] aside, the concept of fitness as applied to individual organisms has become forced and contrived. Before Hamilton's revolution, our world was peopled by individual organisms working single-mindedly to keep themselves alive and to have children. In those days it was natural to measure success in this undertaking at the level of the individual organism. Hamilton changed all that but unfortunately, instead of following his ideas through to their logical conclusion and sweeping the individual organism from its pedestal as notional agent of maximization, he exerted his genius in devising a means of rescuing the individual. He could have persisted in saying: gene survival is what matters; let us examine what a gene would have to do in order to propagate copies of itself. Instead he, in effect, said: gene survival is what matters; what is the minimum change we have to make to our old view of what individuals must do, in order that we may cling on to our idea of the individual as the unit of action? The result—inclusive fitness—was technically correct, but complicated and easy to misunderstand. I shall avoid mentioning fitness again in this book, which I trust will make for easier reading. The next three chapters develop the theory of the extended phenotype itself.

11

THE GENETICAL EVOLUTION OF ANIMAL ARTEFACTS

What do we really mean by the phenotypic effect of a gene? A smattering of molecular biology may suggest one kind of answer. Each gene codes for the synthesis of one protein chain. In a proximal sense that protein is its phenotypic effect. More distal effects like eye colour or behaviour are, in their turn, effects of the protein functioning as an enzyme. Such a simple account does not, however, bear much searching analysis. The 'effect' of any would-be cause can be given meaning only in terms of a comparison, even if only an implied comparison, with at least one alternative cause. It is strictly incomplete to speak of blue eyes as 'the effect' of a given gene G_1. If we say such a thing, we really imply the potential existence of at least one alternative allele, call it G_2, and at least one alternative phenotype, P2, in this case, say, brown eyes. Implicitly we are making a statement about a relation between a pair of genes {*G1*, *G2*} and a pair of distinguishable phenotypes {P1, P2}, in an environment which either is constant or varies in a non-systematic way so that its contribution randomizes out. 'Environment', in that last clause, is taken to include all the genes at other loci that must be present in order for P1 or P2 to be expressed. Our statement is that there is a statistical tendency for individuals with *G1* to be more likely than individuals with *G2* to show P1 (rather than P2). Of course there is no need to demand that P1 should *always* be associated with *G1*, nor that *G1* should always lead to P1: in the real world outside logic textbooks, the simple concepts of 'necessary' and 'sufficient' must usually be replaced by statistical equivalents.

Such an insistence that phenotypes are not caused by genes, but only phenotypic differences caused by gene differences (Jensen 1961; Hinde 1975) may seem to weaken the concept of genetic determination to the point where it ceases to be interesting. This is far from the case, at least if the subject of our interest is natural selection, because natural selection too is concerned with differences (Chapter 2). Natural selection is the process by which some alleles out-propagate their alternatives, and the instruments by which they achieve this are their phenotypic effects. It follows that phenotypic effects can always be thought of as *relative* to alternative phenotypic effects.

It is customary to speak as if differences always mean differences between individual bodies or other discrete 'vehicles'. The purpose of the next three chapters is to show that we can emancipate the concept of the phenotypic difference from that of the discrete vehicle altogether, and this is the meaning of the title 'extended phenotype'. I shall show that the ordinary logic of genetic terminology leads inevitably to the conclusion that genes can be said to have *extended* phenotypic effects, effects which need not be expressed at the level of any particular vehicle. Following an earlier paper (Dawkins 1978a) I shall take a step-by-step approach to the extended phenotype, beginning with conventional examples of 'ordinary' phenotypic effects and gradually extending the concept of the phenotype outwards so that the continuity is easy to accept. The idea of the genetic determination of animal artefacts is a didactically useful intermediate example, and this will be the main topic of this chapter.

But first, consider a gene *A* whose immediate molecular effect is the synthesis of a black protein which directly colours the skin of an animal black. Then the gene's only proximal effect, in the molecular biologist's simple sense, is the synthesis of this one black protein. But is *A* a gene 'for being black'? The point I want to make is that, as a matter of definition, that depends on how the population varies. Assume that *A* has an allele *A*´, which fails to synthesize the black pigment, so that individuals homozygous for *A*´ tend to be white. In this case *A* is truly a gene 'for' being black, in the sense in which I wish to use the phrase. But it may alternatively be that all the variation in skin colour that actually occurs in the population is due to variation at a quite different locus, *B*. *B*'s immediate biochemical effect is the synthesis of a protein which is not a black pigment, but which acts as an enzyme, one of whose indirect effects (in comparison with its allele *B*´), at some distant remove, is the facilitation of the synthesis by *A* of the black pigment in skin cells.

To be sure, *A*, the gene whose protein product is the black pigment, is necessary in order for an individual to be black: so are thousands of other genes, if only because they are necessary to make the individual exist at all. But I shall not call *A* a gene for blackness unless some of the variation in the population is due to lack of *A*. If all individuals, without fail, have *A*, and the only reason individuals are not black is that they have *B*´ rather than *B*, we shall say that *B*, but not *A*, is a gene for blackness. If there is variation at both loci affecting blackness, we shall refer to both *A* and *B* as genes for blackness. The point that is relevant here is that both *A* and *B* are potentially entitled to be called genes for blackness, depending on the alternatives that exist in the population. The fact that the causal chain linking *A* to the production of the black pigment molecule is short, while that for *B* is long and tortuous, is irrelevant. Most gene effects seen by whole animal biologists, and all those seen by ethologists, are long and tortuous.

A geneticist colleague has argued that there are virtually no behaviour-genetic traits, because all those so-far discovered have turned out to be 'byproducts' of more fundamental morphological or physiological effects. But what on earth does he think *any* genetic trait is, morphological, physiological, or behavioural, if not a 'byproduct' of something more fundamental? If we think the matter through we find that all genetic effects are 'byproducts' except protein molecules.

Returning to the black skin example, it is even possible that the chain of causation linking a gene such as *B* to its black-skinned phenotype might involve a behavioural link. Suppose that A can synthesize black pigment only in the presence of sunlight, and suppose that *B* works by making individuals seek sunlight, in comparison with *B*′ which makes them seek shade. *B* individuals will then tend to be blacker than *B*′ individuals, because they spend more time in the sun. *B* is still, by existing terminological convention, a gene 'for blackness', no less than it would be if its causal chain involved internal biochemistry only, rather than an 'external' behavioural loop. Indeed, a geneticist in the pure sense of the word need not care about the detailed pathway from gene to phenotypic effect. Strictly speaking, a geneticist who concerns himself with these interesting matters is temporarily wearing the hat of an embryologist. The pure geneticist is concerned with end products, and in particular with differences between alleles in their effects on end products. Natural selection's concerns are precisely the same, for natural selection 'works on outcomes' (Lehrman 1970). The interim conclusion is that we are already accustomed to phenotypic effects being attached to their genes by long and devious chains of causal connection; therefore further extensions of the concept of phenotype should not overstretch our credulity. This chapter takes the first step towards such further extension, by looking at animal artefacts as examples of the phenotypic expression of genes.

The fascinating subject of animal artefacts is reviewed by Hansell (1984). He shows that artefacts provide useful case studies for several principles of general ethological importance. This chapter uses the example of artefacts in the service of explaining another principle, that of the extended phenotype. Consider a hypothetical species of caddis-fly whose larvae build houses out of stones which they select from those available on the bottom of the stream. We might observe that the population contains two rather distinct colours of house, dark and light. By breeding experiments we establish that the characters 'dark house' and 'light house' breed true in some simple Mendelian fashion, say with dark house dominant to light house. In principle it ought to be possible to discover, by analysing recombination data, where the genes for house colour sit on the chromosomes. This is, of course, hypothetical. I do not know of any genetic work on caddis houses, and it would be difficult to do because adults are difficult to breed in captivity (M. H. Hansell, personal communication). But my point is that, if the practical difficulties could be overcome, nobody would be very surprised if house colour did turn out to be a simple Mendelian character in accordance with my thought experiment. (Actually, colour is a slightly unfortunate example to have chosen, since caddis vision is poor and they almost certainly ignore visual cues in choosing stones. Rather than use a more realistic example like stone shape (Hansell), I stay with colour for the sake of the analogy with the black pigment discussed above.)

The interesting sequel is this. House colour is determined by the colour of the stones chosen from the stream bed by the larva, not by the biochemical synthesis of a black pigment. The genes determining house colour must work via the behavioural mechanism that chooses stones, perhaps via the eyes. So much would be agreed by any ethologist. All that this chapter adds is a logical point: once we have accepted that there are genes for building behaviour, the rules of existing terminology imply that the artefact itself should be treated as part of the phenotypic expression of genes in the animal. The stones are outside the body of the organism, yet logically such a gene is a gene 'for' house colour, in exactly as strong a sense as the hypothetical gene *B* was for skin colour. And *B* was indeed a gene for skin colour, even though it worked by mediating sun-seeking behaviour, in exactly as strong a sense as a gene 'for' albinism is called a gene for skin colour. The logic is identical in all three cases. We have taken the first step in extending the concept of a gene's phenotypic effect outside the individual body. It was not a difficult step to take, because we had already softened up our resistance by realizing that even normal 'internal' phenotypic effects may lie at the end of long, ramified, and indirect, causal chains. Let us now step out a little further.

The house of a caddis is strictly not a part of its cellular body, but it does fit snugly round the body. If the body is regarded as a gene vehicle, or survival machine, it is easy to see the stone house as a kind of extra protective wall, in a functional sense the outer part of the vehicle. It just happens to be made of stone rather than chitin. Now consider a spider sitting at the centre of her web. If she is regarded as a gene vehicle, her web is not a part of that vehicle in quite the same obvious sense as a caddis house, since when she turns round the web does not turn with her. But the distinction is clearly a frivolous one. In a very real sense her web is a temporary functional extension of her body, a huge extension of the effective catchment area of her predatory organs.

Once again, I know of no genetic analysis of spider web morphology, but there is nothing difficult in principle about imagining such an analysis. It is known that individual spiders have consistent idiosyncrasies which are repeated in web after web. One female *Zygiella-x-notata*, for instance, was seen to build more than 100 webs, all lacking a particular concentric ring (Witt, Reed & Peakall 1968). Nobody familiar with the literature on behaviour genetics (e.g. Manning 1971) would be surprised if the observed idiosyncrasies of individual spiders turned out to have a genetic basis. Indeed, our belief that spiders' webs have evolved their efficient shape through natural selection necessarily commits us to a belief that, at least in the past, web variation must have been under genetic influence (Chapter 2). As in the case of the caddis houses, the genes must have worked via building behaviour, before that in embryonic development perhaps via neuroanatomy, before that perhaps via cell membrane biochemistry. By whatever embryological routes the genes may work in detail, the small extra step from behaviour to web is not any more difficult to conceive of than the many causal steps which preceded the behavioural effect, and which lie buried in the labyrinth of neuroembryology.

Nobody has any trouble understanding the idea of genetic control of morphological differences. Nowadays few people have trouble understanding that there is, in principle, no difference between genetic control of morphology and genetic control of behaviour, and we are unlikely to be misled by unfortunate statements such as 'Strictly speaking, it is the brain (rather than the behaviour) that is genetically inherited'. The point here is, of course, that if there is any sense in which the brain is inherited, behaviour may be inherited in exactly the same sense. If we object to calling behaviour inherited, as some do on tenable grounds, then we must, to be consistent, object to calling brains inherited too. And if we do decide to allow that both morphology and behaviour may be inherited, we cannot reasonably at the same time object to calling caddis house colour and spider web shape inherited. The extra step from behaviour to extended phenotype, in this case the stone house or the web, is as conceptually negligible as the step from morphology to behaviour.

From the viewpoint of this book an animal artefact, like any other phenotypic product whose variation is influenced by a gene, can be regarded as a phenotypic tool by which that gene could potentially lever itself into the next generation. A gene may so lever itself by adorning the tail of a male bird of paradise with a sexually attractive blue feather, or by causing a male bower bird to paint his bower with pigment crushed in his bill out of blue berries. The details may be different in the two cases but the effect, from the gene's point of view, is the same. Genes that achieve sexually attractive phenotypic effects when compared with their alleles are favoured, and it is trivial whether those phenotypic effects are 'conventional' or 'extended'. This is underlined by the interesting observation that bower bird species with especially splendid bowers tend to have relatively drab plumage, while those species with relatively bright plumage tend to build less elaborate and spectacular bowers (Gilliard 1963). It is as though some species have shifted part of the burden of adaptation from bodily phenotype to extended phenotype.

So far the phenotypic effects we have been considering have extended only a few yards away from the initiating genes, but in principle there is no reason why the phenotypic levers of gene power should not reach out for miles. A beaver dam is built close to the lodge, but the effect of the dam may be to flood an area thousands of square metres in extent. As to the advantage of the pond from the beaver's point of view, the best guess seems to be that it increases the distance the beaver can travel by water, which is safer than travelling by land, and easier for transporting wood. A beaver that lives by a stream quickly exhausts the supply of food trees lying along the stream bank within a reasonable distance. By building a dam across the stream the beaver creates a large shoreline which is available for safe and easy foraging without the beaver having to make long and difficult journeys overland. If this interpretation is right, the lake may be regarded as a huge extended phenotype, extending the foraging range of the beaver in a way which is somewhat analogous to the web of the spider. As in the case of the spider web, nobody has done a genetic study of beaver dams, but we really do not need to in order to convince ourselves of the rightness of regarding the dam, and the lake, as part of the phenotypic expression of beaver genes. It is enough that we accept that beaver dams must have evolved by Darwinian natural selection: this can only have come about if dams have varied under the control of genes (Chapter 2).

Just by talking about a few examples of animal artefacts, then, we have pushed the conceptual range of the gene's phenotype out to many miles. But now we come up against a complication. A beaver dam is usually the work of more than one individual. Mated pairs routinely work together, and successive generations of a family may inherit responsibility for the upkeep and extension of a 'traditional' dam-complex comprising a staircase of half a dozen dams stepping downstream, and maybe several 'canals' as well. Now it was easy to argue that a caddis house, or a spider web, was the extended phenotype of the genes of the single individual that built it. But what are we to make of an artefact that is the joint production of a pair of animals or a family? Worse, consider the mound built by a colony of compass termites, a tombstone-shaped slab, one of a vista of similar monoliths all oriented precisely north–south, and rising to a height that dwarfs its builders as a mile-high skyscraper would dwarf a man (von Frisch 1975). It is built by perhaps a million termites, separated by time into cohorts, like medieval masons who could work a lifetime on one cathedral and never meet their colleagues that would complete it. A partisan of the individual as the unit of selection might pardonably ask exactly *whose* extended phenotype the termite mound is supposed to be.

If this consideration seems to complicate the idea of the extended phenotype beyond all reason, I can only point out that exactly the same problem has always arisen with 'conventional' phenotypes. We are thoroughly accustomed to the idea that a given phenotypic entity, an organ say, or a behaviour pattern, is influenced by a large number of genes whose effects interact in additive or more complex ways. Human height at a given age is affected by genes at many loci, interacting with each other and with dietary and other environmental effects. The height of a termite mound at a given mound-age is, no doubt, also controlled by many environmental factors and many genes, adding to or modifying each others' effects. It is incidental that in the case of the termite mound the *proximal* theatre of within-body gene effects happens to be distributed among the cells of a large number of worker bodies.

If we are going to worry about proximal effects, the genes influencing my height act in ways that are distributed among many separate cells. My body is full of genes, which happen to be identically distributed among my many somatic cells. Each gene exerts its effects at the cellular level, only a minority of genes expressing themselves in any one cell. The summed effects of all these effects on cells, together with similar effects from the environment, may be measured as my total height. Similarly, a termite mound is full of genes. These genes, too, are distributed among the nuclei of a large number of cells. It happens that the cells are not contained in quite such a compact single unit as the cells in my body, but even here the difference is not very great. Termites move around relative to each other more than human organs do, but it is not unknown for human cells to travel rapidly about the body in pursuance of their errands, for instance phagocytes hunting down and engulfing microscopic parasites. A more important difference (in the case of a termite mound, though not in the case of a coral reef built by a clone of individuals) is that the cells in the termite mound are gathered into genetically heterogeneous packages: each individual termite is a clone of cells but a different clone from all other individuals in the nest. This, however, is only a relative complication. Fundamentally what is going on is that genes, compared with their alleles, exert quantitative, mutually interacting, mutually modifying, effects on a shared phenotype, the mound. They do this proximally by controlling the chemistry of cells in worker bodies, and hence worker behaviour. The principle is the same, whether the cells happen to be organized into one large homogeneous clone, as in the human body, or into a heterogeneous collection of clones, as in the termite mound. I postpone until later the complicating point that a termite body itself is a 'colony', with a substantial fraction of its genetic replicators contained in symbiotic protozoa or bacteria.

What, then, would a genetics of termite mounds look like? Suppose we were to do a population survey of compass mounds in the Australian steppe, scoring a trait such as colour, basal length/width ratio, or some internal structural feature—for termite mounds are like bodies with a complex 'organ' structure. How could we do a genetic study of such group-manufactured phenotypes? We need not hope to find normal Mendelian inheritance with simple dominance. An obvious complication, as already mentioned, is that the genotypes of the individuals working on any one mound are not identical. For most of the life of an average colony, however, all the workers are full siblings, the children of the primary royal pair of alates who founded the colony. Like their parents the workers are diploid. We may presume that the king's two sets of genes and the queen's two sets of genes are permuted throughout the several million worker bodies. The 'genotype' of the aggregate of workers may therefore be regarded, in a sense, as a single *tetraploid* genotype consisting of all the genes which the original founding pair contributed. It is not quite as simple as that, for various reasons; for instance secondary reproductives often arise in older colonies and these may take on the full reproductive role if one of the original royal pair dies. This means that the workers building the later parts of a mound may not be full siblings of those that began the task, but their nephews and nieces (probably inbred and rather uniform, incidentally—Hamilton 1972; Bartz 1979). These later reproductives still draw their genes from the 'tetraploid' set introduced by the original royal pair, but their progeny will permute a particular subset of those original genes. One of the things a 'mound-geneticist' might look out for, then, is a sudden change in details of mound-building after replacement of a primary reproductive by a secondary reproductive.

Ignoring the problem introduced by secondary reproductives, let us confine our hypothetical genetic study to younger colonies whose workers consist entirely of full siblings. Some characters in which mounds vary may turn out to be largely controlled at one locus, while others will be polygenically controlled at many loci. This is no different from ordinary diploid genetics, but our new quasi-tetraploid genetics now introduces some complications. Suppose the behavioural mechanism involved in choosing the colour of the mud used in building varies genetically. (Again, colour is chosen for continuity with earlier thought experiments, although again it would be more realistic to avoid a visual trait, since termites make little use of vision. If necessary, we may suppose the choice to be made chemically, mud colour being incidentally correlated with the chemical cues. This is instructive, for it again emphasizes the fact that our way of labelling a phenotypic trait is a matter of arbitrary convenience.) For simplicity, assume that mud choice is influenced by the diploid genotype of the individual worker doing the choosing, in a simple one locus Mendelian fashion with choice of dark mud dominant over choice of light mud. Then a mound built by a colony containing some dark-preferring workers and some light preferring workers will consist of a mixture of dark and light muds and will presumably be intermediate in overall colour. Of course such simple genetic assumptions are highly improbable. They are equivalent to the simplifying assumptions we ordinarily make when explaining elementary conventional genetics, and I make them here to explain analogously the principles of how a science of 'extended genetics' might work.

Using these assumptions, then, we can write down the expected extended phenotypes, considering mud colour only, resulting from crosses between the various possible founding pair genotypes. For instance, all colonies founded by a heterozygous king and a heterozygous queen will contain dark-building and light-building workers in the ratio 3:1. The resulting extended phenotype will be a mound built of three parts dark mud and one part light mud, therefore nearly, but not completely, dark in colour. If choice of mud colour is influenced by polygenes at many loci, the 'tetraploid genotype' of the colony may be expected to influence the extended phenotype, perhaps in an additive way. The colony's immense size will lead to its acting as a statistical averaging device, making the mound as a whole become the extended phenotypic expression of the genes of the royal pair, manifested via the behaviour of several million workers each containing a different diploid sample of those genes.

Mud colour was an easy character for us to choose, because mud itself blends in a simple additive manner: mix dark and light mud, and you get khaki mud. It was therefore easy for us to deduce the result of assuming that each worker goes its own way, choosing mud of its own preferred colour (or chemical associated with colour), as determined by its own diploid genotype. But what can we say about a characteristic of overall mound morphology, say basal width/length ratio? In itself, this is not a character that any single worker can determine. Each single worker must be obeying behavioural rules, the result of which, summed over thousands of individuals, is the production of a mound of regular shape and stated dimensions. Once again, the difficulty is one we have met before, in the embryonic development of an ordinary diploid multicellular body. Embryologists are still wrestling with problems of this kind, and very formidable they are. There do appear to be some close analogies with termite mound development. For instance, conventional embryologists frequently appeal to the concept of the chemical gradient, while in *Macrotermes* there is evidence that the shape and size of the royal cell is determined by a pheromonal gradient around the body of the queen (Bruinsma & Leuthold 1977). Each cell in a developing embryo behaves as if it 'knows' where it is in the body, and it grows to have a form and physiology appropriate to that part of the body (Wolpert 1970).

Sometimes the effects of a mutation are easy to interpret at the cellular level. For instance, a mutation that affects skin pigmentation has a rather obvious local effect on each skin cell. But other mutations affect complex characters in drastic ways. Well-known examples are the 'homeotic' mutants of *Drosophila*, such as the one that grows a complete and well-formed leg in the socket where an antenna ought to be. For a change in one single gene to wreak such a major, yet orderly, change in the phenotype, it must make its lesion rather high in a hierarchical chain of command. By analogy, if a single infantryman goes off his head he alone runs amok; but if a general loses his reason an entire army behaves crazily on a grand scale—invades an ally instead of an enemy, say—while each individual soldier in that army is obeying orders normally and sensibly, and his individual behaviour as he puts one foot in front of the other will be indistinguishable from that of a soldier in an army with a sane general.

Presumably an individual termite working on a little corner of a big mound is in a similar position to a cell in a developing embryo, or a single soldier tirelessly obeying orders whose purpose in the larger scheme of things he does not understand. Nowhere in the single termite's nervous system is there anything remotely equivalent to a complete image of what the finished mound will look like (Wilson 1971, p. 228). Each worker is equipped with a small toolkit of behavioural rules, and he/she is probably stimulated to choose an item of behaviour by local stimuli emanating from the work already accomplished, no matter whether he/she or other workers accomplished it—stimuli emanating from the present state of the nest in the worker's immediate vicinity ('stigmergie', Grassé 1959). For my purposes it doesn't matter exactly what the behavioural rules are, but they would be something like: 'If you come upon a heap of mud with a certain pheromone on it, put another dollop of mud on the top.' The important point about such rules is that they have a purely local effect. The grand design of the whole mound emerges only as the summed consequences of thousands of obeyings of micro-rules (Hansell 1984). Particular interest attaches to the local rules that are responsible for determining global properties such as the base length of the compass mound. How do the individual workers on the ground 'know' that they have reached the boundary of the ground plan? Perhaps in something like the same way as the cells at the boundary of a liver 'know' that they are not in the middle of the liver. In any case, whatever the local behavioural rules may be that determine the overall shape and size of a termite mound, they are presumably subject to genetic variation in the population at large. It is entirely plausible, indeed almost inevitable, that both the shape and size of compass termite mounds have evolved by natural selection, just like any feature of bodily morphology. This can only have come about through the selection of mutations acting at the local level on the building behaviour of individual worker termites.

Now our special problem arises, which would not arise in the ordinary embryogenesis of a multicellular body, nor in the example of mixing light and dark muds. Unlike the cells of a multicellular body, the workers are not genetically identical. In the example of the dark and light mud, it was easy to suppose that a genetically heterogeneous work-force would simply construct a mound of mixed mud. But a work-force which was genetically heterogeneous with respect to one of the behavioural rules that affected overall mound shape might produce curious results. By analogy with our simple Mendelian model of mud selection, a colony might contain workers favouring two different rules for determining the boundary of the mound, say in the ratio three to one. It is pleasing to imagine that such a bimodal colony might produce a mound with a strange double wall and a moat between! More probably, the rules obeyed by individuals would include provision for the minority to allow themselves to be overruled by majority decisions, so that only one clean-cut wall would emerge. This could work in a similar way to the 'democratic' choice of a new nest site in honeybee swarms, observed by Lindauer (1961).

Scout bees leave the swarm hanging in a tree, and prospect for suitable new permanent sites such as hollow trees. Each scout returns and dances on the surface of the swarm, using the well-known von Frisch code to indicate the direction and distance of the prospective site that she has just investigated. The 'vigour' of the dance indicates the individual scout's assessment of the virtues of the site. New bees are recruited to go out and examine it for themselves, and if they 'approve' they too dance 'in its favour' on returning. After some hours, then, the scouts have formed themselves into a few 'parties', each one 'advocating' a different nest site. Finally, minority 'opinions' become even smaller minorities, as allegiances are transferred to majority dances. When an overwhelming majority has been achieved for one site, the whole swarm takes off and flies there to set up home.

Lindauer observed this procedure for nineteen different swarms, and in only two of these cases was a consensus not soon reached. I quote his account of one of these:

> In the first case two groups of messengers had got into competition; one group announced a nesting place to the northwest, the other to the northeast. Neither of the two wished to yield. The swarm then finally flew off and I could scarcely believe my eyes—it sought to divide itself. The one half wanted to fly to the northwest, the other to the northeast. Apparently each group of scouting bees wanted to abduct the swarm to the nesting place of its own choice. But that was naturally not possible, for one group was always without the queen, and there resulted a remarkable tug of war in the air, once 100 meters to the northwest, then again 150 meters to the northeast, until finally after half an hour the swarm gathered together at the old location. Immediately both groups began again with their soliciting dances, and it was not until the next day that the northeast group finally yielded; they ended their dance and

> thus an agreement was reached on the nesting place in the northwest [Lindauer 1961, p. 43].

There is no suggestion here that the two subgroups of bees were genetically different, though they may have been. What matters for the point I am making is that each individual follows local behavioural rules, the combined effect of which normally gives rise to coordinated swarm behaviour. These evidently include rules for resolving 'disputes' in favour of the majority. Disagreements over the preferred location for the outer wall of a termite mound might be just as serious for colony survival as disagreement over nesting sites among Lindauer's bees (colony survival matters, because of its effects on the survival of the genes causing individuals to resolve disputes). As a working hypothesis we might expect that disputes resulting from genetic heterogeneity in termites would be resolved by similar rules. In this way the extended phenotype could take up a discrete and regular shape, despite being built by genetically heterogeneous workers.

The analysis of artefacts given in this chapter seems, at first sight, vulnerable to *reductio ad absurdum*. Isn't there a sense, it may be asked, in which every effect that an animal has upon the world is an extended phenotype? What about the footprints left in the mud by an oystercatcher, the paths worn through the grass by sheep, the luxuriant tussock that marks the site of a last year's cowpat? A pigeon's nest is an artefact without a doubt, but in gathering the sticks the bird also changes the appearance of the ground where they had lain. If the nest is called extended phenotype, why shouldn't we so call the bare patch of ground where the sticks used to lie?

To answer this we must recall the fundamental reason why we are interested in the phenotypic expression of genes in the first place. Of all the many possible reasons, the one which concerns us in this book is as follows. We are fundamentally interested in natural selection, therefore in the differential survival of replicating entities such as genes. Genes are favoured or disfavoured relative to their alleles as a consequence of their phenotypic effects upon the world. Some of these phenotypic effects may be incidental consequences of others, and have no bearing on the survival chances, one way or the other, of the genes concerned. A genetic mutation that changes the shape of an oystercatcher's foot will doubtless thereby influence the oystercatcher's success in propagating it. It may, for instance, slightly reduce the bird's risk of sinking into mud, while at the same time it slightly slows him down when he is running on firm ground. Such effects are likely to be of direct relevance to natural selection. But the mutation will also have an effect on the shape of the footprints left behind in the mud—arguably an extended phenotypic effect. If, as is perfectly likely, this has no influence on the success of the gene concerned (Williams 1966, pp. 12–13), it is of no interest to the student of natural selection, and there is no point in bothering to discuss it under the heading of the extended phenotype, though it would be formally correct to do so. If, on the other hand, the changed footprint did influence survival of the oystercatcher, say by making it harder for predators to track the bird, I would want to regard it as part of the extended phenotype of the gene. Phenotypic effects of genes, whether at the level of intracellular biochemistry, gross bodily morphology, or extended phenotype, are potentially devices by which genes lever themselves into the next generation, or barriers to their doing so. Incidental side-effects are not always effective as tools or barriers, and we do not bother to regard them as phenotypic expressions of genes, either at the conventional or the extended phenotype level.

It is unfortunate that this chapter has had to be rather hypothetical. There have been only a few studies of the genetics of building behaviour in any animal (e.g. Dilger 1962), but there is no reason to think that 'artefact genetics' will be any different, in principle, from behaviour genetics generally (Hansell 1984). The idea of the extended phenotype is still sufficiently unfamiliar that it might not immediately occur to a geneticist to study termite mounds as a phenotype, even if it were practically easy to do so—and it wouldn't be easy. Yet we must acknowledge at least the theoretical validity of such a branch of genetics if we are to countenance the Darwinian evolution of beaver dams and termite mounds. And who can doubt that, if termite mounds fossilized plentifully, we would see graded evolutionary series with trends as smooth (or as punctuated!) as any that we find in vertebrate skeletal palaeontology (Schmidt 1955; Hansell 1984)?

Permit me one further speculation to lead us into the next chapter. I have spoken as if the genes inside a termite mound were all enclosed in the nuclei of cells of termite bodies. The 'embryological' forces bearing on the extended phenotype have been assumed to originate from the genes of individual termites. Yet the chapter on arms races and manipulation should have alerted us to another way of looking at it. If all the DNA could be distilled out of a termite mound, perhaps as much as one-quarter of it would not have originated from termite nuclei at all. Some such proportion of the body weight of each individual termite is typically made up of symbiotic cellulose-digesting microorganisms in the gut—flagellates or bacteria. The symbionts are obligately dependent on the termites, and the termites on them. The proximal phenotypic power of the symbiont genes is exerted via protein synthesis in symbiont cytoplasm. But just as termite genes reach out beyond the cells that enclose them and manipulate the development of whole termite bodies and hence of the mound, is it not almost inevitable that the symbiont genes will have been selected to exert phenotypic power on their surroundings? And will this not include exerting phenotypic power on termite cells and hence bodies, on termite behaviour and even termite mounds? Along these lines, could the evolution of eusociality in the Isoptera be explained as an adaptation of the microscopic symbionts rather than of the termites themselves?

This chapter has explored the idea of the extended phenotype, first of genes in a single individual, then of genes from different but closely related individuals, members of a kin-group. The logic of the argument now seems to compel us to contemplate the possibility of an extended phenotype's being jointly manipulated, not necessarily cooperatively, by genes from distantly related individuals, individuals of different species, even different kingdoms. This is the direction in which our next outward step must take us.

12

HOST PHENOTYPES OF PARASITE GENES

Let us briefly take stock of where we have reached in our outward march. The phenotypic expression of a gene can extend outside the cell in which the genes exert their immediate biochemical influence, to affect gross features of a whole multicellular body. This is commonplace, and we are conventionally used to the idea of a gene's phenotypic expression being extended this far.

In the previous chapter we took the small further step of extending the phenotype to artefacts, built by individual behaviour which is subject to genetic variation, for instance caddis houses. Next we saw that an extended phenotype can be built under the joint influence of genes in more than one individual body. Beaver dams and termite mounds are collectively built by the behavioural efforts of more than one individual. A genetic mutation in one individual beaver could show itself in phenotypic change in the shared artefact. If the phenotypic change in the artefact had an influence on the success of replication of the new gene, natural selection would act, positively or negatively, to change the probability of similar artefacts existing in the future. The gene's extended phenotypic effect, say an increase in the height of the dam, affects its chances of survival in precisely the same sense as in the case of a gene with a normal phenotypic effect, such as an increase in the length of the tail. The fact that the dam is the shared product of the building behaviour of several beavers does not alter the principle: genes that tend to make beavers build high dams will themselves, on average, tend to reap the benefits (or costs) of high dams, even though every dam may be jointly built by several beavers. If two beavers working on the same dam have different genes for dam height, the resulting extended phenotype will reflect the interaction between the genes, in the same way as bodies reflect gene interactions. There could be extended genetic analogues of epistasis, of modifier genes, even of dominance and recessiveness.

Finally, at the end of the chapter, we saw that genes 'sharing' a given extended phenotypic trait might come from different species, even different phyla and different kingdoms. This chapter will develop two further ideas. One is that phenotypes that extend outside the body do not have to be inanimate artefacts: they can themselves be built of living tissue. The other idea is that wherever there are 'shared' genetic influences on an extended phenotype, the shared influences may be in conflict with each other rather than cooperative. The relationships we shall be concerned with are those of parasites and their hosts. I shall show that it is logically sensible to regard parasite genes as having phenotypic expression in host bodies and behaviour.

The caddis larva rides inside the stone house that it built. It therefore seems appropriate to regard the house as the outer wall of the gene vehicle, the casing of the survival machine. It is even easier to regard the shell of a snail as part of the phenotypic expression of snail genes since, although the shell is inorganic and 'dead', its chemical substance was directly secreted by snail cells. Variations in, say, shell thickness would be called genetic if genes in snail cells affected shell thickness. Otherwise they would be called 'environmental'. But there are reports of snails with trematode parasites having thicker shells than unparasitized snails (Cheng 1973). From the point of view of snail genetics, this aspect of shell variation is under 'environmental' control—the fluke is part of the environment of the snail—but from the point of view of fluke genetics it might well be under genetic control: it might, indeed, be an evolved adaptation of the fluke. It is admittedly also possible that the thickened shell is a pathological response of the snail, a dull byproduct of infection. But let me explore the possibility that it is a fluke adaptation, because it is an interesting idea to use in further discussion.

If we consider snail shell variation as, in part, phenotypic expression of snail genes, we might recognize an optimum shell thickness in the following sense. Selection presumably penalizes snail genes that make shells too thick, as well as those that make shells too thin. Thin shells provide inadequate protection. Genes for too-thin shells therefore endanger their germ-line copies, which are thus not favoured by natural selection. Shells that are too thick presumably protect their snails (and the enclosed germ-line genes for extra thickness) superlatively, but the extra cost of making a thick shell detracts from the snail's success in some other way. In the economy of the body, resources that are consumed in making extra-thick shells, and in carrying the extra weight around, might better have been diverted into making, say, larger gonads. Continuing with the hypothetical example, therefore, genes for extra-thick shells will tend to induce in their bodies some compensating disadvantage such as relatively small gonads, and they will therefore not be passed on to the next generation so effectively. Even if there is, in fact, no trade-off between shell thickness and gonad size, there is bound to be some kind of analogous trade-off, and a compromise will be reached at an intermediate thickness. Genes that tend to make snail shells either too thick or too thin will not prosper in the snail gene-pool.

But this whole argument presupposes that the only genes that have power over variation in shell thickness are snail genes. What if some of the causal factors that are, by definition, environmental from the snail's point of view, turn out to be genetic from some other point of view, say that of the fluke? Suppose we adopt the suggestion made above that some fluke genes are capable, through an influence on snail physiology, of exerting an effect on snail shell thickness. If shell thickness influences the replication success of such fluke genes, natural selection is bound to work on their frequencies relative to their alleles in the fluke gene-pool. Changes in snail shell thickness may be regarded, then, at least in part, as potential adaptations for the benefit of fluke genes.

Now, the optimum shell thickness from the point of view of the fluke genes is hardly likely to be the same as the optimum from the point of view of snail genes. For instance, snail genes will be selected for their beneficial effects on snail reproduction as well as snail survival, but (except under special circumstances which we shall come on to) fluke genes may value snail survival but they will not value snail reproduction at all. In the inevitable trade-off between the demands of snail survival and snail reproduction, therefore, snail genes will be selected to produce an optimal compromise, while fluke genes will be selected to devalue snail reproduction to the advantage of snail survival, and hence to thicken the shell. A thickening of the shell in parasitized snails is, it will be remembered, the observed phenomenon with which we began.

It may be objected here that, although a fluke has no direct stake in the reproduction of its own snail host, it does have a stake in there being a new generation of snails at large. This is true, but we must be very careful before we use the fact to predict that selection would favour fluke adaptations to enhance snail reproduction. The question we have to ask is this. Given that the fluke gene-pool was dominated by genes that aided snail reproduction at the expense of snail survival, would selection favour a selfish fluke gene that sacrificed the reproduction of its particular snail host, even parasitically castrated the snail, in the interest of prolonging the life of that host, and hence of promoting its own survival and reproduction? Except under special circumstances the answer is surely yes; such a rare gene would invade the fluke gene-pool, since it could exploit the free supply of new snails encouraged by the public-spirited majority of the fluke population. In other words, favouring snail reproduction at the expense of snail survival would not be a fluke ESS. Fluke genes that manage to shift the snail's investment of resources away from reproduction and into survival will tend to be favoured in the fluke gene-pool. It is entirely plausible, therefore, that the extra thickness of shells observed in parasitized snails is a fluke adaptation.

On this hypothesis, the shell phenotype is a shared phenotype, influenced by fluke genes as well as by snail genes, just as the beaver dam is a phenotype shared by genes in more than one individual beaver. According to the hypothesis, there are two optimum thicknesses of snail shells: a relatively thick fluke optimum, and a somewhat thinner snail optimum. The observed thickness in parasitized snails will probably be somewhere between the two optima, since snail genes and fluke genes are both in a position to exert power, and they are exerting their power in opposite directions.

As for parasite-free snails, it might be expected that their shells would have the snail-optimal thickness, since there are no fluke genes to exert power. This is too simple, however. If the population at large has a high incidence of fluke infestation, the gene-pool will probably contain genes that tend to compensate for the thickening effect of fluke genes. This would lead to uninfected snails having over-compensating phenotypes, shells that are thinner even than the snail optimum. I therefore predict that shell thicknesses in fluke-free areas should be intermediate between those of infected snails and uninfected snails in fluke-infested areas. I do not know of any evidence bearing on this prediction, but it would be interesting to look. Note that this prediction does not depend on any *ad hoc* assumption about snails 'winning' or flukes 'winning'. It assumes that both snail genes and fluke genes exert *some* power over the snail phenotype. The prediction will follow regardless of the quantitative details of that power.

Flukes live inside snail shells in a sense that is not too far removed from the sense in which snails live inside snail shells, and caddis larvae live inside their stone houses. Having accepted the idea that the form and colour of a caddis house might constitute phenotypic expression of caddis genes, it is not difficult to accept the idea of the form and colour of a snail shell being phenotypic expression of genes in a fluke inside the snail. If we could fancifully imagine a fluke gene and a snail gene intelligently discussing with a caddis gene the problems of making a hard outer wall for protection, I doubt if the conversation would make any reference to the fact that the fluke was a parasite while the caddis and the snail were not. The rival merits of secreting calcium carbonate, recommended by the fluke and the snail genes, versus picking up stones, preferred by the caddis gene, would be discussed. There might be some reference to the fact that a convenient and economical way of secreting calcium carbonate involves the use of a snail. But, from the gene's eye viewpoint, I suspect that the concept of parasitism would be treated as irrelevant. All three genes might regard themselves as parasitic, or alternatively as using comparable levers of power to influence their respective worlds so as to survive. The living cells of the snail would be regarded by the snail gene and the fluke gene as useful objects to be manipulated in the outside world, in exactly the same sense as the stones on the bottom of the stream would be regarded by the caddis gene.

By discussing inorganic snail *shells*, I have retained continuity with the caddis houses and other non-living artefacts of the previous chapter, thereby pursuing my policy of sustaining credulity by extending the concept of the phenotype gradually by insensible degrees. But now it is time to grasp the living snail firmly by the horns. Flukes of the genus *Leucochloridium* invade the horns of snails where they can be seen through the skin, conspicuously pulsating. This tends to make birds, who are the next host in the life cycle of the fluke, bite off the tentacles mistaking them, Wickler (1968) suggests, for insects. What is interesting here is that the flukes seem also to manipulate the *behaviour* of the snails. Whether it is because the snail's eyes are at the ends of the horns, or whether through some more indirect physiological route, the fluke manages to change the snail's behaviour with respect to light. The normal negative phototaxis is replaced in infected snails by positive light-seeking. This carries them up to open sites where they are presumably more likely to be eaten by birds, and this benefits the fluke.

Once again, if this is to be regarded as a parasitic adaptation, and indeed it is widely so regarded (Wickler 1968; Holmes & Bethel 1972), we are forced to postulate the sometime existence of genes in the parasite gene-pool that influenced the behaviour of hosts, since all Darwinian adaptations evolved by the selection of genes. By definition such genes were genes 'for' snail behaviour, and the snail behaviour has to be regarded as part of the phenotypic expression of fluke genes.

The genes in one organism's cells, then, can have extended phenotypic influence on the living body of another organism; in this case a parasite's genes find phenotypic expression in the behaviour of its host. The parasitology literature is full of interesting examples which are now usually interpreted as adaptive manipulation of hosts by parasites (e.g. Holmes & Bethel 1972; Love 1980). It has not, to be sure, always been fashionable among parasitologists to make such interpretations explicit. For instance, an important review of parasitic castration in Crustacea (Reinhard 1956) is packed with detailed information and speculation on the precise physiological routes by which parasites castrate their hosts, but is almost devoid of discussion on why they might have been selected to do so, or whether, instead, castration is simply a fortuitous byproduct of parasitization. It may be an interesting indication of changing scientific fashions that a more recent review (Baudoin 1975) extensively considers the functional significance of parasitic castration from the individual parasite's point of view. Baudoin concludes: 'The main theses of this paper are (1) that parasitic castration can be viewed as a parasite's adaptation and (2) that advantages derived from this adaptation are a result of reduction in host reproductive effort; which in turn gives rise to increased host survivorship, increased host growth and/or increased energy available to the parasite thereby increasing the parasite's Darwinian fitness.' This is, of course, very much the argument I have just followed in discussing parasite-induced thickening of snail shells. Again, a belief that parasitic castration is a parasite adaptation logically implies that there must be, or at least must have been, parasite genes 'for' changes in host physiology. The symptoms of parasitic castration, sex-change, increased size, or whatever they may be, are properly regarded as extended phenotypic manifestations of parasite genes.

The alternative to Baudoin's interpretation is that changes in host physiology and behaviour are not parasite adaptations, but simply dull pathological byproducts of infection. Consider the parasitic barnacle (though in its adult stage it looks more like a fungus) *Sacculina*. It might be said that *Sacculina* does not directly benefit from the castration of its crab host, but simply sucks nutriment from all through the host's body, and as a side-effect of its devouring the tissues of the gonads the crab shows the symptoms of castration. In support of the adaptation hypothesis, however, Baudoin points to those cases in which parasites achieve the castration by synthesis of host hormones, surely a specific adaptation rather than a boring byproduct. Even in those cases where the castration is initially caused as a byproduct of the devouring of gonadal tissue, my suspicion is that selection would subsequently act on parasites to modify the details of their physiological effects on hosts, modify them in a way favourable to the parasite's well-being. *Sacculina* presumably has some options over which parts of the crab's body its root system invades first. Natural selection is surely likely to favour genes in *Sacculina* that cause it to invade gonadal tissue before invading vital organs on which the crab's survival depends. Applying this kind of argument at a more detailed level, since gonadal destruction has multiple and complex effects on crab physiology, anatomy, and behaviour, it is entirely reasonable to guess that selection would act on parasites to fine-tune their castration technique so as to increase their benefit from the initially fortuitous consequences of castration. I believe many modern parasitologists would agree with this feeling (P. O. Lawrence, personal communication). All that I am adding, then, is the logical point that the common belief that parasitic castration is an adaptation implies that the modified host phenotype is part of the extended phenotype of parasite genes.

Parasites often stunt the growth of their hosts, and it is easy to see this as a boring byproduct of infection. More interest therefore attaches to those rarer cases in which parasites enhance host growth, and I have already mentioned the case of the thickened snail shells. Cheng (1973, p. 22) begins his account of such cases with a revealing sentence: 'Although one generally considers parasites to be detrimental to their hosts and cause the loss of energy and poor health, instances are known where the occurrence of parasites actually induces enhanced growth of the host.' But Cheng is here sounding like a medical doctor rather than like a Darwinian biologist. If 'detriment' is defined in terms of reproductive success rather than survival and 'health', it is probable that growth-enhancement is indeed detrimental to the host, for the reasons given in my discussion of snail shells. Natural selection has presumably favoured an optimal host size, and if a parasite causes the host to deviate from this size *in either direction* it is probably harming the host's reproductive success, even if it is at the same time promoting host survival. All the examples of growth-enhancement given by Cheng can be easily understood as parasite-induced switches of resources away from investment in host reproduction, which is of no interest to the parasite, to the growth and survival of the host's own body, which is of great interest to the parasite (again we must beware of the group-selectionist plea that the existence of a new generation of hosts is of importance to the parasite *species*).

Mice infected by larvae of the tapeworm *Spirometra mansanoides* grow faster than uninfected mice. It has been shown that the tapeworms achieve this by secreting a substance resembling a mouse growth hormone. More dramatically, beetle larvae of the genus *Tribolium*, when infected by the sporozoan *Nosema*, usually fail to metamorphose into adults. Instead they continue to grow through as many as six extra larval moults, ending up as giant larvae weighing more than twice as much as non-parasitized controls. Evidence suggests that this major shift in beetle priorities, from reproduction to individual growth, is due to the synthesis of juvenile hormone, or its close analogue, by the protozoan parasite. Again this is of interest because, as already suggested for the case of parasitic castration of crustacea, it makes the boring byproduct theory all but untenable. Juvenile hormones are special molecules ordinarily synthesized by insects, not protozoa. The synthesis of an insect hormone by a protozoan parasite has to be regarded as a specific and rather elaborate adaptation. The evolution of the capacity to synthesize juvenile hormone by *Nosema* must have come about through the selection of genes in the *Nosema* gene-pool. The phenotypic effect of these genes, which led to their survival in the *Nosema* gene-pool, was an extended phenotypic effect, an effect which manifested itself in beetle bodies.

Once again the problem of individual benefit versus group benefit arises, and in an acute form. A protozoan is so small in comparison with a beetle larva that a single protozoan, on its own, could not muster a sufficient dose of hormone to affect the beetle. Hormone manufacture must be a group effort by large numbers of individual protozoa. It benefits all individual parasites in the beetle, but it must also cost each individual something to add his tiny contribution to the group chemical effort. If the individual protozoa were genetically heterogeneous, consider what would happen. Assume a majority of protozoa cooperating in synthesizing hormone. An individual with a rare gene that made him opt out of the group effort would save himself the cost of synthesis. Such a saving would be of immediate benefit to him, and to the selfish gene that made him opt out. The loss of his contribution to the group synthesis would hurt his rivals just as much as it hurt him. In any case the loss to the group's productivity would be very small, though it would represent a major saving to him. Therefore, except under special conditions, taking part in a cooperative group synthesis together with genetic rivals is not an evolutionarily stable strategy. We must therefore predict that all the *Nosema* in a given beetle will be found to be close relatives, probably an identical clone. I don't know of any direct evidence here, but the expectation is in line with the typical sporozoan life cycle.

Baudoin properly emphasizes the analogous point in connection with parasitic castration. He has a section entitled 'Kinship of castrators in the same individual host', in which he says 'Parasitic castration is almost invariably produced either by single parasites or by their immediate offspring...parasitic castration is usually produced either by single genotypes or by very closely related genotypes. Metacercarial infections in snails are exceptions...In these cases, however, parasitic castration may be incidental.' Baudoin is fully aware of the significance of these facts: 'the genetic relationship of castrators within individual hosts is such that natural selection at the level of individual genotypes can account for the observed effects'.

Many fascinating examples of parasites manipulating the behaviour of their hosts can be given. For nematomorph larvae who need to break out of their insect hosts and get into water where they live as adults, 'a major difficulty in the parasite's life is the return to water. It is, therefore, of particular interest that the parasite appears to affect the behaviour of its host, and "encourages" it to return to water. The mechanism by which this is achieved is obscure, but there are sufficient isolated reports to certify that the parasite does influence its host, and often suicidally for the host.... One of the more dramatic reports describes an infected bee flying over a pool and, when about six feet over it, diving straight into the water. Immediately on impact the gordian worm burst out and swam into the water, the maimed bee being left to die' (Croll 1966).

Parasites that have a life cycle involving an intermediate host, from which they have to move to a definitive host, often manipulate the behaviour of the intermediate host to make it more likely to be eaten by the definitive host. We have already seen an example of this, the case of *Leucochloridium* in snail tentacles. Holmes and Bethel (1972) have reviewed many examples, and they have themselves provided us with one of the most thoroughly researched cases (Bethel & Holmes 1973). They studied two species of acanthocephalan worms, *Polymorphus paradoxus* and *P. marilis*. Both use a freshwater 'shrimp' (really an amphipod), *Gammarus lacustris*, as an intermediate host, and both use ducks as the definitive host. *P. paradoxus*, however, specializes in the mallard, which is a surface-dabbling duck, while *P. marilis* specializes in diving ducks. Ideally then, *P. paradoxus* might benefit by making its shrimps swim to the surface, where they are likely to be eaten by mallards, while *P. marilis* might benefit by making its shrimps avoid the surface.

Uninfected *Gammarus lacustris* tend to avoid light, and stay close to the lake bottom. Bethel and Holmes noticed striking differences in the behaviour of shrimps infected with cystacanths of *P. paradoxus*. They stayed close to the surface, and clung tenaciously to surface plants and even to the hairs on the researchers' legs. This surface-hugging behaviour would presumably make them vulnerable to predation by dabbling mallards, and also by muskrats which are an alternative definitive host of *P. paradoxus*. Bethel and Holmes believe that the habit of clinging to weeds makes infected shrimps particularly vulnerable to muskrats, who gather floating vegetation and take it home to feed off it.

Laboratory tests confirmed that shrimps infected with cystacanths of *P. paradoxus* seek the lighted half of a tank, and also positively approach the source of light. This is the opposite of the behaviour shown by uninfected shrimps. Incidentally, it is not that the infected shrimps were generally sick and floated passively to the surface, as in the comparable case of Crowden and

Broom's (1980) fish. These shrimps fed actively, often leaving the surface layer to do so, but when they captured a morsel of food they promptly took it to the surface to eat it whereas a normal shimp would have taken it to the bottom. And when startled in mid water, instead of diving to the bottom as a normal shrimp would, they headed for the surface.

Shrimps infected by cystacanths of the other species, *P. marilis*, however, do not hug the surface. In laboratory tests they admittedly sought the lighted half of an aquarium in preference to the dark half, but they did not orient positively towards the source of light: they distributed themselves randomly in the lighted half, rather than at the surface. When startled, they went to the bottom rather than to the surface. Bethel and Holmes suggest that the two species of parasite modify the behaviour of their intermediate host in different ways, calculated to make the shrimps more vulnerable to predation by their definitive hosts, surface-feeding and diving predators, respectively.

A later paper (Bethel & Holmes 1977) provides partial confirmation of this hypothesis. Captive mallards and muskrats in the laboratory took shrimps infected with *P. paradoxus* at a higher rate than they took uninfected shrimps. Shrimps infected with *P. marilis*, however, were not taken any more often than uninfected ones, either by mallards or by muskrats. It would obviously be desirable to do the reciprocal experiment with a diving predator, predicting that shrimps infected with cystacanths of *P. marilis* would, in this case, be relatively more vulnerable. This experiment does not appear to have been done.

Let us provisionally accept Bethel and Holmes's hypothesis, and rephrase it in the language of the extended phenotype. The altered behaviour of the shrimp is regarded as an adaptation on the part of the acanthocephalan parasite. If this has come about through natural selection, there must have been genetic variation 'for' shrimp behaviour in the worm gene-pool; otherwise, there would have been nothing for natural selection to work on. We may, therefore, talk of worm genes having phenotypic expression in shrimp bodies, in just the same sense as we are accustomed to talking of human genes having phenotypic expression in human bodies.

The case of the fluke ('brainworm') *Dicrocoelium dendriticum* is often quoted as another elegant example of a parasite manipulating an intermediate host to increase its chances of ending up in its definitive host (Wickler 1976; Love 1980). The definitive host is an ungulate such as a sheep, and the intermediate hosts are first a snail and then an ant. The normal life cycle calls for the ant to be accidentally eaten by the sheep. It seems that the fluke cercaria achieves this in a way analogous to that of the *Leucochloridium* mentioned above. By burrowing into the suboesophageal ganglion, the aptly named 'brainworm' changes the ant's behaviour. Whereas an uninfected ant would normally retreat into its nest when it became cold, infected ants climb to the top of grass stems, clamp their jaws in the plant and remain immobile as if asleep. Here they are vulnerable to being eaten by the worm's definitive host. The infected ant, like a normal ant, does retreat down the grass stem to avoid death from the midday heat—which would be bad for the parasite—but it returns to its aerial resting position in the cool of the afternoon (Love 1980). Wickler (1976) says that of the approximately fifty cercariae that infect a given ant, only one burrows into the brain and it dies in the process: 'It sacrifices itself for the benefit of the other cercariae.' Understandably, Wickler therefore predicts that the group of cercariae in an ant will be found to be a polyembryonic clone.

An even more elaborate example is the case of crown gall, one of the few known plant cancers (Kerr 1978; Schell *et al.* 1979). Exceptionally for a cancer it is induced by a bacterium, *Agrobacterium.*

These bacteria induce the cancer in the plant only if they themselves contain a Ti plasmid, a small ring of extrachromosomal DNA. The Ti plasmid may be regarded as an autonomous replicator (Chapter 9) although, like any other DNA replicator, it cannot succeed apart from the cellular machinery put together under the influence of other DNA replicators, in this case those of the host. Ti genes are transferred from bacterial to plant cells, and the infected plant cells are induced to multiply uncontrollably, which is why the condition is called a cancer. The Ti genes also cause the plant cells to synthesize large quantities of substances called opines, which the plants would not normally make, and which they cannot use. The interesting point is that, in an environment rich in opines, Ti-infected bacteria survive and reproduce much better than non-infected bacteria. This is because the Ti plasmid provides the bacterium with a set of genes which enable the bacterium to use the opines as a source of energy and chemicals. The Ti plasmids could almost be regarded as practising artificial selection in favour of infected bacteria, hence in favour of copies of themselves. The opines also function as bacterial 'aphrodisiacs', as Kerr puts it, promoting bacterial conjugation and therefore plasmid transfer.

Kerr (1978) concludes: 'It is a very elegant example of biological evolution; it even demonstrates apparent altruism in bacterial genes.... The DNA which is transferred from bacterium to plant has no future; it dies when the plant cell dies. However, by altering the plant cell to produce an opine, it ensures (a) preferential selection of the same DNA in bacterial cells and (b) transfer of that DNA to other bacterial cells. It demonstrates evolution at the level of genes, not organisms, which may only be gene carriers.' (Such statements are, of course, music to my ears, but I hope Kerr will forgive my publicly wondering at the gratuitous caution of 'may only be' gene carriers. It is a bit like saying 'The eyes may be the windows of the soul' or 'O, my Luve may be like a red red rose'. It may have been the work of an editorial hand!) Kerr continues: 'In naturally-induced crown galls on many (but not all) hosts, very few bacteria survive in the gall.... At first sight, it would appear that pathogenicity is conferring no biological advantage. It is only when one considers opine production by the host and its effect on bacteria living at the surface of the gall, that the strong selective advantage of genes for pathogenicity becomes clear.'

Mayr (1963, pp. 196–7) discusses the phenomenon of plants making galls to house insects, in terms so favourable to my thesis that I can quote him verbatim almost without comment:

> Why...should a plant make the gall such a perfect domicile for an insect that is its enemy? Actually we are dealing here with two selection pressures. On the one hand, selection works on a population of gall insects and favors those whose gall-inducing chemicals stimulate the production of galls giving maximum protection to the young larva. This, obviously, is a matter of life or death for the gall insect and thus constitutes a very high selection pressure. The opposing selection pressure on the plant is in most cases quite small because having a few galls will depress viability of the plant host only very slightly. The 'compromise' in this case is all in favor of the gall insect. Too high a density of the gall insect is usually prevented by density-dependent factors not related to the plant host.

Mayr is here invoking the equivalent of the 'life/dinner principle' to explain why the plant does not fight back against the remarkable manipulation by the insect. It is necessary for me to add only this. If Mayr is right that the gall is an adaptation for the benefit of the insect and not the plant, it can have evolved only through the natural selection of genes in the insect gene-pool. Logically, we have to regard these as genes with phenotypic expression in plant tissue, in the same sense as some other gene of the insect, say one for eye colour, can be said to have phenotypic expression in insect tissue.

Colleagues with whom I discuss the doctrine of the extended phenotype repeatedly come up with the same entertaining speculations. Is it just an accident that we sneeze when getting a cold, or could it be a result of manipulation by viruses to increase their chances of infecting another host? Do any venereal diseases increase the libido, even if only by inducing an itch, like extract of Spanish fly? Do the behavioural symptoms of rabies infection increase the chance of the virus being passed on (Bacon & Macdonald 1980)? 'When a dog gets rabies, its temper changes quickly. It is often more affectionate for a day or two, and given to licking its human contacts, a dangerous practice, for the virus is already in its saliva. Soon it grows restless and wanders off, ready to bite anyone that gets in its way' (*Encyclopaedia Britannica* 1977). Even non-carnivorous animals are driven by the rabies virus to vicious biting, and there are recorded cases of humans contracting the disease from the bites of normally harmless fruit-eating bats. Apart from the obvious power of biting to spread a saliva-borne virus, 'restless wandering' might very well serve to spread the virus more effectively (Hamilton & May 1977). That the widespread availability of cheap air travel has had a dramatic impact on the spread of human disease is obvious: dare we wonder whether the phrase 'travel bug' might have a more than metaphorical significance?

The reader will probably, like me, find such speculations far-fetched. They are intended only as light-hearted illustrations of the *kind* of thing that might go on (see also Ewald, 1980, who draws attention to the medical significance of this kind of thinking). All I really need to establish is that in *some* examples host symptoms can properly be regarded as parasitic adaptation; say, for instance, the Peter Pan syndrome in *Tribolium* induced by protozoan-synthesized juvenile hormone. Given such an admitted parasite adaptation, the conclusion I wish to draw is not really disputable. If host behaviour or physiology is a parasite adaptation, there must be (have been) parasite genes 'for' modifying the host, and the host modifications are therefore part of the phenotypic expression of those parasite genes. The extended phenotype reaches out of the body in whose cells the genes lie, reaches out to the living tissues of other organisms.

The relationship of *Sacculina* gene to crab body is not in principle different from the relationship of caddis gene to stone, nor indeed different from the relationship of human gene to human skin. This is the first of the points that I intended to establish in this chapter. It has the corollary, which I have already emphasized in other terms in Chapter 4, that the behaviour of an individual may not always be interpretable as designed to maximize its own genetic welfare: it may be maximizing somebody else's genetic welfare, in this case that of a parasite inside it. In the next chapter we shall go further, and see that some of the attributes of an individual may be regarded as phenotypic expression of genes in other individuals who need not necessarily be parasites inside.

The second point of this present chapter is that the genes that bear upon any given extended phenotypic trait may be in conflict rather than in concert with one another. I could talk in terms of any of the examples given above, but I shall stick to one, the case of the snail shell thickened by the influence of a fluke. To recapitulate that story in slightly different terms, a student of snail genetics and a student of fluke genetics might each look at the same phenotypic variation, variation in snail shell thickness. The snail geneticist would partition the variance between a genetic and an environmental component, by correlating the thickness of shells in parent snails and their offspring. The fluke geneticist would independently partition the same observed variance into a genetic and an environmental component, in his case by correlating the shell thickness of snails containing particular flukes with the shell thickness of snails containing the offspring of those same flukes. As far as the snail geneticist is concerned, the fluke contribution is part of what he calls the 'environmental' variation. Reciprocally, for the fluke geneticist, variation due to snail genes is 'environmental' variation.

An 'extended geneticist' would acknowledge both sources of genetic variation. He would have to worry about the form of their interaction—is it additive, multiplicative, 'epistatic', etc.?—but in principle such worries are already familiar to both the snail geneticist and the fluke geneticist. Within any organism, different genes influence the same phenotypic traits, and the form of the interaction is a problem with respect to genes within one normal genome just as much as it is for genes in an 'extended' genome. Interactions between the effects of snail gene and fluke gene are, in principle, no different from interactions between the effects of one snail gene and another snail gene.

And yet, it may be said, is there not a rather important difference? Snail gene may interact with snail gene in additive, multiplicative, or any other ways, but do they not both have the same interests at heart? Both have been selected in the past because they worked for the same end, the survival and reproduction of the snails that bore them. Fluke gene and fluke gene, too, are working for the same end, the reproductive success of the fluke. But snail gene and fluke gene do not have the same interests at heart. One is selected to promote snail reproduction, the other to promote fluke reproduction.

There is truth in the protestations of the last paragraph, but it is important to be clear about exactly where that truth lies. It is not that there is some obvious trade-union spirit uniting fluke genes against a rival union of snail genes. To persist with this harmless anthropomorphism, each gene is fighting only its alleles at the same locus, and it will 'unite' with genes at other loci only insofar as doing so assists it in its selfish war against its own alleles. A fluke gene may 'unite' with other fluke genes in this way but, equally, if it was convenient to do so, it might unite with particular snail genes. And if it remains true that snail genes are in practice selected to work together with each other and against an opposing gang of fluke genes, the reason is only that snail genes tend to gain from the same events in the world as do other snail genes. Fluke genes stand to gain from other events. And the real reason why snail genes stand to gain from the same events as each other, while fluke genes stand to gain from a different set of events, is simply this: all snail genes share the same route into the next generation—snail gametes. All fluke genes, on the other hand, must use a different route, fluke cercariae, to get into the next generation. It is this fact alone that 'unites' snail genes against fluke genes and vice versa. If it were the case that the parasite genes passed out of the host's body inside the host's gametes, things might be very different. The interests of host genes and parasite genes might not be quite identical, but they would then be very much closer than in the case of fluke and snail.

It follows from the extended phenotype view of life, then, that crucial importance attaches to the means by which parasites propagate their genes out of a given host into a new host. If the parasite's means of genetic exit from the host's body is the same as the host's, namely the host's gametes or spores, there will be relatively little conflict between the 'interests' of parasite and host genes. For example, both would 'agree' about the optimum thickness of the host's shell. Both will be selected to work not only for host survival but for host reproduction as well, with all that that entails. This might include host success in courtship and even, if the parasites aspire to be 'inherited' by the host's own offspring, host success in parental care. Under such circumstances the interests of parasite and host would be likely to coincide to such an extent that it might become difficult for us to detect that a separate parasite existed at all. It is clearly of great interest for parasitologists and 'symbiologists' to study such very intimate parasites or symbionts, symbionts with an interest in the success of their host's gametes as well as in the survival of their host's body. Some lichens might be promising examples, and so might bacterial endosymbionts of insects which are transmitted transovarially and in some cases seem to influence the host sex ratio (Peleg & Norris 1972).

Mitochondria, chloroplasts, and other cell organelles with their own replicating DNA might also be good candidates for study in this connection. A fascinating account of cell organelles and microorganisms seen as semi-autonomous symbionts inhabiting a cellular ecology is given in the proceedings of a symposium entitled *The Cell as a Habitat*, edited by Richmond and Smith (1979). The closing words of Smith's introductory chapter are particularly memorable and apposite: 'In non-living habitats, an organism either exists or it does not. In the cell habitat, an invading organism can progressively lose pieces of itself, slowly blending into the general background, its former existence betrayed only by some relic. Indeed, one is reminded of Alice in Wonderland's encounter with the Cheshire Cat. As she watched it, "it vanished quite slowly, beginning with the tail, and ending with the grin, which remained some time after the rest of it had gone"' (Smith 1979). Margulis (1976) gives an interesting survey of all the degrees of vanishing of the grin.

Richmond's (1979) chapter, too, is very congenial to the present thesis: 'It is conventional to regard cells as units of biological function. Another view, particularly apposite to this symposium, is that the cell is the minimal unit capable of replicating DNA.... Such a concept places DNA at the centre of biology. Thus DNA is not regarded simply as a hereditary means of ensuring the long term survival of the organisms of which it forms part. Rather, it stresses that the primary role of cells is to maximize the amount and diversity of DNA in the biosphere'. This last remark, incidentally, is unfortunate. Maximizing the amount and diversity of DNA in the biosphere is the concern of nobody and nothing. Rather, each small piece of DNA is selected for its power to maximize its *own* survival and replication. Richmond goes on: 'If the cell is considered as a unit for the replication of DNA, it follows that DNA additional to that required for the duplication of the cell may also be carried; molecular parasitism, symbiosis and mutualism may occur at the DNA level, as it does at higher organizational levels in biology.' We have arrived back at the concept of 'selfish DNA' which was a subject of Chapter 9.

It is interesting to speculate on whether mitochondria, chloroplasts, and other DNA-bearing organelles originated from parasitic prokaryotes (Margulis 1970, 1981). But, important as that question is as a matter of history, it does not bear, one way or the other, on my present concern. I am here interested in whether mitochondrial DNA is likely to work for the same phenotypic ends as nuclear DNA, or whether it is likely to be in conflict with it. This should depend not on the historic origins of mitochondria but on their present method of propagating their DNA. Mitochondrial genes are passed out of one metazoan body into a body of the next generation in egg cytoplasm. An optimal female phenotype from the point of view of the female's own nuclear genes is likely to be very much the same as an optimal female phenotype from the point of view of her mitochondrial DNA. Both have an interest in her successfully surviving, reproducing, and rearing offspring. At least, that is true as far as female offspring are concerned. Mitochondria presumably have no 'wish' for their bodies to have sons: a male body represents the end of the line as far as mitochondrial descent is concerned. All mitochondria in existence have spent the vast majority of their ancestral careers in female bodies, and they might tend to have what it takes to persist in inhabiting female bodies. In birds the interests of mitochondrial DNA will be closely similar to those of Y-chromosomal DNA, and slightly divergent from those of autosomal and X-chromosomal DNA. And if mitochondrial DNA could exert phenotypic power in the egg of a mammal, it is perhaps not too fanciful to imagine it frantically fighting off the kiss of death of Y-bearing sperms (Eberhard 1980; Cosmides & Tooby 1981). But in any case, if the interests of mitochondrial DNA and nuclear DNA are not always identical, they are very close, certainly much closer than the interests of fluke DNA and snail DNA.

The message of the present section is this. The fact that snail genes conflict with fluke genes more than they conflict with other snail genes at different loci is not the obvious foregone conclusion it might appear to be. It results simply from the fact that any two genes in a snail nucleus are obliged to use the same exit route from the present body into the future. Both have the same stake in the success of the present snail in manufacturing gametes, getting them fertilized, and securing the survival and reproduction of the offspring so formed. Fluke genes conflict with snail genes in their influence on the shared phenotype, simply because their destiny is shared for only a short part of the future: their common cause is limited to the life of the present host body, and does not carry over into the gametes and offspring of the present host.

The role of mitochondria in the argument is to exemplify cases where parasite and host genes share the same gametic destiny, at least in part. If nuclear genes do not conflict with nuclear genes at other loci, it is only because meiosis is even-handed: meiosis does not normally favour some loci over others, nor some alleles over others, but scrupulously puts one gene at random from each diploid pair in every gamete. Of course there are instructive exceptions, and for my thesis they are sufficiently important to have dominated the two chapters on 'outlaws' and 'selfish DNA'. There, as here, an important message is that replicating entities will tend to work against each other to the extent that they employ different methods of egress from vehicle to vehicle.

Returning to the main subject of the present chapter, parasitic and symbiotic relationships can be classified in various ways for different purposes. The classifications developed by parasitologists and doctors are no doubt useful for their purposes, but I want to develop a particular classification based on the concept of gene power. It should be remembered that, from this point of view, the normal relationship between different genes in the same nucleus, even on the same chromosome, is just one extreme on the continuum of parasitic or symbiotic relationships.

The first dimension of my classification has already been stressed. It concerns the degree of similarity or difference in the methods of egress from hosts, and propagation of host genes and parasite genes. At one extreme will be parasites that use the host's propagules for their own reproduction. For such parasites an optimum host phenotype from the parasite's point of view is likely to coincide with the optimum from the point of view of the host's own genes. This is not to say that the host genes would not 'prefer' to be rid of the parasite altogether. But both have an interest in mass-producing the same propagules, and both have an interest in developing a phenotype that is good for mass-producing those same propagules: the right beak length, wing shape, courtship behaviour, clutch size, etc., down to minute details of all aspects of the phenotype.

At the other extreme will be parasites whose genes are passed on not in the host's reproductive propagules but, say, in the host's exhaled breath, or in the host's dead body. In these cases the optimum host phenotype from the parasite genes' point of view is likely to be very different from the optimum host phenotype from the host genes' point of view. The phenotype which emerges will be a compromise. This, then, is one dimension of classification of host–parasite relationships. I shall call it the dimension of 'propagule overlap'.

A second dimension of classification concerns the time of action of parasite genes during host development. A gene, whether a host gene or a parasite gene, can exert a more fundamental influence on the final host phenotype if it acts early in the development of the host embryo than if it acts late. A radical change such as the development of two heads could be achieved by a single mutation (in host or parasite genome) provided the mutation acted sufficiently early in the embryonic development of the host. A late-acting mutant (again, in host or parasite genome)—a mutant that does not begin to act until the host body has reached adulthood—is likely to have only a small effect, since the general architecture of the body will, by then, have been laid down. Therefore a parasite that enters its host after the host has reached adulthood is less likely to have a radical effect on the host's phenotype than a parasite that gets in early. There are notable exceptions, however, such as the parasitic castration of crustacea already mentioned.

My third dimension of classification of host–parasite relations concerns the continuum from what may be called close intimacy to action at a distance. All genes exert power primarily by serving as templates for the synthesis of proteins. The locus of primary gene power is, therefore, the cell, in particular the cytoplasm surrounding the nucleus in which the gene sits. Messenger RNA streams through the nuclear membrane and mediates genetic control over cytoplasmic biochemistry. The phenotypic expression of a gene is then, in the first place, its influence on cytoplasmic biochemistry. In its turn, this influences the form and structure of the whole cell, and the nature of its chemical and physical interactions with neighbouring cells. This affects the build-up of multicellular tissues, and in turn the differentiation of a variety of tissues in the developing body. Finally emerge the attributes of the whole organism that gross anatomists and ethologists identify at their level as phenotypic expressions of genes.

Where parasite genes exert shared power with host genes over the same host phenotypic characteristic, the confluence of the two powers may occur at any stage in the chain just described. Snail genes, and the genes of the fluke that parasitizes the snail, exert their power separately from each other at the cellular and even the tissue level. They influence the cytoplasmic chemistry of their respective cells separately, because they do not share cells. They influence tissue formation separately, because snail tissues are not intimately infiltrated by fluke tissues in the way that, say, the algal and fungal tissues of a lichen are intimate. Snail genes and fluke genes influence the development of organ systems, indeed of whole organisms, separately, because all the fluke cells are gathered together in one mass rather than being interspersed among snail cells. If fluke genes influence snail shell thickness, they do so by first collaborating with other fluke genes to make a whole fluke.

Other parasites and symbionts more intimately infiltrate the systems of the host. At the extreme are the plasmids and other fragments of DNA which, as we saw in Chapter 9, literally insert themselves in the host chromosomes. It is impossible to imagine a more intimate parasite. 'Selfish DNA' itself is not more intimate, and indeed we may never know how many of our genes, whether 'junk' or 'useful', originated as inserted plasmids. It seems to follow from the thesis of this book that there is no important distinction between our 'own' genes and parasitic or symbiotic insertion sequences. Whether they conflict or cooperate will depend not on their historical origins but on the circumstances from which they stand to gain now.

Viruses have their own protein jacket, but they insert their DNA into the host's cell. They are therefore in a position to influence the cellular chemistry of the host at an intimate level, if not quite such an intimate level as an insertion sequence in the host chromosome. Intracellular parasites in the cytoplasm, too, may be presumed to be in a position to exert considerable power over host phenotypes.

Some parasites do not infiltrate the host at the cellular level, but at the tissue level. Examples are *Sacculina*, and many fungal and plant parasites, where parasite cells and host cells are distinct, but where the parasite invades the host's tissues by means of an intricate and finely divided root system. The separate cells of parasitic bacteria and protozoa may infiltrate the host tissues with similarly comprehensive intimacy. To a slightly lesser extent than a cell parasite, such a 'tissue parasite' is in a strong position to influence organ development and gross phenotypic form and behaviour. Other internal parasites, such as the flukes we have been discussing, do not mix their tissues with those of the host, but keep their tissues to themselves and exert power only at the level of the whole organism.

But we have not yet reached the end of our continuum of proximity. Not all parasites live physically inside their hosts. They may even seldom come into contact with their hosts. A cuckoo is a parasite in very much the same way as a fluke. Both are whole-organism parasites rather than tissue parasites or cell parasites. If fluke genes can be said to have phenotypic expression in a snail's body, there is no sensible reason why cuckoo genes should not be said to have phenotypic expression in a reed warbler's body. The difference is a practical one, and a rather smaller one than the difference between, say, a cellular parasite and a tissue parasite. The practical difference is that the cuckoo does not live inside the reed warbler's body, so has less opportunity for manipulating the host's internal biochemistry. It has to rely on other media for its manipulation, for instance sound waves and light waves. As discussed in Chapter 4, it uses a supernormally bright gape to inject its control into the reed warbler's nervous system via the eyes. It uses an especially loud begging cry to control the reed warbler's nervous system via the ears. Cuckoo genes, in exerting their developmental power over host phenotypes, have to rely on action at a distance.

The concept of genetic action at a distance pushes our idea of the extended phenotype out to its logical culmination. That is where we must go in the next chapter.

13

ACTION AT A DISTANCE

Snail shells coil either to the right or to the left. Usually all individuals in one species coil the same way, but a few polymorphic species are to be found. In the Pacific island land snail *Partula suturalis* some local populations are right-handed, others are left-handed, and others are mixed in various proportions. It is therefore possible to study the genetics of directionality of coiling (Murray & Clarke 1966). When snails from right-handed populations were crossed with snails from left-handed populations, every offspring coiled the same way as its 'mother' (the parent that provided the egg: the snails are hermaphrodites). This might be thought to indicate a non-genetic maternal influence. But when Murray and Clarke crossed F1 snails with each other they obtained a curious result. All the progeny were left-handed, regardless of the direction of coiling of either parent. Their interpretation of the results is that coiling is genetically determined, with left-handedness dominant to right-handedness, but that an animal's phenotype is controlled not by its own genotype but by its mother's genotype. Thus the F1 individuals displayed the phenotypes dictated by their mothers' genotypes, although all contained the same heterozygous genotypes since they were produced by mating two pure strains. Similarly, the F2 progeny of F1 matings all displayed the phenotype appropriate to an F1 genotype—left-handed since that is dominant and the F1 genotype was heterozygous. The underlying genotypes of the F2 generation presumably segregated in classic 3:1 Mendelian fashion, but this did not show itself in their phenotypes. It would have shown itself in the phenotypes of their progeny.

Note that it is the mother's genotype, not her phenotype, which controls her offspring's phenotype. The F1 individuals themselves were left-handed or right-handed in equal proportion, yet all had the same heterozygous genotype, and all therefore produced left-handed offspring. A similar effect had been obtained earlier in the freshwater snail *Limnaea peregra*, though in that case right-handedness was dominant. Other such 'maternal effects' have long been known to geneticists. As Ford (1975) put it, 'We have here simple Mendelian inheritance the expression of which is constantly delayed one generation.' The phenomenon perhaps arises when the embryological event determining the phenotypic trait occurs so early in development as to be influenced by maternal messenger RNA from the egg cytoplasm, before the zygote has begun to manufacture its own messenger RNA. The direction of coiling in snails is determined by the initial direction of spiral cleavage, which occurs before the embryo's own DNA has begun to work (Cohen 1977).

This kind of effect provides a special opportunity for the kind of maternal manipulation of offspring that we discussed in Chapter 4. More generally, it is a special example of genetic 'action at a distance'. It illustrates, in a particularly clear and simple manner, that the power of a gene may extend beyond the boundaries of the body in whose cells it sits (Haldane 1932b). We cannot hope that all genetic action at a distance will reveal itself in so elegant a Mendelian manner as in the case of the snails. Just as, in conventional genetics, the Mendelian major genes of the schoolroom are the tip of the iceberg of reality, so we may make conjectures about a polygenic 'extended genetics', a genetics in which action at a distance is rife but in which the effects of the genes are complex and interacting, and therefore difficult to sort out. Again as in conventional genetics, we do not necessarily have to do genetic experiments in order to infer the presence of genetic influence on variation. Once we have satisfied ourselves that a given characteristic is a Darwinian adaptation this, in itself, is tantamount to satisfying ourselves that variation in that character must at one time have had a genetic basis. If it had not, selection could not have preserved the advantageous adaptation in the population.

One phenomenon that looks like an adaptation and which, in some sense, involves action at a distance is the 'Bruce Effect'. A female mouse who has just been inseminated by one male has her pregnancy blocked by exposure to chemical influence from a second male. The effect seems also to occur in a variety of species of mice and voles in nature. Schwagmeyer (1980) considers three main hypotheses of the adaptive significance of the Bruce Effect, but for the sake of argument I shall not, here, advocate the hypothesis that Schwagmeyer attributes to me—that the Bruce Effect represents a kind of female adaptation. Instead I shall look at it from the male point of view, and simply assume that the second male benefits himself by preventing the female's pregnancy, thereby eliminating the offspring of a male rival, while at the same time bringing the female quickly into oestrus so that he can mate with her himself.

I have expressed the hypothesis in the language of Chapter 4, the language of individual manipulation. But it can equally well be expressed in the language of the extended phenotype and genetic action at a distance. Genes in male mice have phenotypic expression in female bodies, in just the same sense as genes in mother snails have phenotypic expression in the bodies of their children. In the snail case, the medium of the action at a distance was assumed to be maternal messenger RNA. In the mouse case it is apparently a male pheromone. My thesis is that the difference between the two cases is not a fundamental one.

Consider how an 'extended geneticist' might talk about the genetical evolution of the Bruce Effect. A mutant gene arose which, when present in the body of a male mouse, had phenotypic expression in the bodies of female mice with whom he came in contact. The route of action of the gene on its final phenotype was long and complex, but not noticeably more so than routes of genetic action within bodies customarily are. In conventional within-body genetics, the chain of causation leading from gene to observed phenotype may have many links. The first link is always RNA, the second is protein. A biochemist may detect the phenotype that interests him at this second link stage. Physiologists or anatomists will not pick up the phenotype that interests them until more stages have been passed. They will not concern themselves with the details of these earlier links in the chain, but will take them for granted. Whole-organism geneticists find it sufficient to do breeding experiments looking only at what, for them, is the final link in the chain, eye colour, crinkliness of hair, or whatever it is. The behaviour geneticist looks at an even more distant link—waltzing in mice, creeping-through mania in sticklebacks, hygiene in honeybees, etc. He arbitrarily chooses to regard a behaviour pattern as the end link in the chain, but he knows that the abnormal behaviour of a mutant is caused by, say, abnormal neuroanatomy, or abnormal endocrine physiology. He knows that he could have looked with a microscope at the nervous system in order to detect his mutants, but he preferred to look at behaviour instead (Brenner 1974). He made an arbitrary decision to regard observed behaviour as the end link in the chain of causation.

Whichever link in the chain a geneticist chooses to regard as the 'phenotype' of interest, he knows that the decision was an arbitrary one. He might have chosen an earlier stage, and he might have chosen a later one. So, a student of the genetics of the Bruce Effect could assay male pheromones biochemically in order to detect the variation upon which to base his genetic study. Or he could look further back in the chain, ultimately to the immediate polypeptide products of the genes concerned. Or he could look later in the chain.

What is the next later link in the chain after the male pheromone? It is outside the male body. The chain of causation extends across a gap into the female body. It goes through a number of stages in the female body, and once again our geneticist does not have to bother himself with the details. He chooses, for convenience, to end his conceptual chain at the point where the gene causes pregnancy blockage in females. That is the phenotypic gene-product which he finds most easy to assay, and it is the phenotype which is of direct interest to him as a student of adaptation in nature. Abortion in female mice, according to this hypothesis, is a phenotypic effect of a gene in male mice.

How, then, would the 'extended geneticist' visualize the evolution of the Bruce Effect? The mutant gene which, when present in males, has the phenotypic effect in female bodies of causing them to abort, is favoured by natural selection over its alleles. It is favoured because it tends to be carried in the bodies of the offspring which the female bears after blocking her previous pregnancy. But, following the habit of Chapter 4, we now guess that females would be unlikely to submit to such manipulation without resistance, and that a kind of arms race might develop. In the language of individual advantage, selection would favour mutant females that resisted the pheromonal manipulation of the males.

How would the 'extended geneticist' think about this resistance? By invoking the concept of the modifier gene.

Once again, we turn first to conventional within-body genetics to remind ourselves of a principle, then carry that principle over into the realm of extended genetics. In within-body genetics we are quite used to the idea of more than one gene affecting variation in any given phenotypic character. Sometimes it is convenient to designate one locus as having the 'major' effect on the character, the others having 'modifying' effects. At other times no one locus predominates over the others sufficiently to be called major. All the genes may be thought of as modifying the effects of each other. In the chapter on 'Outlaws and Modifiers', we saw that two loci bearing on the same phenotypic character may be subject to conflicting selection pressures. The end result may be stalemate, compromise, or outright victory for one side or the other. The point is that conventional within-body genetics is already accustomed to thinking of the natural selection of genes at different loci bearing upon the same phenotypic character but in opposite directions.

Apply the lesson in the extended genetics domain. The phenotypic trait of interest is abortion in female mice. The genes bearing upon it no doubt include a set of genes in the female's own body, and also another set of genes in the male's body. In the case of the male genes the links in the chain of causation include pheromonal action at a distance, and this may make the influence of the male genes seem very indirect. But the causal links in the case of the female genes are likely to be nearly as indirect, albeit they are confined inside her body. Probably they make use of various chemical secretions flowing in her bloodstream, whereas the male genes make use, in addition, of chemical secretions flowing in the air. The point is that both sets of genes, by long and indirect causal links, bear upon the same phenotypic character, abortion in the female, and either set of genes may be regarded as modifiers of the other set, just as some genes within each set may be regarded as modifiers of others within the same set.

Male genes influence the female phenotype. Female genes influence the female phenotype, and also modify the influence of male genes. For all we know, female genes influence the male phenotype in counter-manipulation, in which case we expect the selection of modifiers among genes in males.

This whole story could have been told in the language of Chapter 4, the language of individual manipulation. The language of extended genetics is not demonstrably more correct. It is a different way of saying the same thing. The Necker Cube has flipped. Readers must decide for themselves whether they like the new view better than the old. I suggest that the way the extended geneticist tells the story of the Bruce Effect is more elegant and parsimonious than the way the conventional geneticist would have told it. Both geneticists potentially have to contend with a formidably long and complex chain of causation, leading from gene to phenotype. Both admit that their choice of which link in the chain to designate as the phenotypic character of interest—earlier links being consigned to the embryologist—is arbitrary. The conventional geneticist makes the further arbitrary decision to cut off all chains at the point where they reach the outer wall of the body.

Genes affect proteins, and proteins affect X which affects Y which affects Z which...affects the phenotypic character of interest. But the conventional geneticist defines 'phenotypic effect' in such a way that X, Y, and Z must all be confined inside one individual body wall. The extended geneticist recognizes that this cut-off is arbitrary, and he is quite happy to allow his X, Y, and Z to leap the gap between one individual body and another. The conventional geneticist takes in his stride the bridging of gaps between cells within bodies. Human red blood cells, for instance, have no nuclei, and must express the phenotypes of genes in other cells. So why should we not, when the occasion warrants it, conceive of the bridging of gaps between cells in different bodies? And when will the occasion warrant it? Whenever we find it convenient, and this will tend to be in any of those cases where, in conventional language, one organism appears to be manipulating another. The extended geneticist would, in fact, be quite happy to rewrite the whole of Chapter 4, fixing his gaze on the new face of the Necker Cube. I shall spare the reader any such rewriting, although it would be an interesting task to undertake. I shall not pile example on example of genetic action at a distance, but instead will discuss the concept, and problems that it raises, more generally.

In the chapter on arms races and manipulation I said that an organism's limbs might be adapted to work for the genes of another organism, and I added that this idea could not be made fully meaningful until later in the book. I meant that it could be made meaningful in terms of genetic action at a distance. So, what does it mean to say that a female's muscles work for a male's genes, or that a parent's limbs work for its offspring's genes, or that a reed warbler's limbs work for a cuckoo's genes? It will be remembered that the 'central theorem' of the selfish organism claims that an animal's behaviour tends to maximize its own (inclusive) fitness. We saw that to talk of an individual behaving so as to maximize its inclusive fitness is equivalent to talking of the gene or genes 'for' that behaviour pattern maximizing their survival. We have now also seen that, in precisely the same sense as it is ever possible to talk of a gene 'for' a behaviour pattern, it is possible to talk of a gene, in one organism, 'for' a behaviour pattern (or other phenotypic characteristic) in another organism. Putting these three things together we arrive at our own 'central theorem' of the extended phenotype: *An animal's behaviour tends to maximize the survival of the genes 'for' that behaviour, whether or not those genes happen to be in the body of the particular animal performing it.*

And how far afield can the phenotype extend? Is there any limit to action at a distance, a sharp cut-off, an inverse square law? The farthest action at a distance I can think of is a matter of several miles, the distance separating the extreme margins of a beaver lake from the genes for whose survival it is an adaptation. If beaver lakes could fossilize, we would presumably see a trend towards increased lake size if we arranged the fossils in chronological order. The increase in size was doubtless an adaptation produced by natural selection, in which case we have to infer that the evolutionary trend came about by allele replacement. In the terms of the extended phenotype, alleles for larger lakes replaced alleles for smaller lakes. In the same terms, beavers can be said to carry within themselves genes whose phenotypic expression extends many miles away from the genes themselves.

Why not hundreds of miles, thousands of miles? Could an ectoparasite which stayed behind in England inject a swallow with a drug which affected that swallow's behaviour on its arrival in Africa, and could the consequence in Africa be usefully regarded as the phenotypic expression of parasite genes in England? The logic of the extended phenotype might seem to favour the idea, but I think in practice it is unlikely, at least if we are talking about phenotypic expression as *adaptation*. I see a crucial practical difference from the case of the beaver dam. A gene in a beaver which, when compared with its alleles, causes a larger lake to come into existence, can directly benefit itself by means of its lake. Alleles causing smaller lakes are less likely to survive, as a direct consequence of their smaller phenotypes. It is, however, hard to see how a gene in an English ectoparasite could benefit itself, at the expense of its alleles in England, as a direct result of its African phenotypic expression. Africa is probably too far away for the consequences of the gene's action to feed back and affect the welfare of the gene itself.

By the same token, beyond a certain size of beaver lakes, it would become hard to regard further increases in size as adaptations. The reason is that, beyond a certain size, other beavers than the builders of the dam are just as likely to benefit from each increase in size as the dam-builders themselves. A big lake benefits all the beavers in the area, whether they created it or whether they just found it and exploited it. Similarly, even if a gene in an English animal could exert some phenotypic effect on Africa which directly benefited the survival of the gene's 'own' animal, other English animals of the same kind would almost certainly benefit just as much. We must not forget that natural selection is all about *relative* success.

It is admittedly possible to speak of a gene as having a particular phenotypic expression, even when its own survival is not influenced by that phenotypic expression. In this sense, then, a gene in England might indeed have phenotypic expression in a remote continent where its consequences do not feed back upon its own success in the English gene-pool. But I have already argued that in the world of the extended phenotype this is not a profitable way of speaking. I used the example of footprints in mud as phenotypic expression of genes for foot shape, and I gave my intention of using extended phenotype language only when the character concerned might conceivably influence, positively or negatively, the replication success of the gene or genes concerned.

It is not plausible, but it helps to make the point if I construct a thought experiment in which it would indeed be useful to speak of a gene as having phenotypic expression extending to another continent. Swallows return, each year, to exactly the same nest. It follows that an ectoparasite, waiting dormant in a swallow's nest in England, can expect to see the very same swallow both before and after the swallow's journey to Africa. If the parasite could engineer some change in the swallow's behaviour in Africa, it might indeed reap the consequences on the swallow's return to England. Suppose, for instance, that the parasite needs a rare trace element which is not found in England, but which occurs in the fat of a particular African fly. Swallows normally have no preference for this fly, but the parasite, by injecting a drug into the swallow before it leaves for Africa, so changes its dietary preferences as to increase the likelihood of its eating specimens of this fly. When the swallow returns to England, its body contains enough of the trace element to benefit the individual parasite (or its children) waiting in the original nest, benefit them at the expense of rivals within the parasite species. Only in circumstances such as these would I wish to speak of a gene in one continent as having phenotypic expression in another continent.

There is a risk, which I had better forestall, that such talk of adaptation on a global scale may call to the reader's mind the fashionable image of the ecological 'web', of which the most extreme manifestation is the 'Gaia' hypothesis of Lovelock (1979). My web of interlocking extended phenotypic influences bears a superficial resemblance to the webs of mutual dependence and symbiosis that bulk so largely in the pop-ecology literature (e.g. *The Ecologist*) and in Lovelock's book. The comparison could hardly be more misleading. Since Lovelock's Gaia hypothesis has been enthusiastically espoused by no less a scientist then Margulis (1981), and extravagantly praised by Mellanby (1979) as the work of a genius, it cannot be ignored, and I must digress in order categorically to disclaim any connection with the extended phenotype.

Lovelock rightly regards homeostatic self-regulation as one of the characteristic activities of living organisms, and this leads him to the daring hypothesis that the whole Earth is equivalent to a single living organism. Whereas Thomas's (1974) likening of the world to a living cell can be accepted as a throwaway poetic line, Lovelock clearly takes his Earth/organism comparison seriously enough to devote a whole book to it. He really means it. His explanations of the nature of the atmosphere are representative of his ideas. The Earth has much more oxygen than is typical of comparable planets. It has long been widely suggested that green plants are probably almost entirely responsible for this high oxygen content. Most people would regard oxygen production as a byproduct of plant activity, and a fortunate one for those of us who need to breathe oxygen (presumably, too, we have been selected to breathe oxygen partly because there is so much of it about). Lovelock goes further, and regards oxygen production by plants as an adaptation on the part of the Earth/organism or 'Gaia' (named after the Greek Earth goddess): plants produce oxygen *because* it benefits life as a whole.

He uses the same kind of argument for other gases that occur in small amounts:

> What, then, is the purpose of methane and how does it relate to oxygen? One obvious function is to maintain the integrity of the anaerobic zones of its origin [p. 73].
>
> Another puzzling atmospheric gas is nitrous oxide.... We may be sure that the efficient biosphere is unlikely to squander the energy required in making this odd gas unless it has some useful function. Two possible uses come to mind... [p. 74].
>
> Another nitrogenous gas made in large volumes in the soil and the sea and released to the air is ammonia.... As with methane, the biosphere uses a great deal of energy in producing ammonia, which is now entirely of biological origin. Its function is almost certainly to control the acidity of the environment... [p. 77].

The fatal flaw in Lovelock's hypothesis would have instantly occurred to him if he had wondered about the level of natural selection process which would be required in order to produce the Earth's supposed adaptations. Homeostatic adaptations in individual bodies evolve because individuals with improved homeostatic apparatus pass on their genes more effectively than individuals with inferior homeostatic apparatuses. For the analogy to apply strictly, there would have to have been a set of rival Gaias, presumably on different planets. Biospheres which did not develop efficient homeostatic regulation of their planetary atmospheres tended to go extinct. The Universe would have to be full of dead planets whose homeostatic regulation systems had failed, with, dotted around, a handful of successful, well-regulated planets of which Earth is one. Even this improbable scenario is not sufficient to lead to the evolution of planetary adaptations of the kind Lovelock proposes. In addition we would have to postulate some kind of reproduction, whereby successful planets spawned copies of their life forms on new planets.

I am not, of course, suggesting that Lovelock believes it happened like that. He would surely consider the idea of interplanetary selection as ludicrous as I do. Obviously he simply did not see his hypothesis as entailing the hidden assumptions that I think it entails. He might dispute that it does entail those assumptions, and maintain that Gaia could evolve her global adaptations by the ordinary processes of Darwinian selection acting within the one planet. I very much doubt that a model of such a selection process could be made to work: it would have all the notorious difficulties of 'group selection'. For instance, if plants are supposed to make oxygen for the good of the biosphere, imagine a mutant plant which saved itself the costs of oxygen manufacture. Obviously it would outreproduce its more public-spirited colleagues, and genes for public-spiritedness would soon disappear.

It is no use protesting that oxygen manufacture need not have costs: if it did not have costs, the most parsimonious explanation of oxygen production in plants would be the one the scientific world accepts anyway, that oxygen is a byproduct of something the plants do for their own selfish good. I do not deny that somebody may, one day, produce a workable model of the evolution of Gaia (possibly along the lines of 'Model 2' below), although I personally doubt it. But if Lovelock has such a model in mind he does not mention it. Indeed, he gives no indication that there is a difficult problem here.

The Gaia hypothesis is an extreme form of what, for old times' sake although it is now rather unfair, I shall continue to call the 'BBC Theorem'. The British Broadcasting Corporation is rightly praised for the excellence of its nature photography, and it usually strings these admirable visual images together with a serious commentary. Things are changing now, but for years the dominant message of these commentaries was one that had been elevated almost to the status of a religion by pop 'ecology'. There was something called the 'balance of nature', an exquisitely fashioned machine in which plants, herbivores, carnivores, parasites, and scavengers each played their appointed role for the good of all. The only thing that threatened this delicate ecological china shop was the insensitive bull of human progress, the bulldozer of ..., etc. The world needs the patient, toiling dung beetles and other scavengers, but for whose selfless efforts as the sanitary workers of the world ..., etc. Herbivores need their predators, but for whom their populations would soar out of control and threaten them with extinction, just as man's population will unless ..., etc. The BBC Theorem is often expressed in terms of the poetry of webs and networks. The whole world is a fine-meshed network of interrelationships, a web of connections which it has taken thousands of years to build up, and woe betide mankind if we tear it down ..., etc.

There is, no doubt, much merit in the moralistic exhortations that seem to flow from the BBC Theorem, but that does not mean its theoretical basis is sound. Its weakness is the one I have already exposed in the Gaia hypothesis. A network of relationships there may be, but it is made up of small, self-interested components. Entities that pay the costs of furthering the well-being of the ecosystem as a whole will tend to reproduce themselves less successfully than rivals that exploit their public-spirited colleagues, and contribute nothing to the general welfare. Hardin (1968) summed the problem up in his memorable phrase 'The tragedy of the commons', and more recently (Hardin 1978) in the aphorism, 'Nice guys finish last'.

I have dealt with the BBC Theorem and the Gaia hypothesis, because of the danger that my own language of the extended phenotype and action at a distance may sound like some of the more exuberantly extended networks and webs of the TV 'ecologists'. To emphasize the difference, then, let me borrow the rhetoric of webs and networks, but use it in a very different way, to explain the idea of the extended phenotype and of genetic action at a distance.

Loci in germ-line chromosomes are hotly contested pieces of property. The contestants are allelomorphic replicators. Most of the replicators in the world have won their place in it by defeating all available alternative alleles. The weapons with which they won, and the weapons with which their rivals lost, are their respective phenotypic consequences. These phenotypic consequences are conventionally thought of as being restricted to a small field around the replicator itself, its boundaries being defined by the body wall of the individual organism in whose cells the replicator sits. But the nature of the causal influence of gene on phenotype is such that it makes no sense to think of the field of influence as being limited in this arbitrary way, any more than it makes sense to think of it as limited to intracellular biochemistry. We must think of each replicator as the centre of a field of influence on the world at large. Causal influence radiates out from the replicator, but its power does not decay with distance according to any simple mathematical law. It travels wherever it can, far or near, along available avenues, avenues of intracellular biochemistry, of intercellular chemical and physical interaction, of gross bodily form and physiology. Through a variety of physical and chemical media it radiates out beyond the individual body to touch objects in the world outside, inanimate artefacts and even other living organisms.

Just as every gene is the centre of a radiating field of influence on the world, so every phenotypic character is the centre of converging influences from many genes, both within and outside the body of the individual organism. The whole biosphere—recognize the superficial affinity with the BBC Theorem—the whole world of plant and animal matter is criss-crossed with an intricate network of fields of genetic influence, a web of phenotypic power. I can almost hear the television commentary: 'Imagine being shrunk to the size of a mitochondrion stationed at a convenient vantage point outside the nuclear membrane of a human zygote. Watch the molecules of messenger RNA as they stream, by the millions, out into the cytoplasm on their errands of phenotypic power-play. Now grow to the size of a cell in the developing limb-bud of a chick embryo. Feel the wafts of chemical inducers as they roll down the gentle slopes of their axial gradients! Now grow again to your full size, and stand in the middle of a wood, at dawn in spring. Birdsong surges round you. Male syrinxes pour out sound, and all round the wood female ovaries swell. Here the influence travels in the form of pressure waves in the open air, rather than molecules in cytoplasm, but the principle is the same. At all three levels of the Lilliputian/Brobdingnagian thought experiment, you are privileged to stand in the midst of uncountable interlocking fields of replicator power.'

The reader will gather that it was the *message* of the BBC Theorem that I wanted to criticize, not its rhetoric! Nevertheless, a more restrained brand of eloquence is often more effective. A master of restrained eloquence in biological writing is Ernst Mayr. His chapter on 'The unity of the genotype' (Mayr 1963) is often held up to me in conversation as deeply antithetical to my replicator-based viewpoint. Since, on the contrary, I find myself enthusiastically endorsing almost every word of that chapter, something must have been misunderstood, somewhere.

Much the same could be said of Wright's (1980) equally eloquent article on 'Genic and organismic selection', which purports to be a repudiation of the genic selection view that I hold, yet almost none of which I find myself disagreeing with. I think this is a valuable paper, even though its ostensible purpose is to attack the view that 'with respect to natural selection . . . it is the gene, not the individual or group, that is the unit'. Wright concludes that 'The likelihood of organismic, instead of merely genic, selection goes far toward meeting one of the most serious objections to the theory of natural selection encountered by Darwin.' He attributes the 'genic selection' view to Williams, Maynard Smith, and me, and traces it back to R. A. Fisher, I think correctly. All of which might lead him to be somewhat bemused by the following accolade from Medawar (1981): 'The most important single innovation in the modern synthesis was however the new conception that a population that was deemed to undergo evolution could best be thought of as a population of fundamental replicating units—of genes—rather than as a population of individual animals or of cells. Sewall Wright . . . was a principal innovator in this new way of thinking—a priority for which R. A. Fisher, an important but lesser figure, never forgave him'.

In the rest of this chapter, I hope to show that the version of 'genic selectionism' that can be attacked as naively atomistic and reductionistic is a straw man; that it is not the view that I am advocating; and that if genes are correctly understood as being selected *for their capacity to cooperate* with other genes in the gene-pool, we arrive at a theory of genic selection which Wright and Mayr will recognize as fully compatible with their own views. Not only compatible but, I would claim, a truer and a clearer expression of their views. I shall quote key passages from the summary of Mayr's chapter (pp. 295–6), showing how they may be adapted to the world of the extended phenotype.

> The phenotype is the product of the harmonious interaction of all genes. The genotype is a 'physiological team' in which a gene can make a maximum contribution to fitness by elaborating its chemical 'gene product' in the needed quantity and at the time when it is needed in development [Mayr 1963].

An *extended* phenotypic character is the product of the interaction of many genes whose influence impinges from both inside and outside the organism. The interaction is not necessarily harmonious—but then nor are gene interactions within bodies necessarily harmonious, as we saw in Chapter 8. The genes whose influences converge on a particular phenotypic character are a 'physiological team' only in a special and subtle sense, and this is true of the conventional within-body interactions to which Mayr refers, as well as of extended interactions.

I have previously tried to convey that special sense with the metaphor of a rowing crew (Dawkins 1976a, pp. 91–2), and with the metaphor of cooperation between myopic and normal-sighted people (Dawkins 1980, pp. 22–4). The principle might also be labelled the Jack Sprat principle. Two individuals with complementary appetites, say for fat and lean, or with complementary skills, say in growing wheat and milling it, form naturally harmonious partnerships, and it is possible to regard a partnership as a higher-order unit. The interesting question is how such harmonious units come about. I want to make a general distinction between two models of selective processes, both of which could, in theory, lead to harmonious cooperation and complementarity.

The first model invokes selection at the level of the higher-order units: in a metapopulation of higher-order units, harmonious units are favoured against disharmonious units. It was a version of this first model that I suggested was implicit in the Gaia hypothesis—selection among planets in that case. Coming down to earth, the first model might suggest that groups of animals whose members complement one anothers' skills, say groups containing both farmers and millers, survive better than groups of farmers alone, or groups of millers alone. The second model is the one that I find more plausible. It does not need to postulate a metapopulation of groups. It is related to what population geneticists call frequency-dependent selection. Selection goes on at the lower level, the level of the component parts of a harmonious complex. Components within a population are favoured by selection if they happen to interact harmoniously with the other components that happen to be frequent in the population. In a population dominated by millers, individual farmers prosper, while in a population dominated by farmers it pays to be a miller.

Both kinds of model lead to a result which Mayr would call harmonious and cooperative. But I am afraid that the contemplation of harmony too often leads biologists to think automatically in terms of the first of the two models, and to forget the plausibility of the second. This is true of genes within a body just as it is true of farmers and millers in a community. The genotype may be a 'physiological team', but we do not have to believe that that team was necessarily selected as a harmonious unit in comparison with less harmonious rival units. Rather, each gene was selected because it prospered in its environment, and its environment necessarily included the other genes which were simultaneously prospering in the gene-pool. Genes with complementary 'skills' prosper in each others' presence.

What does complementariness mean for genes? Two genes may be said to be complementary if the survival of each, relative to its alleles, is enhanced when the other is abundant in the population. The most obvious reason for such mutual assistance stems from the two genes performing a mutually complementary function within individual bodies that they happen to share. The synthesis of chemical substances of biological importance often depends upon a chain of steps in a biochemical pathway, each one mediated by a particular enzyme. The usefulness of any one of these enzymes is conditional upon the presence of the other enzymes in the chain. A gene-pool which is rich in genes for all enzymes in a given chain except one may set up a selection pressure in favour of the gene for the missing link in the chain. If there are alternative pathways to the same biochemical end product, selection may favour either pathway (but not both) depending upon initial conditions. Rather than regard the alternative pathways as units between which selection chooses (Model 1), it is better to think as follows (Model 2): selection will favour a gene that makes a given enzyme, to the extent that genes for making the other enzymes in its pathway are already abundant in the gene-pool.

But we do not have to stay at a biochemical level. Imagine a species of moth with stripes on the wings which resemble grooves in tree bark. Some individuals have transverse stripes while others, in a different area, have longitudinal stripes, the difference being determined at a single genetic locus. Clearly a moth will be well camouflaged only if it points itself in the right direction when sitting on tree bark (Sargent 1969b). Suppose some moths sit vertically and others horizontally, the behavioural difference being controlled at a second locus. An observer finds that, fortunately, all the moths in one area have longitudinal stripes and sit vertically, while in another area all the moths have transverse stripes and sit horizontally. We might say, then, that in both areas there is 'harmonious cooperation' between the genes for stripe orientation and the genes determining the orientation of sitting. How has this harmony come about?

Again we invoke our two models. Model 1 says that disharmonious gene combinations—transverse stripes with vertical resting behaviour, or longitudinal stripes with horizontal resting behaviour—died out, leaving only harmonious gene combinations. Model 1 invokes selection among *combinations* of genes. Model 2, on the other hand, invokes selection at the lower level of the gene. If, for whatever reason, the gene-pool in a given area happens to be already dominated by genes for transverse stripes, this will automatically set up a selection pressure at the behavioural locus in favour of genes for sitting horizontally. This in turn will set up a selection pressure to increase the predominance of transverse stripe genes at the striping locus which will, in turn, reinforce the selection in favour of sitting horizontally. The population will therefore rapidly converge on the evolutionarily stable combination, transverse stripes/sit horizontally. Conversely, a different set of starting conditions would lead the population to converge on the other evolutionarily stable state, longitudinal stripes/sit vertically. Any given combination of starting frequencies at the two loci will converge, after selection, on one or other of the two stable states.

Model 1 is applicable only if the pairs or sets of cooperating genes are especially likely to find themselves together in bodies, for instance if they are closely linked into a 'supergene' on one chromosome. They might indeed be so linked (Ford 1975), but Model 2 is particularly interesting because it enables us to visualize the evolution of harmonious gene complexes without such linkage. In Model 2 the cooperating genes may be on different chromosomes, and frequency-dependent selection will still lead to populations being dominated by genes that interact harmoniously with the other genes in the population, as a result of evolution to one or another evolutionarily stable state (Lawlor & Maynard Smith 1976). In principle the same kind of reasoning is applicable to sets of three loci (suppose stripes on the hindwing were controlled at a different locus from stripes on the forewing), four loci... *n* loci. If we try to model the interactions in detail the mathematics become difficult, but that does not matter for the point I want to make. All I am saying is that there are two general ways in which harmonious cooperation can come about. One way is for harmonious complexes to be favoured by selection over dis-harmonious complexes. The other is for the separate *parts* of complexes to be favoured in the presence, in the population, of other parts with which they happen to harmonize.

Having used Model 2 for the kind of within-body gene harmony that Mayr had in mind, we now generalize it to between-body, 'extended', gene interactions. We are going to be talking about genetic *inter*action at a distance, rather than the phenotypic action at a distance which was the subject of the earlier part of this chapter. This is easy to do, because frequency-dependent selection has classically been applied to between-body interactions, from Fisher's (1930a) theory of the sex ratio on. Why do populations have a balanced sex ratio? Model 1 would suggest that it is because populations with an unbalanced sex ratio have gone extinct. Fisher's own hypothesis is, of course, a version of Model 2. If a population happens to have an unbalanced sex ratio, selection *within* that population favours genes that tend to restore the balance. There is no need to postulate a meta-population of populations, as in the case of Model 1.

Other examples of frequency-dependent advantage are well known to geneticists (e.g. Clarke 1979), and I have previously discussed their relevance to the controversy over 'harmonious cooperation' (Dawkins 1980, pp. 22–4). The point I want to stress here is that, from the point of view of each replicating entity, its relationships of harmony, cooperation, and complementariness *within* genomes are not, in principle, different from relationships between genes in different genomes. The gene for sitting vertically on tree-trunks is favoured in a gene-pool which happens to be rich in genes for longitudinal stripes, and vice versa. Here, as in the biochemical example of the chain of enzymes, the cooperation takes place within bodies: the significance of the fact that the gene-pool is rich in genes for longitudinal stripes is that any given gene at the locus determining sitting behaviour is therefore statistically likely to be in a longitudinally striped body. I suggest that we should think primarily of genes as being selected against the background of the other genes that happen to be frequent in the gene-pool, and only secondarily make a distinction between whether the salient between-gene interactions happen to occur within bodies or between them.

Wickler (1968), in his fascinating review of animal mimicry, points out that individuals sometimes appear to cooperate in achieving mimetic resemblance. He recounts an observation by Koenig of what appeared to be a sea anemone in an aquarium tank. The following day there were two anemones, each half the size of the original one, and the day after that the original large anemone had apparently reconstituted itself. The impossibility of this finally led Koenig to investigate in detail, and he discovered that the 'anemone' was in fact a fake, put together by numerous cooperating annelid worms. Each worm represented one tentacle, and they grouped themselves into a circle in the sand. Fish seemed to be fooled by the deception just as Koenig originally was, for they gave the fake anemone the same kind of wide berth they would a real one. Each individual worm presumably gained protection from fish predators, by joining in the cooperative mimicry ring. I suggest that it is not helpful to speak of *groups* of worms that form rings as being selected over groups that do not. Rather, ring-joining individuals are favoured in populations of ring-joiners.

In various insect species, each individual mimics one flower of a many-flowered inflorescence, and therefore a cooperating crowd of them is needed before a whole inflorescence can be convincingly mimicked. 'In East Africa it is possible to find a particular plant with extremely beautiful inflorescences... the individual flowers are about half a centimetre in height, look rather like broom flowers, and are arranged around a vertical stem like the flowers of the lupin. Experienced botanists have taken this plant for *Tinnaea* or *Sesamopteris* and found themselves suddenly holding a bare stem after plucking the "flower". The flower had not fallen off—it had flown away! The "flower" consists of cicadas, either *Ityraea gregorii* or *Oyarina nigritarsus*' (Wickler 1968, p. 61).

In order to develop my argument I need to make certain detailed assumptions. Since the details of the selection pressures bearing on these particular cicada species are not known, it will be safest if I invent a hypothetical cicada which basically practises the same group-mimicry trick as *Ityraea* and *Oyarina*. I assume that my species occurs in two colour morphs, pink and blue, and that the two morphs mimic two different colour varieties of lupin. Pink and blue lupins are assumed to be equally abundant over the whole range of the cicada species, but in any one local area all the cicadas are either pink or blue. 'Cooperation' occurs, in that individuals cluster together near the tips of plant stems and together resemble lupin inflorescences. It is 'harmonious' in that mixed colour clusters do not occur: I assume that a mixed colour cluster is especially likely to be spotted by predators as a fake, since real lupins do not have two-tone inflorescences.

Here is how the harmony might have come about through Model 2's frequency-dependent selection. In any given area, historical accident determined that there was an initial majority in favour of one colour type or the other. In an area that happened to be dominated by pink cicadas, blue ones were penalized. In an area that happened to be dominated by blue cicadas, pink ones were penalized. In both cases, simply being in the minority was disfavoured, because a member of the minority type was, by the laws of chance, more likely than a member of the majority type to find itself participating in a mixed cluster. At the gene level we may say that pink genes are favoured in a gene-pool dominated by pink genes, and blue genes are favoured in a gene-pool dominated by blue genes.

We now invent another insect, say a caterpillar, which is large enough to mimic a whole lupin inflorescence, instead of a single flower. Each segment of the caterpillar mimics a different flower of the inflorescence. The colour of each segment is controlled at a different locus, the alternatives being pink and blue. A caterpillar that is all blue or all pink is more successful than one that is a mixture of colours because, once again, predators have learned that mixed colour lupins do not occur. There is no theoretical reason why two-tone caterpillars should not occur, but suppose that, as a result of selection, they do not: in any one area the local caterpillars are either all pink or all blue. We have 'harmonious cooperation' again.

How might the harmonious cooperation come about? By definition, Model 1 would be applicable only if the genes responsible for the coloration of the different segments were linked tightly in a supergene. Multi-coloured supergenes would be penalized at the expense of pure pink and pure blue supergenes. In this hypothetical species, however, the relevant genes are widely spread on different chromosomes, and we have to apply Model 2. In any given local area, once one colour starts to predominate at a majority of loci, selection works to increase the frequency of that colour at all loci. In a particular area, if all the loci save one are dominated by pink genes, the odd locus which is dominated by blue genes will soon be brought into line by selection. As in the case of the hypothetical cicadas, historical accidents in different local areas automatically set up selection pressures in favour of one or another of two evolutionarily stable states.

The point of this thought experiment is that Model 2 is equally applicable both between and within individuals. In both the caterpillar and the cicada case, pink genes are favoured in gene-pools already dominated by pink genes, and blue genes are favoured in gene-pools already dominated by blue genes. In the caterpillars the reason is that each gene benefits if it shares a *body* with other genes producing the same colour as itself. In the cicadas the reason is that each gene benefits if the body it is in meets *another* body bearing a gene producing the same colour as itself. In the caterpillar example, the cooperating genes occupy different loci in the same individual. In the cicada example the cooperating genes occupy the same locus in different individuals. My purpose is to close the conceptual gap between these two kinds of gene interaction, by showing that genetic interaction at a distance is not, in principle, different from genetic interaction within one body.

To resume my series of quotations from Mayr:

> The result of the coadapting selection is a harmoniously integrated gene complex. The coaction of the genes may occur

> at many levels, that of the chromosome, nucleus, cell, tissue, organ, and whole organism.

The reader will by now have no difficulty in guessing how Mayr's list is to be extended. Coaction among genes in different organisms is not fundamentally different from coaction among genes in the same organism. Each gene works in a world of phenotypic consequences of other genes. Some of those other genes will be members of the same genome. Others will be members of the same gene-pool operating through other bodies. Yet others may be members of different gene-pools, different species, different phyla.

> The nature of the functional mechanisms of physiological interaction are [*sic*] only of minor interest to the evolutionist, whose main concern is the viability of the ultimate product, the phenotype.

Mayr hits the nail on the head again, but his 'phenotype' is not the ultimate: it can be extended outside the individual body.

> Many devices tend to maintain the *status quo* of gene pools, quantitatively and qualitatively. The lower limit of genetic diversity is determined by the frequent advantage of heterozygosity.... The upper limit is determined by the fact that only those genes can be incorporated that are able to 'coadapt' harmoniously. No gene has a fixed selective value; the same gene may confer high fitness on one genetic background and be virtually lethal on another.

Excellent, but remember that 'genetic background' can include genes in other organisms as well as genes within the same organism.

> The result of the close interdependence of all genes in a gene pool is tight cohesion. No gene frequency can be changed, nor

> any gene be added to the gene pool, without an effect on the genotype as a whole, and thus indirectly on the selective value of other genes.

Mayr himself has now subtly shifted to talking about a coadapted gene-*pool*, rather than a coadapted individual genome. This is a great step in the right direction, but we must still take one further step. Mayr is here talking about interactions between all the genes in one gene-pool, regardless of the bodies they happen to be sitting in. The doctrine of the extended phenotype ultimately requires us to acknowledge the same kind of interactions among genes of different gene-pools, different phyla, different kingdoms.

Consider again the ways in which a pair of genes in the same gene-pool can interact, more specifically the ways in which the frequency of each in the gene-pool can affect the survival prospects of the other. The first way, and the one which I suspect Mayr had mainly in mind, is through sharing the same body. The survival prospects of gene *A* are influenced by the frequency in the population of gene *B*, because *B*'s frequency influences the probability that *A* will find itself sharing a body with *B*. The interaction between the loci determining moth stripe direction and sitting direction was an example of this. So was the hypothetical lupin-mimicking caterpillar. So is a pair of genes coding for enzymes that are necessary for successive stages in a particular pathway synthesizing a useful substance. Call this type of gene interaction 'within-body' interaction.

The second way in which the frequency of a gene *B* in a population can affect the survival prospects of a gene *A* is 'between-body' interaction. The vital influence here is on the probability that any body in which *A* sits will meet another body in which *B* sits. My hypothetical cicadas provided an example of this. So does Fisher's sex-ratio theory. As I have emphasized, it has been one of my purposes in this chapter to minimize the distinction between the two kinds of gene interaction, within-body and between-body interaction.

But now consider interactions between genes in different gene-pools, different species. It will be seen that there is rather little distinction between a cross-species gene interaction and a within-species between-individual gene interaction. In neither case do the interacting genes share a body. In both cases the survival prospects of each may depend on the frequency, in its own gene-pool, of the other gene. Let me illustrate the point using the lupin thought experiment again. Suppose there is a species of beetle which is polymorphic like the cicadas. In some areas it turns out that the pink morphs of both species, cicadas and beetles, predominate, while in other areas the blue morphs of both species predominate. The two species differ in body size. They 'cooperate' in faking inflorescences, the smaller-bodied cicadas tending to sit near the tips of stems, where small flowers might be expected, the larger beetles tending to sit nearer the base of each fake inflorescence. A joint beetle/cicada 'inflorescence' fools birds more effectively than either a pure beetle or a pure cicada one.

Model 2's frequency-dependent selection will tend to lead to the evolution of one of two evolutionarily stable states, just as before, except that two species are now involved. If historical accident leads to one local area being dominated by pink morphs (regardless of species), selection within both species will favour pink morphs over blue; and vice versa. If the beetle species was relatively recently introduced into areas already colonized by the cicada species, the direction of selection within the beetle species will depend on the colour of the locally predominant morph of cicadas. Thus there will be frequency-dependent interaction between genes in two different gene-pools, the gene-pools of two non-interbreeding species. In faking the inflorescence of a lupin, cicadas might cooperate with spiders or snails just as effectively as with beetles or with cicadas of another species. Model 2 works across species and across phyla, as well as across individuals and within individuals.

Across kingdoms, too. Consider the interaction between flax (*Linum usitissimum*) and the rust fungus *Melampsora lini*, although this is an antagonistic rather than a cooperative interaction. 'There is essentially a one-to-one matching in which a specific allele in the flax confers resistance to a specific allele in the rust. This "gene-for-gene" system has since been found in numerous other plant species.... Models of gene-for-gene interactions are not formulated in terms of ecological parameters because of the specific nature of the genetic systems. It is one case in which the genetic interactions between species can be understood without reference to the phenotypes. A model of a gene-for-gene system would necessarily have between-species frequency dependence' (Slatkin & Maynard Smith 1979, pp. 255–6).

In this chapter, as in others, I have used hypothetical thought experiments to aid clear explanation. In case they are found too far-fetched, let me turn to Wickler again for an example of a real cicada which does something at least as far-fetched as anything I have invented. *Ityraea nigrocincta*, like *I. gregorii*, practises cooperative mimicry of lupin-like inflorescences, but it 'possesses a further peculiarity in that both sexes have two morphs, a green form and a yellow form. These two morphs may squat together, and the green forms tend to sit at the top of the stem, especially on vertical stems, with the yellow forms below. The result is an extremely convincing "inflorescence", because the true flowers of inflorescences often open progressively from base to apex, so that green buds are still present at the tip when the base is covered with open flowers' (Wickler 1968).

These three chapters have extended the concept of phenotypic expression of genes by easy stages. We began with the recognition that even within a body there are many degrees of distance of gene control over phenotypes. For a nuclear gene to control the shape of the cell in which it sits is presumably simpler than to control the shape of some other cell, or of the whole body in which the cell sits. Yet we conventionally lump the three together and call them all genetic control of phenotype. My thesis has been that the slight further conceptual step outside the immediate body is a comparatively minor one. Nevertheless it is an unfamiliar one, and I tried to develop the idea in stages, working through inanimate artefacts to internal parasites controlling their hosts' behaviour. From internal parasites we moved via cuckoos to action at a distance. In theory, genetic action at a distance could include almost all interactions between individuals of the same or different species. The living world can be seen as a network of interlocking fields of replicator power.

It is hard for me to imagine the kind of mathematics that the understanding of the details will eventually demand. I have a dim vision of phenotypic characters in an evolutionary space being tugged in different directions by replicators under selection. It is of the essence of my approach that the replicators tugging on any given phenotypic feature will include some from outside the body as well as those inside it. Some will obviously be tugging harder than others, so the arrows of force will have varying magnitude as well as direction. Presumably the theory of arms races—the rare-enemy effect, the life/dinner principle, etc.—will have a prominent role to play in the assignment of these magnitudes. Sheer physical proximity will probably play a role: genes seem likely, other things being equal, to exert more power over nearby phenotypic characters than over distant ones. As an important special case of this, cells are likely to be quantitatively more heavily influenced by genes inside them than by genes inside other cells. The same will go for bodies. But these will be quantitative effects, to be weighed in the balance with other considerations from arms race theory. Sometimes, say because of the rare-enemy effect, genes in other bodies may exert more power than the body's 'own' genes, over particular aspects of its phenotype. My hunch is that almost all phenotypic characters will turn out to bear the marks of compromise between internal and external replicator forces.

The idea of conflict and compromise between many selection pressures bearing on a given phenotypic character is, of course, familiar from ordinary biology. We often speak of, say, the size of a bird's tail as a compromise between the needs of aerodynamics and the needs of sexual attractiveness. I do not know what kind of mathematics are considered suitable for describing this kind of within-body conflict and compromise, but whatever they are, they should be generalized to cope with the analogous problems of genetic action at a distance and extended phenotypes.

But I have not the wings to fly in mathematical spaces. There must be a verbal message for those that study animals in the field. What difference will the doctrine of the extended phenotype make to how we actually see animals? Most serious field biologists now subscribe to the theorem, largely due to Hamilton, that animals are expected to behave as if maximizing the survival chances of all the genes inside them. I have amended this to a new central theorem of the extended phenotype: An animal's behaviour tends to maximize the survival of the genes 'for' that behaviour, whether or not those genes happen to be in the body of the particular animal performing the behaviour. The two theorems would amount to the same thing if animal phenotypes were always under the unadulterated control of their own genotypes, and uninfluenced by the genes of other organisms. In advance of a mathematical theory to handle the quantitative interactions between conflicting pressures, perhaps the simplest qualitative conclusion is that the behaviour we are looking at may be, at least partly, an adaptation for the preservation of some other animal's or plant's genes. It may therefore be positively maladaptive for the organism performing the behaviour.

Once when I tried to persuade a colleague of this—he is a staunch believer in the power of Darwinian selection, and a good field investigator of it—he thought that I was making an anti-adaptation point. He warned me that time and again people had written off some quirk of animal behaviour or morphology as functionless or maladaptive, only to discover that they had not fully understood it. He was right. But the point I am making is different. When I say here that a behaviour pattern is maladaptive, I only mean it is maladaptive for the *individual animal* performing it. I am suggesting that the individual performing the behaviour is not the entity for whose benefit the behaviour is an adaptation. Adaptations benefit the genetic replicators responsible for them, and only incidentally the individual organisms involved.

This could have been the end of the book. We have extended the phenotype out as far as it can go. The past three chapters build to a climax of a sort, and we might have been content with this as consummation. But I prefer to end on an upbeat, to begin the arousing of a tentative new curiosity. I confessed at the outset to being an advocate, and an easy way for any advocate to prepare the ground for his case is to attack the alternative. Before advocating the doctrine of the extended phenotype of an active germ-line replicator, therefore, I tried to undermine the reader's confidence in the individual organism as the unit of adaptive benefit. But, now that we have discussed the extended phenotype itself, it is time to reopen the question of the organism's existence and obvious importance in the hierarchy of life, and see whether we see it any clearer in the light of the extended phenotype. Given that life did not *have* to be packaged into discrete organisms, and allowing that organisms are not always totally discrete, why, nevertheless, did active germ-line replicators so conspicuously opt for the organismal way of doing things?

REDISCOVERING THE ORGANISM

Having devoted most of this book to playing down the importance of the individual organism, and to building up an alternative image of a turmoil of selfish replicators, battling for their own survival at the expense of their alleles, reaching unimpeded through individual body walls as though those walls were transparent, interacting with the world and with each other without regard to organismal boundaries, we now hesitate. There really *is* something pretty impressive about individual organisms. If we actually could wear spectacles that made bodies transparent and displayed only DNA, the distribution of DNA that we would see in the world would be overwhelmingly non-random. If cell nuclei glowed like stars and all else was invisible, multicellular bodies would show up as close-packed galaxies with cavernous space between them. A million billion glowing pinpricks move in unison with each other and out of step with all the members of other such galaxies.

The organism is a physically discrete machine, usually walled off from other such machines. It has an internal organization, often of staggering complexity, and it displays to a high degree the quality that Julian Huxley (1912) labelled 'individuality'—literally indivisibility—the quality of being sufficiently heterogeneous in form to be rendered non-functional if cut in half. Genetically speaking, too, the individual organism is usually a clearly definable unit, whose cells have the same genes as each other but different genes from the cells of other organisms. To an immunologist the individual organism has a special kind of 'uniqueness' (Medawar 1957), in that it will readily accept grafts from other parts of its own body, but not from other bodies. To the ethologist—and this is really an aspect of Huxley's indivisibility—the organism is a unit of behavioural action in a much stronger sense than, say, two organisms, or a limb of an organism. The organism has one coordinated central nervous system. It takes 'decisions' (Dawkins & Dawkins 1973) as a unit. All its limbs conspire harmoniously together to achieve one end at a time. On those occasions when two or more organisms try to coordinate their efforts, say when a pride of lions cooperatively stalks prey, the feats of coordination among individuals are feeble compared with the intricate orchestration, with high spatial and temporal precision, of the hundreds of muscles within each individual. Even a starfish, whose tube-feet enjoy a measure of autonomy and may tear the animal in two if the circum-oral nerve ring has been surgically cut, looks like a single entity, and in nature behaves as if it had a single purpose.

I am grateful to Dr J. P. Hailman for not withholding from me the sarcastic reaction of a colleague to the paper that was a brief trial-run for this book (Dawkins 1978a): 'Richard Dawkins has rediscovered the organism.' The irony was not lost on me, but there are wheels within wheels. We agree that there is something special about the individual organism as a level in the hierarchy of life, but it is not something obvious, to be accepted without question. My hope is that this book has revealed that there is a second side to the Necker Cube. But Necker Cubes have a habit of flipping back again to their original orientation, and then continuing to alternate. Whatever it is that is special about the individual organism as a unit of life, we should at least see it more clearly for having viewed the other side of the Necker Cube, for having trained our eyes to see through body walls into the world of replicators, and out and beyond to their extended phenotypes.

So, what is it that is special about the individual organism? Given that life can be viewed as consisting of replicators with their extended phenotypic tools of survival, why in practice have replicators chosen to group themselves together by the hundreds of thousands in cells, and why have they influenced those cells to clone themselves by the millions of billions in organisms?

One kind of answer is suggested by the logic of complex systems. Simon (1962) has written a stimulating essay on 'The architecture of complexity', which suggests (using a parable of two watchmakers called Tempus and Hora, which has become well known) a general functional reason why complex organization of any kind, biological or artificial, tends to be organized into nested hierarchies of repeated subunits. I have developed his argument in the ethological context, concluding that the evolution of statistically 'improbable assemblies proceeds more rapidly if there is a succession of intermediate stable sub-assemblies. Since the argument can be applied to each sub-assembly, it follows that highly complex systems which exist in the world are likely to have a hierarchical architecture' (Dawkins 1976b). In the present context the hierarchy consists of genes within cells and cells within organisms. Margulis (1981) makes a persuasive and fascinating case for an old idea that the hierarchy contains yet another intermediate level: eukaryotic 'cells' are themselves, in a sense, multicellular clusters, symbiotic unions of entities such as mitochondria, plastids, and cilia, which are homologous to, and descended from, prokaryotic cells. I will not pursue the matter further here. Simon's point is a very general one, and we need a more specific answer to the question of why replicators chose to organize their phenotypes into functional units, especially at the two levels of the cell and the multicellular organism.

In order to ask questions about why the world is the way it is, we have to imagine how it might have been. We invent possible worlds in which life might have been organized differently, and ask what would have happened if it had been. What instructive alternatives to the way life is, then, can we imagine? First, in order to see why replicating molecules ganged up in cells, we imagine a world in which there are replicating molecules floating freely in the sea. There are different varieties of replicator, and they are competing with each other for space and for the chemical resources needed to build copies of themselves, but they are not grouped together in chromosomes or nuclei. Each solitary replicator exerts phenotypic power to make copies of itself, and selection favours those with the most effective phenotypic power. It is easy to believe that this form of life would not be evolutionarily stable. It would be invaded by mutant replicators that 'gang up'. Certain replicators would have chemical effects that complement those of other replicators, complement them in the sense that when the two chemical effects are put together replication of both is facilitated (Model 2 in the previous chapter). I have already used the example of genes coding for enzymes that catalyse successive stages of a biochemical chain reaction. The same principle may be applied to larger groups of mutually complementary replicating molecules. Indeed, earthly biochemistry suggests that the minimal unit of replication, except possibly in the food-rich environment of a total parasite, is about fifty cistrons (Margulis 1981). It makes no difference to the argument whether new genes arise by duplication of old ones and remain close together, or whether previously independent genes positively come together. We can still discuss the evolutionary stability of the state of being 'ganged up'.

Ganging up of genes together into cells, then, is easily understood, but why did cells gang together into multicellular clones? In this case it might be thought that we do not have to invent thought experiments, because unicellular, or acellular, organisms abound on our world. These, however, are all small, and it may be useful to imagine a possible world in which there exist large and complex unicellular or mononucleate organisms. Could there be a form of life in which one single set of genes, enthroned in a single central nucleus, directed the biochemistry of a macroscopic body with complex organs, either a single gigantic 'cell', or a multicellular body in which all but one of the cells lacked their own private copies of the genome? I think such a form of life could only exist if its embryology followed principles very different from those with which we are familiar. In the embryologies that we know, only a minority of genes are 'turned on' in any one type of differentiating tissue at any one time (Gurdon 1974). It is admittedly a weak argument at this stage, but if there were only one set of genes in the entire body, it is not easy to see how the appropriate gene products could be conveyed to the various parts of the differentiating body at the appropriate times.

But why does there have to be a *complete* set of genes in every cell of a developing body? It is surely easy to imagine a form of life in which parts of the genome are hived off during differentiation, so that a given type of tissue, liver tissue or kidney tissue say, has only the genes that it needs. Only the germ-line cells, it would seem, really need to preserve the entire genome. It may be that the reason is simply that there is no easy way, physically, to hive off parts of the genome. It is not, after all, as though the genes needed in any particular differentiated region of the developing body are all segregated on one chromosome. We could, I suppose, now ask ourselves why *this* had to be the case. Given that it *is* the case, full division of the entire genome at every cell division may simply be the easiest and most economical way of doing things. However, in the light of my parable (Chapter 9) of the rosy-spectacled Martian and the need for cynicism, the reader may be tempted to speculate further. Could it be that the total, rather than partial, duplication of the genome in mitosis is an adaptation by some genes to keep themselves in a position to oversee and thwart would-be outlaws among their colleagues? Personally I doubt it, not because the idea is inherently farfetched but because it is hard to see how a gene in the liver, say, could stand to gain from being an outlaw and manipulating the liver in a way that would be to the detriment of genes in the kidney or the spleen. Following the logic of the chapter on parasites, the interests of 'liver genes' and 'kidney genes' would overlap because they share the same germ-line and the same gametic route out of the present body.

I have not provided a rigorous definition of the organism. It is, indeed, arguable that the organism is a concept of dubious utility, precisely because it is so difficult to define satisfactorily. From the immunological or genetic points of view, a pair of monozygotic twins would have to count as a single organism, yet clearly they would not so qualify from the point of view of the physiologist, the ethologist, or Huxley's indivisibility criterion. What is 'the individual' in a colonial siphonophore or bryozoan? Botanists have good reason to be less fond of the phrase 'individual organism' than zoologists are: 'The individual fruit fly, flour beetle, rabbit, flatworm or elephant is a population at the cellular but not at any higher level. Starvation does not change the number of legs, hearts or livers of an animal but the effect of stress on a plant is to alter both the rate of formation of new leaves and the rate of death of old ones: a plant may react to stress by varying the number of its parts' (Harper 1977, pp. 20–1). To Harper, as a plant population biologist, the leaf may be a more salient 'individual' than 'the plant', since the plant is a straggling, vague entity for whom reproduction may be hard to distinguish from what a zoologist would happily call 'growth'. Harper feels obliged to coin two new terms for different kinds of 'individual' in botany. 'The "ramet" is the unit of clonal growth, the module that may often follow an independent existence if severed from the parent plant.' Sometimes, as in strawberries, the ramet is the unit that we ordinarily call a 'plant'. In other cases such as white clover the ramet may be the single leaf. The 'genet', on the other hand, is the unit which springs from one single-celled zygote, the 'individual' in the sense of a zoologist whose animals reproduce sexually.

Janzen (1977) faces up to the same difficulty, suggesting that a clone of dandelions should be regarded as one 'evolutionary individual' (Harper's genet), equivalent to a single tree although spread out along the ground rather than raised up in the air on a trunk, and although divided up into separate physical 'plants' (Harper's ramets). According to this view, there may be as few as four individual dandelions competing with each other for the territory of the whole of North America. Janzen sees a clone of aphids in the same way. His paper has no literature citations at all, but the view is not a new one. It goes back at least as far as 1854, when T. H. Huxley 'treated each life cycle as an individual, with all the products from sexual event to sexual event being a single unit. He even treated an asexual lineage of aphids as an individual' (Ghiselin 1981). There is merit in this way of thinking, but I shall show that it leaves out something important.

One way to re-express the Huxley/Janzen argument is as follows. The germ-line of a typical organism, say a human, goes through a sequence of perhaps a few dozen mitotic divisions between each meiosis. Employing Chapter 5's 'backwards' way of looking at the 'past experience of a gene', any given gene in a living human has a history of cell divisions as follows: meiosis mitosis mitosis...mitosis meiosis. In every successive body, in parallel with the mitotic division of the germ-line, other mitotic divisions have furnished the germ-line with a large clone of 'helper' cells, grouped together into a body in which the germ-line is housed. In every generation the germ-line is funnelled down into a one-celled 'bottleneck' (a gamete followed by a zygote), then it fans out into a many-celled body, then it is funnelled down into a new bottleneck, etc. (Bonner 1974).

The many-celled body is a machine for the production of single-celled propagules. Large bodies, like elephants, are best seen as heavy plant and machinery, a temporary resource drain, invested in so as to improve later propagule production (Southwood 1976). In a sense the germ-line would 'like' to reduce capital investment in heavy machinery, reduce the number of cell divisions in the growth part of the cycle, so as to reduce the interval between recurrence of the reproduction part of the cycle. But this recurrence interval has an optimal length which is different for different ways of life. Genes that caused elephants to reproduce when too young and small would propagate themselves less efficiently than alleles tending to produce an optimal recurrence interval. The optimal recurrence interval for genes that happen to find themselves in elephant gene-pools is much longer than the optimal recurrence interval for genes in mouse gene-pools. In the elephant case, more capital investment is required to be laid down before returns on investment are sought. A protozoan largely dispenses with the growth phase of the cycle altogether, and its cell divisions are all 'reproductive' cell divisions.

It follows from this way of looking at organisms that the end product, the 'goal' of the growth phase of the cycle, is reproduction. The mitotic cell divisions which put together an elephant are all directed to the end of finally propagating viable gametes which succeed in perpetuating the germ-line. Now, holding this in mind, look at aphids. During the summer, asexual females go through repeated generations of asexual reproduction culminating in a single sexual generation which restarts the cycle. Clearly, by analogy with the elephant, it is easy to follow Janzen in seeing the summer asexual generations as all directed towards the final end of sexual reproduction in autumn. Asexual reproduction, according to this view, is not really reproduction at all. It is growth, just like the growth of a single elephant's body. For Janzen the whole clone of female aphids is a single evolutionary individual because it is the product of a single sexual fusion. It is an unusual individual in that it happens to be split up into a number of physically separate units, but so what? Each of those physical units contains its own fragment of germ-line, but then so does the left ovary and the right ovary of a female elephant. The fragments of germ-line in the aphid case are separated by thin air, while the two ovaries of an elephant are separated by guts but, again, so what?

Persuasive as this line of argument is, I have already mentioned that I think it misses an important point. It is right to regard most mitotic cell divisions as 'growth', 'aimed' at the final goal of reproduction, and it is right to regard the individual organism as the product of one reproductive event, but Janzen is wrong to equate the reproduction/growth distinction with the sexual/asexual distinction. There is, to be sure, an important distinction lurking here, but it is not the distinction between sex and non-sex, nor is it the distinction between meiosis and mitosis.

The distinction that I wish to emphasize is that between germ-line cell division (reproduction), and somatic or 'dead-end' cell division (growth). A germ-line cell division is one where the genes being duplicated have a chance of being the ancestors of an indefinitely long line of descendants, where the genes are, in fact, true germ-line replicators in the sense of Chapter 5. A germ-line cell division may be a mitosis or a meiosis. If we simply look at a cell division under a microscope, there may be no way of telling whether it is a germ-line division or not. Both germ-line and somatic cell divisions may be mitoses of identical appearance.

If we look at a gene in any cell in a living organism and trace its history backwards in evolutionary time, the most recent few cell divisions of its 'experience' may be somatic, but once we reach a germ-line cell division in our backwards march, all previous ones in the gene's history must be germ-line divisions. Germ-line cell divisions may be thought of as proceeding forwards in evolutionary time, while somatic cell divisions are proceeding sideways. Somatic cell divisions are used to make mortal tissues, organs, and instruments whose 'purpose' is the promoting of germ-line cell divisions. The world is populated by genes which have survived in germ-lines as a consequence of aid that they received from their exact duplicates in somatic cells. Growth comes about through the propagation of dead-end somatic cells, while reproduction is the means of the propagation of germ-line cells.

Harper (1977) makes a distinction between reproduction and growth in plants, which will normally amount to the same as my distinction between germ-line and somatic cell division: 'The distinction made here between "reproduction" and "growth" is that reproduction involves the formation of a new individual from a single cell: this is usually (though not always e.g. apomicts) a zygote. In this process a new individual is "reproduced" by the information that is coded in the cell. Growth, in contrast, results from the development of organized meristems' (Harper 1977, p. 27 fn.). What matters here is whether there really is an important biological distinction between growth and reproduction which is not the same as the distinction between mitosis and meiosis + sex. Is there really a crucial difference between 'reproducing' to make two aphids on the one hand, and 'growing' to make one aphid twice as large on the other? Janzen would presumably say no. Harper would presumably say yes. I agree with Harper, but I would not have been able to justify my position until I had read J. T. Bonner's (1974) inspiring book *On Development*. The justification is best made with the aid of thought experiments.

Imagine a primitive plant consisting of a flat, pad-like thallus, floating on the surface of the sea, absorbing nutrients through its lower surface and sunlight through its upper surface. Instead of 'reproducing' (i.e. sending off single-celled propagules to grow elsewhere), it simply grows at its margins, spreading into an ever larger circular green carpet, like a monstrous lily pad several miles across and still growing. Maybe older parts of the thallus eventually die, so that it consists of an expanding ring rather than a filled circle like a true lily pad. Perhaps also, from time to time, chunks of the thallus split off, like icefloes shearing away from the pack ice, and separate chunks drift to different parts of the ocean. Even if we assume this kind of fission, I shall show that it is not reproduction in an interesting sense.

Now consider a similar kind of plant which differs in one crucial respect. It stops growing when it attains a diameter of 1 foot, and reproduces instead. It manufactures single-celled propagules, either sexually or asexually, and sheds them into the air where they may be carried a long way on the wind. When one of these propagules lands on the water surface it becomes a new thallus, which grows until it is 1 foot wide, then reproduces again. I shall call the two species G (for growth) and R (for reproduction), respectively.

Following the logic of Janzen's paper, we should see a crucial difference between the two species only if the 'reproduction' of the second species, R, is sexual. If it is asexual, the propagules shed into the air being mitotic products genetically identical to the cells of the parent thallus, there is, for Janzen, no important difference between the two species. The separate 'individuals' in R are no more genetically distinct than different regions of the thallus in G might be. In either species, mutation can initiate new clones of cells. There is no particular reason why, in R, mutations should occur during propagule formation any more than during thallus growth. R is simply a more fragmented version of G, just as a clone of dandelions is like a fragmented tree. My purpose in making the thought experiment, however, was to disclose an important difference between the two hypothetical species, representing the difference between growth and reproduction, even when reproduction is asexual.

G just grows, while R grows and reproduces in alternation. Why is the distinction important? The answer cannot be a genetic one in any simple sense because, as we have seen, mutations are just as likely to initiate genetic change during growth-mitosis as during reproduction-mitosis. I suggest that the important distinction between the two species is that a lineage of R is capable of evolving complex adaptations in a way that G is not. The reasoning goes as follows.

Consider again the past history of a gene, in this case a gene sitting in a cell of R. It has had a history of passing repeatedly from one 'vehicle' to another similar vehicle. Each of its successive bodies began as a single-celled propagule, then grew through a fixed cycle, then passed the gene on into a new single-celled propagule and hence a new multicellular body. Its history has been a cyclical one, and now here is the point. Since each of this long series of successive bodies developed anew from single-celled beginnings, it is possible for successive bodies to develop slightly differently from their predecessors. Evolution of complex body structure with organs, say a complex apparatus for catching insects like a Venus fly trap, is only possible if there is a cyclically repeating developmental process to evolve. I shall return to this point in a moment.

Meanwhile, compare G. A gene sitting in a young cell at the growing margin of the huge thallus has a history which is not cyclical, or is cyclical only at the cellular level. The ancestor of the present cell was another cell, and the career of the two cells was very similar. Each cell of an R plant, by contrast, has a definite place in the growth sequence. It is either near the centre of the 1-foot thallus, or near the rim, or at some particular place in between. It can therefore differentiate to fill its special role in its appointed place in an organ of the plant. A cell of G has no such specific developmental identity. All cells first appear at the growing margin, and later find themselves enclosed by other, younger cells. There is cyclicity only at the cellular level, which means that in G evolutionary change can take place only at the cellular level. Cells might improve on their predecessors in the cell lineage, developing more complex internal organelle structure, say. But the evolution of organs and adaptations at the multicellular level could not take place, because recurrent, cyclical development of whole groups of cells does not occur. It is, of course, true that in G the cells and their ancestors are in physical contact with other cells, and in this sense form a multicellular 'structure'. But as far as putting together complex multicellular organs is concerned, they might just as well have been free-swimming protozoa.

In order to put together a complex multicellular organ you need a complex developmental sequence. A complex developmental sequence has to have evolved from an earlier developmental sequence which was slightly less complex. There has to be an evolutionary progression of developmental sequences, each one in the series being a slight improvement on its predecessor. G does not have a recurring developmental sequence other than the high-frequency cycle of development at the single-cell level. Therefore it cannot evolve multicellular differentiation and organ-level complexity. To the extent that it can be said to have a multicellular developmental process at all, that development continues non-cyclically through geological time: the species makes no separation between the growth time-scale and the would-be evolution time-scale. The only high-frequency developmental cycle available to it is the cell cycle. R, on the other hand, has a multicellular developmental cycle which is fast compared with evolutionary time. Therefore, as the ages succeed, later developmental cycles can be different from earlier developmental cycles, and multicellular complexity can evolve. We are moving towards a definition of the organism as the unit which is initiated by a new act of *reproduction* via a single-celled developmental 'bottleneck'.

The importance of the difference between growth and reproduction is that each act of reproduction involves a new developmental cycle. Growth simply involves swelling of the existing body. When an aphid gives rise to a new aphid by parthenogenetic reproduction the new aphid, if it is a mutant, may be radically different from its predecessor. When an aphid grows to twice its original size, on the other hand, all its organs and complex structures simply swell. It might be said that somatic mutations could occur within cell lineages of the growing giant aphid. True, but a mutation within a somatic cell line in a heart, say, cannot radically re-organize the structure of the heart. To switch the example to vertebrates, if the present heart is two-chambered, with one auricle feeding one ventricle, new mutations in the mitotic cells at the growing margin of the heart are very unlikely to achieve radical restructuring of the heart so that it comes to have four chambers with a pulmonary circulation kept separate from the rest. In order to put together new complexity, new developmental beginnings are required. A new embryo must start from scratch, without any heart at all. Then a mutation can act on sensitive key points in early development to bring about a new fundamental architecture of the heart. Developmental recycling allows a return 'back to the drawing board' (see below) in every generation.

We began this chapter by wondering why replicators have ganged up into large, multicellular clones called organisms, and we initially gave a rather unsatisfactory answer. A more satisfying answer is now starting to emerge. An organism is the physical unit associated with one single life cycle. Replicators that gang up in multicellular organisms achieve a regularly recycling life history, and complex adaptations to aid their preservation, as they progress through evolutionary time.

Some animals have a life cycle involving more than one distinct body. A butterfly is utterly different from the caterpillar which preceded it. It is hard to imagine a butterfly growing from a caterpillar by slow, within-organ change: caterpillar organ growing into corresponding butterfly organ. Instead, what happens is that the complex organ structure of the caterpillar is largely broken down and the tissues of the caterpillar are used as fuel for the development of a whole new body. The new butterfly body does not quite restart from a single cell, but the principle is the same. It develops a radically new bodily structure from simple, relatively undifferentiated imaginal discs. There is a partial return to the drawing board.

Returning to the growth/reproduction distinction itself, Janzen was not actually wrong. Distinctions can be unimportant for some purposes, while they remain important for other purposes. For discussing certain kinds of ecological or economic questions, there may be no important distinction between growth and asexual reproduction. A sisterhood of aphids may indeed be analogous to a single bear. But for other purposes, for discussing the evolutionary putting together of complex organization, the distinction is crucial. A certain type of ecologist may gain illumination from comparing a field full of dandelions with a single tree. But for other purposes it is important to understand the differences, and to see the single dandelion ramet as analogous to the tree.

But Janzen's position is, in any case, a minority one. A more typical biologist might think it perverse of Janzen to regard asexual reproduction in aphids as growth, and equally perverse of Harper and me to regard vegetative propagation by multicellular runners as growth and not reproduction. Our decision is based on the assumption that the runner is a multicellular meristem rather than a single-celled propagule, but why should we regard this as an important point? Again, the answer may be seen in a thought experiment involving two hypothetical species of plants, in this case strawberry-like plants called M and S.

Both the hypothetical strawberry-like species propagate vegetatively, by runner. In both there is a population of what appear to be distinct and recognizable 'plants' connected by a network of runners. In both species, each 'plant' (i.e. ramet) can give rise to more than one daughter plant, so that we have the possibility of exponential growth of the 'population' (or growth of the 'body' depending on your point of view). Even though there is no sex, there can be evolution since mutations will sometimes occur in the mitotic cell divisions (Whitham & Slobodchikoff 1981). Now comes the crucial difference between the two species. In species M (for many, or multicellular, or meristem), the runner is a broad-fronted multicellular meristem. This means that two cells in any one 'plant' may be the mitotic descendants of two different cells in the parent plant. In terms of mitotic descent, a cell may therefore be a closer cousin of a cell in another 'plant' than it is of another cell in its own plant. If mutation has introduced genetic heterogeneity into the cell population, this means that individual plants may be genetic mosaics, with some cells having closer genetic relatives in other plants than in their own. We will see the consequences of this for evolution in a moment. Meanwhile we turn to the other hypothetical species.

Species S (for single) is exactly like M, except that each runner culminates in a single apical cell. This cell acts as the basal mitotic ancestor of all the cells of the new daughter plant. This means that all cells in a given plant are closer cousins to each other than they are to any cells in other plants. If mutation introduces genetic heterogeneity into the population of cells, there will be relatively few mosaic plants. Rather, each plant will tend to be a genetically uniform clone, but it may differ genetically from some other plants, while being genetically identical to yet other plants. There will be a true population of *plants*, each one of which has a genotype characteristic of all its cells. It is therefore possible to conceive of selection, in the sense which I have called 'vehicle selection', acting at the level of the whole plant. Some whole plants may be better than others, because of their superior genotypes.

In species M, especially if the runners are very broad-fronted meristems, a geneticist will not discern a population of plants at all. He will see a population of cells, each with its own genotype. Some cells will be genetically identical, others will have different genotypes. A form of natural selection might go on among cells, but it is hard to imagine selection among 'plants', because 'the plant' is not a unit that can be identified as having its own characteristic genotype. Rather, the whole mass of straggling vegetation will have to be regarded as a population of cells, with cells of any one genotype being untidily peppered across the different 'plants'. The unit which I have called the 'gene vehicle', and which Janzen has called the 'evolutionary individual', will, in such a case, be no larger than the cell. It is *cells* that will be the genetic competitors. Evolution may take the form of improvements in cell structure and physiology, but it is hard to see how it could take the form of improvements in individual plants or their organs.

It might be thought that improvements in organ structure could evolve, if it regularly happened that particular subpopulations of cells, in discrete areas of the plant, were a clone, descended from a single mitotic ancestor. For instance, the runner giving rise to a new 'plant' might be a broad-fronted meristem, but it might still be the case that each leaf sprang from a single cell at its own base. A leaf could therefore be a clone of cells more closely related to each other than to cells anywhere else in the plant. Given the commonness of somatic mutation in plants (Whitham & Slobodchikoff 1981), might one not therefore imagine the evolution of improved complex adaptation at the level of the leaf, if not at the level of the whole plant? A geneticist could now discern a genetically heterogeneous population of leaves, each one made up of genetically homogenous cells, so might not natural selection go on between successful leaves and unsuccessful leaves? It would be tidy if the answer to this question could be yes; that is, if we could assert that vehicle selection will go on at any level in the hierarchy of multicellular units, provided that the cells within a unit tend to be genetically uniform compared with cells in other units at the same level. Unfortunately, however, something has been left out of the reasoning.

It will be remembered that I classified replicators into germ-line replicators and dead-end replicators. Natural selection results in some replicators becoming more numerous at the expense of rival replicators, but this leads to evolutionary change only if the replicators are in germ-lines. A multicellular unit qualifies as a vehicle, in an evolutionarily interesting sense, only if at least some of its cells contain germ-line replicators. Leaves normally do not so qualify, for their nuclei contain only dead-end replicators. Leaf cells synthesize chemical substances which ultimately benefit other cells that do contain germ-line copies of the leaf genes, the genes which gave the leaves their characteristically leafy phenotypes. But we cannot accept the conclusion of the previous paragraph, that inter-leaf vehicle selection, and inter-organ selection generally, could go on if only the cells within an organ were closer mitotic cousins than cells in different organs. Inter-leaf selection could have evolutionary consequences only if leaves directly spawned daughter leaves. Leaves are organs, not organisms. For inter-organ selection to occur, it is necessary that the organs concerned should have their own germ-lines and do their own reproducing, and this they normally do not. Organs are parts of organisms, and reproduction is the prerogative of organisms.

For clarity I have been a bit extreme. There could be a range of intermediates between my two strawberry-like plants. Species M's runner was said to be a broad-fronted meristem, while species S's runner narrowed down to a one-celled bottleneck at the base of each new plant. But what if there was an intermediate species with a two-celled bottleneck at the base of each new plant? There are two main possibilities here. If the pattern of development is such that it is unpredictable which cells in the daughter plant will be descended from which of the two stem cells, the point I have made about developmental bottlenecks will simply be weakened quantitatively: genetic mosaics may occur in the population of plants, but there will still be a statistical tendency for cells to be genetically closer to fellow members of the same plant than to cells in other plants. Therefore we may still talk meaningfully about vehicle selection between plants in a population of plants, but the inter-plant selection pressure may have to be strong to outweigh selection among cells within plants. This is, incidentally, analogous to one of the conditions for 'kin-group selection' (Hamilton 1975a) to work. To make the analogy, we have only to see the plant as a 'group' of cells.

The second possibility arising out of the assumption of a two-celled bottleneck at the base of each plant is that the pattern of development of the species might be such that certain organs of the plant are always the mitotic descendants of a designated one of the two cells. For instance, cells of the root system might develop from a cell in the lower part of the runner, while the rest of the plant developed from the other cell, in the upper part of the runner. If, further, the lower cell is always descended from a root cell in the parent plant, while the upper cell is recruited from an above-ground cell in the parent plant, we would have an interesting situation. Root cells would be closer cousins of other root cells in the population at large than they would be of stalk and leaf cells in their 'own' plant. Mutation would open up the possibility of evolutionary change, but it would be split-level evolution. Genotypes below ground could evolve away from genotypes above ground, irrespective of apparent joint membership in discrete 'plants'. Theoretically we could even have a kind of within-organism 'speciation'.

To recapitulate, the significance of the difference between growth and reproduction is that reproduction permits a new beginning, a new developmental cycle, and a new organism which may be an improvement, in terms of the fundamental organization of complex structure, over its predecessor. Of course it may *not* be an improvement, in which case its genetic basis will be eliminated by natural selection. But growth without reproduction does not even allow the *possibility* of radical change at the organ level, either in the direction of improvement or the reverse. It allows only superficial tinkering. You may divert a developing Bentley into a fully grown Rolls Royce, simply by tinkering with the assembly process at the late point where the radiator is added. But if you want to change a Ford into a Rolls Royce you must start at the drawing board, before the car starts 'growing' on the assembly line at all. The point about recurrent reproduction life cycles, and hence, by implication, the point about organisms, is that they allow repeated returns to the drawing board during evolutionary time.

We must beware here of the heresy of 'biotic' adaptationism (Williams 1966). We have seen that recurrent reproduction life cycles, i.e. 'organisms', make the evolution of complex organs possible. It is all too easy to treat this as a sufficient adaptive explanation for the *existence* of organismal life cycles, on the grounds that complex organs are, in some vague sense, a good idea. A related point is that repeated reproduction is possible only if individuals die (Maynard Smith 1969), but we should not therefore wish to say that individuals die as an adaptation to keep evolution going! The same could be said of mutation: its existence is a necessary precondition for evolution to occur, but it is nevertheless quite likely that natural selection has favoured evolution in the direction of a zero mutation rate—fortunately never attained (Williams 1966). The growth/reproduction/death type of life cycle—the multicellular clonal 'organism' type of life cycle—has had far-reaching consequences and was probably essential for the evolution of adaptive complexity, but this is not tantamount to an adaptive explanation for the existence of this type of life cycle. The Darwinian must begin by seeking immediate benefits to genes promoting this kind of life cycle, at the expense of their alleles. He may go on to acknowledge the possibility of other levels of selection, differential lineage extinction, say. But he must show the same circumspection in this difficult theoretical field as Fisher (1930a), Williams (1975), and Maynard Smith (1978a) brought to the analogous suggestions about sexual reproduction being there because it speeds up evolution.

The organism has the following attributes. It is either a single cell, or if it is multicellular its cells are close genetic kin of each other: they are descended from a single stem cell, which means that they have a more recent common ancestor with each other than with the cells of any other organism. The organism is a unit with a life cycle which, however complicated it may be, repeats the essential characteristics of previous life cycles, and may be an improvement on previous life cycles. The organism either consists of germ-line cells, or it contains germ-line cells as a subset of its own cells, or, as in the case of a sterile social insect worker, it is in a position to work for the welfare of germ-line cells in closely related organisms.

I have not aspired in this final chapter to give a completely satisfying answer to the question of why there are large multicellular organisms. I will be content if I can arouse new curiosity about the question. Instead of accepting that organisms exist and asking how adaptations benefit the organisms displaying them, I have tried to show that the very existence of organisms should be treated as a phenomenon deserving of explanation in its own right. Replicators exist. That is fundamental. Phenotypic manifestations of them, including extended phenotypic manifestations, may be expected to function as tools to keep replicators existing. Organisms are huge and complex assemblages of such tools, assemblages shared by gangs of replicators who in principle need not have gone around together but in fact *do* go around together and share a common interest in the survival and reproduction of the organism. As well as drawing attention to the phenomenon of the organism as one that needs explanation, I have tried in this last chapter to sketch the general direction in which we might proceed in seeking an explanation. It is only a preliminary sketch, but, for what it is worth, I summarize it here.

The replicators that exist tend to be the ones that are good at manipulating the world to their own advantage. In doing this they exploit the opportunities offered by their environments, and an important aspect of the environment of a replicator is other replicators and their phenotypic manifestations. Those replicators are successful whose beneficial phenotypic effects are conditional upon the presence of other replicators which happen to be common. These other replicators are also successful; otherwise, they would not be common. The world therefore tends to become populated by mutually compatible sets of successful replicators, replicators that get on well together. In principle this applies to replicators in different gene-pools, different species, classes, phyla, and kingdoms. But a relationship of specially intimate mutual compatibility has grown up between subsets of replicators that share cell nuclei and, where the existence of sexual reproduction makes the expression meaningful, share gene-pools.

The cell nucleus as a population of uneasily cohabiting replicators is a remarkable phenomenon in itself. Just as remarkable, and quite distinct, is the phenomenon of multicellular cloning, the phenomenon of the multicellular organism. Replicators whose effects interact with those of other replicators to produce multicellular organisms achieve for themselves vehicles with complex organs and behaviour patterns. Complex organs and behaviour patterns are favoured in arms races. The evolution of complex organs and behaviour patterns is possible because the organism is an entity with a recurrent life cycle, each cycle beginning with a single cell. The fact that each cycle restarts in every generation from a single cell permits mutations to achieve radical evolutionary changes by going 'back to the drawing board' of embryological engineering. It also, by concentrating the efforts of all cells in the organism on the welfare of a narrow, shared germ-line, partly removes the 'temptation' for outlaws to work for their own private good at the expense of the other replicators with a stake in the same germ-line. The integrated multicellular organism is a phenomenon which has emerged as a result of natural selection on primitively independent selfish replicators. It has paid replicators to behave gregariously. The phenotypic power by which they ensure their survival is in principle extended and unbounded. In practice the organism has arisen as a partially bounded local concentration, a shared knot of replicator power.

AFTERWORD BY DANIEL DENNETT

Why is a philosopher writing an Afterword for this book? Is *The Extended Phenotype* science or philosophy? It is both; it is science, certainly, but it is also what philosophy should be, and only intermittently is: a scrupulously reasoned argument that opens our eyes to a new perspective, clarifying what had been murky and ill-understood, and *giving us a new way of thinking* about topics we thought we already understood. As Richard Dawkins says at the outset, 'The extended phenotype may not constitute a testable hypothesis in itself, but it so far changes the way we see animals and plants that it may cause us to think of testable hypotheses that we would otherwise never have dreamed of' (p. 2). And what is this new way of thinking? It is not just the 'gene's eye point of view' made famous in Dawkins's 1976 book, *The Selfish Gene*. Building here on that foundation, he shows how our traditional way of thinking about organisms should be replaced by a richer vision in which the boundary between organism and environment first dissolves and then gets (partially) rebuilt on a deeper foundation. 'I shall show that the ordinary logic of genetic terminology leads inevitably to the conclusion that genes can be said to have *extended* phenotypic effects, effects which need not be expressed at the level of any particular vehicle' (p. 298). Dawkins is not declaring revolution; he is using 'the ordinary logic of genetic terminology', to prove a striking implication of the biology already securely in hand, a new 'central theorem': '*An animal's behaviour tends to maximize the survival of the genes "for" that behaviour, whether or not those genes happen to be in the body of the particular animal performing it*' (p. 355). Dawkins's earlier eye-opener, the recommendation that biologists adopt the gene's eye point of view, was also not presented as revolutionary, but rather as clarifying a shift of attention that had already begun to sweep through biology in 1976. There has been so much anxious and misguided criticism of Dawkins's earlier idea that many laypeople and even some biologists may fail to appreciate how bountiful this shift of attention has been. We now know that a genome, such as the human genome, consists of, and depends on, mechanisms of breathtaking deviousness and ingenuity—not just molecular copyists and proof-reading editors, but outlaws and vigilantes to combat them, chaperones and escape artists and protection rackets and addicts and other devious nano-agents, out of whose robotic conflicts and projects emerge the marvels of visible nature. The fruits of this new vision extend far beyond the almost daily headlines about striking new discoveries about one bit of DNA or another. Why and how do we age? Why do we get sick? How does HIV work? How do brains get wired up in the course of embryological development? Can we use parasites instead of poisons to control agricultural pests? Under what conditions is cooperation not just possible, but likely to arise and persist? All these vital questions and many more are illuminated by re-thinking the issues in terms of the processes by which the opportunities for replicators to replicate, and their associated costs and benefits, sort themselves out.

Dawkins, like a philosopher, is primarily concerned with the logic of the explanations we devise to account for these processes and predict their outcomes. But these explanations are scientific explanations, and Dawkins (along with many others) wants to claim that their implications are *scientific results*, not just the tenets of an interesting and defensible philosophy. Since so much is at stake, we need to see if this is good science, and for that, we need to check out the logic down in the trenches, where the data are gathered, where the details matter, where relatively small-scale hypotheses about manageable phenomena can be actually tested. *The Selfish Gene* was written for educated lay readers, and glided over many of the intricacies and technicalities that a proper scientific assessment needs to consider at length. *The Extended Phenotype* was written for the professional biologist, but so graceful and lucid is Dawkins's writing that even outsiders who are prepared to exercise their brains vigorously can follow the arguments, and appreciate the subtlety of the issues.

For the professional philosopher, I cannot resist adding, there is a feast: some of the most masterful, sustained chains of rigorous argument I have ever encountered (check out Chapter 5, and the last four chapters), and a host of ingenious and vivid thought experiments (check out p. 218, and pp. 366–7, among many others). There are even some sidelong but substantial contributions to philosophical controversies undreamt of by Dawkins. For instance, I take the thought experiment about the genetic control of mud-gathering by termites on pp. 307–9 to provide useful insight on theories of intentionality—especially on the debate I have had with Fodor, Dretske, and others about the conditions under which content can be ascribed to mechanisms. In philosophical jargon, pure extensionality reigns in genetics, and this makes any labelling of phenotypic traits 'a matter of arbitrary convenience' but not, for that reason, unmotivated by our interest in drawing attention to the most telling facts about the situation.

For the scientist, there are testable predictions aplenty—about such varied topics as, for instance, wasp copulation strategies (pp. 120–2), sperm size evolution (p. 217), the anti-predator behaviour of moths (pp. 223–4), and the effects of parasites on bees and freshwater shrimps (pp. 329–30). There are also crisp, clear analyses of problems—about the evolution of sex, conditions of intra-genomic conflict (or genomic parasites), and many other initially counter-intuitive topics. His cautionary review of the pitfalls to be avoided in thinking about the green beard effect and its neighbours is an indispensable *vademecum* for anybody who ventures into that confusing territory.

This book has been required reading for any serious student of neo-Darwinian evolutionary theory since it first appeared in 1982, and one of the striking effects of re-reading it today is that it provides a time-lapse snapshot of the glacial pace of criticism. Stephen Jay Gould and Richard Lewontin in the United States, and Steven Rose in the United Kingdom, have long warned the world about the 'genetic determinism' that supposedly rises menacingly out of Dawkins's gene's eye view of biology, and here in Chapter 2 we find *all* their most recent criticisms already handily rebutted. You would think that in almost twenty years his opponents would have come up with some new angle, some new crack into which they could drive a subversive wedge or two, but, as Dawkins remarks in another context in which there has been no evolution, 'there is apparently no available variation for further enhancement' in their thinking. How much more satisfying, when faced with the task of replying to your most vehement critics, to be able simply to republish what you said on the topic many years ago!

What is this dread 'genetic determinism'? Dawkins (p. 15) cites a 1978 definition by Gould: 'If we are programmed to be what we are, then these traits are ineluctable. We may, at best, channel them, but we cannot change them either by will, education, or culture.' But if this is genetic determinism—and I have seen no seriously revised definition from the critics—then Dawkins is no genetic determinist (and neither is E. O. Wilson, or, so far as I know, any well-known sociobiologist or evolutionary psychologist). As Dawkins shows, in an impeccable philosophical analysis, the whole idea of the 'threat' of 'genetic' (or any other kind of) determinism is so ill-thought-out by those who brandish the term that it would be a bad joke if it weren't a scandal. Dawkins doesn't just rebut the charges in Chapter 2; he diagnoses the likely sources of the confusions that make this such a tempting charge, and as he notes, 'There is a wanton eagerness to misunderstand.' Sad to say, he is right.

Not all criticisms of neo-Darwinian thinking are so misbegotten. Adaptationist thinking, the critics say, is seductive; it is all too easy to mistake an unsupported just-so story for a serious evolutionary argument. This is true, and time and again in this book Dawkins deftly exposes tempting lines of argument that run foul of reality in one way or another. (For a few striking examples, see pp. 110–11, 117, 236–7, 401–2.) On p. 58, Dawkins makes the very important point that a change in the environment may not just change the success rate of a phenotypic effect; it may change the phenotypic effect altogether! So much for the standard, and boringly false, charge that the gene's eye point of view *must* ignore or underestimate the contribution of changes (including 'massively contingent' changes) in the selectional environment, but the fact remains that adaptationists often do ignore these (and other) complications, which is why the book fairly bristles with warnings against facile adaptationist reasoning.

The charge of 'reductionism', another standard epithet applied to the gene's eye perspective, is perversely inappropriate when levelled at Dawkins. Far from blinding us to the marvels of higher levels of explanation, the idea of the extended phenotype expands their powers by removing crippling misconceptions. As Dawkins says, it permits us to *rediscover* the organism. Why, if phenotypic effects do not have to honour the boundary between organism and 'outside' world, are there (multicellular) organisms at all? A very good question, and one that one probably wouldn't ask—or ask well—if it weren't for the perspective that Dawkins offers. Each of us walks around each day carrying the DNA of several *thousand* lineages (our parasites, our intestinal flora) in addition to our nuclear (and mitochondrial) DNA, and all these genomes get along pretty well under most circumstances. They are all in the same boat, after all. A herd of antelope, a termite colony, a mating pair of birds and their clutch of eggs, a human society—these groupish entities are no more groupish, in the end, than an individual human being, with its trillion-plus cells, each a descendant of the ma-cell and pa-cell union that started the group's voyage. 'At any level, if a vehicle is destroyed, all the replicators inside it will be destroyed. Natural selection will therefore, at least to some extent, favour replicators that cause their vehicles to resist being destroyed. In principle this could apply to groups of organisms as well as to single organisms, for if a group is destroyed all the genes inside it are destroyed too' (p. 173–4). So are genes all that matter? Not at all. 'There is nothing magic about Darwinian fitness in the genetic sense. There is no law giving it priority as the fundamental quantity that is maximized. . . . A meme has its own opportunities for replication, and its own phenotypic effects, and there is no reason why success in a meme should have any connection whatever with genetic success' (p. 167).

The logic of Darwinian thinking is not just about genes. More and more thinkers are coming to appreciate this: evolutionary economists, evolutionary ethicists, and others in the social sciences and even in the physical sciences and the arts. I take this to be a *philosophical* discovery, and it is undeniably mind-boggling. The book you hold in your hands is one of the best guide-books to this new world of understanding.

REFERENCES

Alcock, J. (1979). *Animal Behavior: an Evolutionary Approach*. Sunderland, Mass.: Sinauer.

Alexander, R. D. (1974). The evolution of social behavior. *Annual Review of Ecology and Systematics* **5**, 325–83.

Alexander, R. D. (1980). *Darwinism and Human Affairs*. London: Pitman.

Alexander, R. D. & Borgia, G. (1978). Group selection, altruism, and the levels of organization of life. *Annual Review of Ecology and Systematics* **9**, 449–74.

Alexander, R. D. & Borgia, G. (1979). On the origin and basis of the male–female phenomenon. In *Sexual Selection and Reproductive Competition in Insects* (eds M. S. Blum & N. A. Blum), pp. 417–40. New York: Academic Press.

Alexander, R. D. & Sherman, P. W. (1977). Local mate competition and parental investment in social insects. *Science* **96**, 494–500.

Allee, W. C., Emerson, A. E., Park, O., Park, T. & Schmidt, K. P. (1949). *Principles of Animal Ecology*. Philadelphia: W. B. Saunders.

Axelrod, R. & Hamilton, W. D. (1981). The evolution of cooperation. *Science* **211**, 1390–6.

Bacon, P. J. & Macdonald, D. W. (1980). To control rabies: vaccinate foxes. *New Scientist* **87**, 640–5.

Baerends, G. P. (1941). Fortpflanzungsverhalten und Orientierung der Grabwespe *Ammophila campestris* Jur. *Tijdschrift voor Entomologie* **84**, 68–275.

Barash, D. P. (1977). *Sociobiology and Behavior*. New York: Elsevier.

Barash, D. P. (1978). *The Whisperings Within*. New York: Harper & Row.

Barash, D. P. (1980). Predictive sociobiology: mate selection in damselfishes and brood defense in white-crowned sparrows. In *Sociobiology: Beyond Nature/Nurture?* (eds G. W. Barlow & J. Silverberg), pp. 209–26. Boulder: Westview Press.

Barlow, H. B. (1961). The coding of sensory messages. In *Current Problems in Animal Behaviour* (eds W. H. Thorpe & O. L. Zangwill), pp. 331–60. Cambridge: Cambridge University Press.

Bartz, S. H. (1979). Evolution of eusociality in termites. *Proceedings of the National Academy of Sciences, U.S.A.* **76**, 5764–8.

Bateson, P. P. G. (1978). Book review: *The Selfish Gene. Animal Behaviour* **26**, 316–18.

Bateson, P. P. G. (1982). Behavioural development and evolutionary processes. In *Current Problems in Sociobiology* (ed. King's College Sociobiology Group), pp. 133–51. Cambridge: Cambridge University Press.

Bateson, P. P. G. (1983) Optimal Outbreeding. In *Mate Choice* (ed. P. Bateson), pp. 257–77. Cambridge: Cambridge University Press.

Baudoin, M. (1975). Host castration as a parasitic strategy. *Evolution* **29**, 335–52.

Beatty, R. A. & Gluecksohn-Waelsch, S. (1972). *The Genetics of the Spermatozoon*. Edinburgh: Department of Genetics of the University.

Bennet-Clark, H. C. (1971). Acoustics of insect song. *Nature* **234**, 255–9.

Benzer, S. (1957). The elementary units of heredity. In *The Chemical Basis of Heredity* (eds W. D. McElroy & B. Glass), pp. 70–93. Baltimore: Johns Hopkins Press.

Bertram, B. C. R. (1978). *Pride of Lions*. London: Dent.

Bethel, W. M. & Holmes, J. C. (1973). Altered evasive behavior and responses to light in amphipods harboring acanthocephalan cystacanths. *Journal of Parasitology* **59**, 945–56.

Bethel, W. M. & Holmes, J. C. (1977). Increased vulnerability of amphipods to predation owing to altered behavior induced by larval acanthocephalans. *Canadian Journal of Zoology* **55**, 110–15.

Bethell, T. (1978). Burning Darwin to save Marx. *Harpers* **257** (Dec.), 31–8 & 91–2.

Bishop, D. T. & Cannings, C. (1978). A generalized war of attrition. *Journal of Theoretical Biology* **70**, 85–124.

Blick, J. (1977). Selection for traits which lower individual reproduction. *Journal of Theoretical Biology* **67**, 597–601.

Boden, M. (1977). *Artificial Intelligence and Natural Man*. Brighton: Harvester Press.

Bodmer, W. F. & Cavalli-Sforza, L. L. (1976). *Genetics, Evolution, and Man*. San Francisco: W. H. Freeman.

Bonner, J. T. (1958). *The Evolution of Development*. Cambridge: Cambridge University Press.

Bonner, J. T. (1974). *On Development*. Cambridge, Mass.: Harvard University Press.

Bonner, J. T. (1980). *The Evolution of Culture in Animals.* Princeton, N.J.: Princeton University Press.

Boorman, S. A. & Levitt, P. R. (1980). *The Genetics of Altruism.* New York: Academic Press.

Brenner, S. (1974). The genetics of *Caenorhabditis elegans. Genetics* **77**, 71–94.

Brent, L., Rayfield, L. S., Chandler, P., Fierz, W., Medawar, P. B. & Simpson, E. (1981). Supposed lamarckian inheritance of immunological tolerance. *Nature* **290**, 508–12.

Brockmann, H. J. (1980). Diversity in the nesting behavior of mud-daubers (*Trypoxylon politum* Say; Sphecidae). *Florida Entomologist* **63**, 53–64.

Brockmann, H. J. & Dawkins, R. (1979). Joint nesting in a digger wasp as an evolutionarily stable preadaptation to social life. *Behaviour* **71**, 203–45.

Brockmann, H. J., Grafen, A. & Dawkins, R. (1979). Evolutionarily stable nesting strategy in a digger wasp. *Journal of Theoretical Biology* **77**, 473–96.

Broda, P. (1979). *Plasmids.* Oxford: W. H. Freeman.

Brown, J. L. (1975). *The Evolution of Behavior.* New York: W. W. Norton.

Brown, J. L. & Brown, E. R. (1981). Extended family system in a communal bird. *Science* **211**, 959–60.

Bruinsma, O. & Leuthold, R. H. (1977). Pheromones involved in the building behaviour of *Macrotermes subhyalinus* (Rambur). *Proceedings of the 8th International Congress of the International Union for the Study of Social Insects,* Wageningen, 257–8.

Burnet, F. M. (1969). *Cellular Immunology.* Melbourne: Melbourne University Press.

Bygott, J. D., Bertram, B. C. R. & Hanby, J. P. (1979). Male lions in large coalitions gain reproductive advantages. *Nature* **282**, 839–41.

Cain, A. J. (1964). The perfection of animals. In *Viewpoints in Biology,* 3 (eds J. D. Carthy & C. L. Duddington), pp. 36–63. London: Butterworths.

Cain, A. J. (1979). Introduction to general discussion. In *The Evolution of Adaptation by Natural Selection* (eds J. Maynard Smith & R. Holliday). *Proceedings of the Royal Society of London,* B **205**, 599–604.

Cairns, J. (1975). Mutation selection and the natural selection of cancer. *Nature* **255**, 197–200.

Cannon, H. G. (1959). *Lamarck and Modern Genetics.* Manchester: Manchester University Press.

Caryl, P. G. (1982). Animal signals: a reply to Hinde. *Animal Behaviour* **30**, 240–4.

Cassidy, J. (1978). Philosophical aspects of the group selection controversy. *Philosophy of Science* **45**, 575–94.

Cavalier-Smith, T. (1978). Nuclear volume control by nucleoskeletal DNA, selection for cell volume and cell growth rate, and the solution of the DNA C-value paradox. *Journal of Cell Science* **34**, 247–78.

Cavalier-Smith, T. (1980). How selfish is DNA? *Nature* **285**, 617–18.

Cavalli-Sforza, L. & Feldman, M. (1973). Cultural versus biological inheritance: phenotypic transmission from parents to children. *Human Genetics* **25**, 618–37.

Cavalli-Sforza, L. & Feldman, M. (1981). *Cultural Transmission and Evolution.* Princeton, N.J.: Princeton University Press.

Charlesworth, B. (1979). Evidence against Fisher's theory of dominance. *Nature* **278**, 848–9.

Charnov, E. L. (1977). An elementary treatment of the genetical theory of kin-selection. *Journal of Theoretical Biology* **66**, 541–50.

Charnov, E. L. (1978). Evolution of eusocial behavior: offspring choice or parental parasitism? *Journal of Theoretical Biology* **75**, 451–65.

Cheng, T. C. (1973). *General Parasitology*. New York: Academic Press.

Clarke, B. C. (1979). The evolution of genetic diversity. *Proceedings of the Royal Society of London*, B **205**, 453–74.

Clegg, M. T. (1978). Dynamics of correlated genetic systems. II. Simulation studies of chromosomal segments under selection. *Theoretical Population Biology* **3**, 1–23.

Cloak, F. T. (1975). Is a cultural ethology possible? *Human Ecology* **3**, 161–82.

Clutton-Brock, T. H. & Harvey, P. H. (1979). Comparison and adaptation. *Proceedings of the Royal Society of London*, B **205**, 547–65.

Clutton-Brock, T. H., Guinness, F. E. & Albon, S. D. (1982). *Red Deer: The Ecology of Two Sexes.* Chicago: Chicago University Press.

Cohen, J. (1977). *Reproduction*. London: Butterworths.

Cohen, S. N. (1976). Transposable genetic elements and plasmid evolution. *Nature* **263**, 731–8.

Cosmides, L. M. & Tooby, J. (1981). Cytoplasmic inheritance and intragenomic conflict. *Journal of Theoretical Biology* **89**, 83–129.

Craig, R. (1980). Sex investment ratios in social Hymenoptera. *American Naturalist* **116**, 311–23.

Crick, F. H. C. (1979). Split genes and RNA splicing. *Science* **204**, 264–71.

Croll, N. A. (1966). *Ecology of Parasites.* Cambridge, Mass.: Harvard University Press.

Crow, J. F. (1979). Genes that violate Mendel's rules. *Scientific American* **240** (2), 104–13.

Crowden, A. E. & Broom, D. M. (1980). Effects of the eyefluke, *Diplostomum spathacaeum,* on the behaviour of dace (*Leuciscus leuciscus*). *Animal Behaviour* **28**, 287–94.

Crozier, R. H. (1970). Coefficients of relationship and the identity by descent of genes in Hymenoptera. *American Naturalist* **104**, 216–17.

Curio, E. (1973). Towards a methodology of teleonomy. *Experientia* **29**, 1045–58.

Daly, M. (1979). Why don't male mammals lactate? *Journal of Theoretical Biology* **78**, 325–45.

Daly, M. (1980). Contentious genes. *Journal of Social and Biological Structures* **3**, 77–81.

Darwin, C. R. (1859). *The Origin of Species.* 1st edn, reprinted 1968. Harmondsworth, Middx: Penguin.

Darwin, C. R. (1866). Letter to A. R. Wallace, dated 5 July. In James Marchant (1916), *Alfred Russel Wallace Letters and Reminiscences,* Vol. 1, pp. 174–6. London: Cassell.

Davies, N. B. (1982). Alternative strategies and competition for scarce resources. In *Current problems in sociobiology* (ed. King's College Sociobiology Group), pp. 363–80. Cambridge: Cambridge University Press.

Dawkins, R. (1968). The ontogeny of a pecking preference in domestic chicks. *Zeitschrift für Tierpsychologie* **25**, 170–86.

Dawkins, R. (1969). Bees are easily distracted. *Science* **165**, 751.

Dawkins, R. (1971). Selective neurone death as a possible memory mechanism. *Nature* **229**, 118–19.

Dawkins, R. (1976a). *The Selfish Gene.* Oxford: Oxford University Press.

Dawkins, R. (1976b). Hierarchical organisation: a candidate principle for ethology. In *Growing Points in Ethology* (eds P. P. G. Bateson & R. A. Hinde), pp. 7–54. Cambridge: Cambridge University Press.

Dawkins, R. (1978a). Replicator selection and the extended phenotype. *Zeitschrift für Tierpsychologie* **47**, 61–76.

Dawkins, R. (1978b). What is the optimon? University of Washington, Seattle, Jessie & John Danz Lecture, unpublished.

Dawkins, R. (1979a). Twelve misunderstandings of kin selection. *Zeitschrift für Tierpsychologie* **51**, 184–200.

Dawkins, R. (1979b). Defining sociobiology. *Nature* **280**, 427–8.

Dawkins, R. (1980). Good strategy or evolutionarily stable strategy? In *Sociobiology: Beyond Nature/Nurture?* (eds G. W. Barlow & J. Silverberg), pp. 331–67. Boulder: Westview Press.

Dawkins, R. (1981). In defence of selfish genes. *Philosophy*, October.

Dawkins, R. (1982). Replicators and vehicles. In *Current Problems in Sociobiology* (ed. King's College Sociobiology Group), pp. 45–64. Cambridge: Cambridge University Press

Dawkins, R. & Brockmann, H. J. (1980). Do digger wasps commit the Concorde fallacy? *Animal Behaviour* **28**, 892–6.

Dawkins, R. & Carlisle, T. R. (1976). Parental investment, mate desertion and a fallacy. *Nature* **262**, 131–3.

Dawkins, R. & Dawkins, M. (1973). Decisions and the uncertainty of behaviour. *Behaviour* **45**, 83–103.

Dawkins, R. & Krebs, J. R. (1978). Animal signals: information or manipulation? In *Behavioural Ecology* (eds J. R. Krebs & N. B. Davies), pp. 282–309. Oxford: Blackwell Scientific Publications.

Dawkins, R. & Krebs, J. R. (1979). Arms races between and within species. *Proceedings of the Royal Society of London*, B **205**, 489–511.

Dilger, W. C. (1962). The behavior of lovebirds. *Scientific American* **206** (1), 89–98.

Doolittle, W. F. & Sapienza, C. (1980). Selfish genes, the phenotype paradigm and genome evolution. *Nature* **284**, 601–3.

Dover, G. (1980). Ignorant DNA? *Nature* **285**, 618–19.

Eaton, R. L. (1978). Why some felids copulate so much: a model for the evolution of copulation frequency. *Carnivore* **1**, 42–51.

Eberhard, W. G. (1980). Evolutionary consequences of intracellular organelle competition. *Quarterly Review of Biology* **55**, 231–49.

Eldredge, N. & Cracraft, J. (1980). *Phylogenetic Patterns and the Evolutionary Process*. New York: Columbia University Press.

Eldredge, N. & Gould, S. J. (1972). Punctuated equilibria: an alternative to phyletic gradualism. In *Models in Paleobiology* (ed. T. J. M. Schopf), pp. 82–115. San Francisco: Freeman Cooper.

Emerson, A. E. (1960). The evolution of adaptation in population systems. In *Evolution after Darwin* (ed. S. Tax), pp. 307–48. Chicago: Chicago University Press.

Evans, C. (1979). *The Mighty Micro*. London: Gollancz.

Ewald, P. W. (1980). Evolutionary biology and the treatment of signs and symptoms of infectious disease. *Journal of Theoretical Biology* **86**, 169–71.

Falconer, D. S. (1960). *Introduction to Quantitative Genetics.* London: Longman.

Fisher, R. A. (1930a). *The Genetical Theory of Natural Selection.* Oxford: Clarendon Press.

Fisher, R. A. (1930b). The distribution of gene ratios for rare mutations. *Proceedings of the Royal Society of Edinburgh* **50**, 204–19.

Fisher, R. A. & Ford, E. B. (1950). The Sewall Wright effect. *Heredity* **4**, 47–9.

Ford, E. B. (1975). *Ecological Genetics.* London: Chapman and Hall.

Fraenkel, G. S. & Gunn, D. L. (1940). *The Orientation of Animals.* Oxford: Oxford University Press.

Frisch, K. von (1967). *A Biologist Remembers.* Oxford: Pergamon Press.

Frisch, K. von (1975). *Animal Architecture.* London: Butterworths.

Futuyma, D. J., Lewontin, R. C., Mayer, G. C., Seger, J. & Stubblefield, J. W. (1981). Macroevolution conference. *Science* **211**, 770.

Ghiselin, M. T. (1974a). *The Economy of Nature and the Evolution of Sex.* Berkeley: University of California Press.

Ghiselin, M. T. (1974b). A radical solution to the species problem. *Systematic Zoology* **23**, 536–44.

Ghiselin, M. T. (1981). Categories, life and thinking. *Behavioral and Brain Sciences* **4**, 269–313.

Gilliard, E. T. (1963). The evolution of bowerbirds. *Scientific American* **209** (2), 38–46.

Gilpin, M. E. (1975). *Group Selection in Predator-Prey Communities.* Princeton, N.J.: Princeton University Press.

Gingerich, P. D. (1976). Paleontology and phylogeny: patterns of evolution at the species level in early Tertiary mammals. *American Journal of Science* **276**, 1–28.

Glover, J. (ed.) (1976). *The Philosophy of Mind.* Oxford: Oxford University Press.

Goodwin, B. C. (1979). Spoken remark in *Theoria to Theory* **13**, 87–107.

Gorczynski, R. M. & Steele, E. J. (1980). Inheritance of acquired immunological tolerance to foreign histocompatibility antigens in mice. *Proceedings of the National Academy of Sciences, U.S.A.* **77**, 2871–5.

Gorczynski, R. M. & Steele, E. J. (1981). Simultaneous yet independent inheritance of somatically acquired tolerance to two distinct H-2 antigenic haplotype determinants in mice. *Nature* **289**, 678–81.

Gould, J. L. (1976). The dance language controversy. *Quarterly Review of Biology* **51**, 211–44.

Gould, S. J. (1977a). *Ontogeny and Phylogeny.* Cambridge, Mass.: Harvard University Press.

Gould, S. J. (1977b). Caring groups and selfish genes. *Natural History* **86** (12), 20–4.

Gould, S. J. (1977c). Eternal metaphors of palaeontology. In *Patterns of Evolution* (ed. A. Hallam), pp. 1–26. Amsterdam: Elsevier.

Gould, S. J. (1978). *Ever Since Darwin*. London: Burnett.

Gould, S. J. (1979). Shades of Lamarck. *Natural History* **88** (8), 22–8.

Gould, S. J. (1980a). The promise of paleobiology as a nomothetic, evolutionary discipline. *Paleobiology* **6**, 96–118.

Gould, S. J. (1980b). Is a new and general theory of evolution emerging? *Paleobiology* **6**, 119–30.

Gould, S. J. & Calloway, C. B. (1980). Clams and brachiopods—ships that pass in the night. *Paleobiology* **6**, 383–96.

Gould, S. J. & Eldredge, N. (1977). Punctuated equilibria: the tempo and mode of evolution reconsidered. *Paleobiology* **3**, 115–51.

Gould, S. J. & Lewontin, R. C. (1979). The spandrels of San Marco and the Panglossian paradigm: a critique of the adaptationist programme. *Proceedings of the Royal Society of London*, B **205**, 581–98.

Grafen, A. (1979). The hawk-dove game played between relatives. *Animal Behaviour* **27**, 905–7.

Grafen, A. (1980). Models of *r* and *d*. *Nature* **284**, 494–5.

Grant, V. (1978). Kin selection: a critique. *Biologisches Zentralblatt* **97**, 385–92.

Grassé, P. P. (1959). La réconstruction du nid et les coordinations interindividuelles chez *Bellicositermes natalensis* et *Cubitermes* sp. La théorie de la stigmergie: essai d'interpretation du comportement des termites constructeurs. *Insectes Sociaux* **6**, 41–80.

Greenberg, L. (1979). Genetic component of bee odor in kin recognition. *Science* **206**, 1095–7.

Greene, P. J. (1978). From genes to memes? *Contemporary Sociology* **7**, 706–9.

Gregory, R. L. (1961). The brain as an engineering problem. In *Current Problems in Animal Behaviour* (eds W. H. Thorpe & O. L. Zangwill), pp. 307–30. Cambridge: Cambridge University Press.

Grey Walter, W. (1953). *The Living Brain*. London: Duckworth.

Grun, P. (1976). *Cytoplasmic Genetics and Evolution*. New York: Columbia University Press.

Gurdon, J. B. (1974). *The Control of Gene Expression in Animal Development*. Oxford: Oxford University Press.

Hailman, J. P. (1977). *Optical Signals*. Bloomington: Indiana University Press.

Haldane, J. B. S. (1932a). *The Causes of Evolution*. London: Longman's Green.

Haldane, J. B. S. (1932b). The time of action of genes, and its bearing on some evolutionary problems. *American Naturalist* **66**, 5–24.

Haldane, J. B. S. (1955). Population genetics. *New Biology* **18**, 34–51.

Hallam, A. (1975). Evolutionary size increase and longevity in Jurassic bivalves and ammonites. *Nature* **258**, 493–6.

Hallam, A. (1978). How rare is phyletic gradualism and what is its evolutionary significance? *Paleobiology* **4**, 16–25.

Hamilton, W. D. (1963). The evolution of altruistic behavior. *American Naturalist* **97**, 31–3.

Hamilton, W. D. (1964a). The genetical evolution of social behaviour. I. *Journal of Theoretical Biology* **7**, 1–16.

Hamilton, W. D. (1964b). The genetical evolution of social behaviour. II. *Journal of Theoretical Biology* **7**, 17–32.

Hamilton, W. D. (1967). Extraordinary sex ratios. *Science* **156**, 477–88.

Hamilton, W. D. (1970). Selfish and spiteful behaviour in an evolutionary model. *Nature* **228**, 1218–20.

Hamilton, W. D. (1971a). Selection of selfish and altruistic behavior in some extreme models. In *Man and Beast: Comparative Social Behavior* (eds J. F. Eisenberg & W. S. Dillon), pp. 59–91. Washington, D.C.: Smithsonian Institution.

Hamilton, W. D. (1971b). Addendum. In *Group selection* (ed. G. C. Williams), pp. 87–9. Chicago: Aldine, Atherton.

Hamilton, W. D. (1972). Altruism and related phenomena, mainly in social insects. *Annual Review of Ecology and Systematics* **3**, 193–232.

Hamilton, W. D. (1975a) Innate social aptitudes of man: an approach from evolutionary genetics. In *Biosocial Anthropology* (ed. R. Fox), pp. 133–55. London: Malaby Press.

Hamilton, W. D. (1975b). Gamblers since life began: barnacles, aphids, elms. *Quarterly Review of Biology* **50**, 175–80.

Hamilton, W. D. (1977). The play by nature. *Science* **196**, 757–9.

Hamilton, W. D. & May R. M. (1977). Dispersal in stable habitats. *Nature* **269**, 578–81.

Hamilton, W. J. & Orians, G. H. (1965). Evolution of brood parasitism in altricial birds. *Condor* **67**, 361–82.

Hansell, M. H. (1984). *Animal Architecture and Building Behaviour*. London: Longman.

Hardin, G. (1968). The tragedy of the commons. *Science* **162**, 1243–8.

Hardin, G. (1978). Nice guys finish last. In *Sociobiology and Human Nature* (eds M. S. Gregory et al.). pp. 183–94. San Francisco: Jossey-Bass.

Hardy, A. C. (1954). Escape from specialization. In *Evolution as a Process* (eds J. S. Huxley, A. C. Hardy & E. B. Ford), pp. 122–40. London: Allen & Unwin.

Harley, C. B. (1981). Learning the evolutionarily stable strategy. *Journal of Theoretical Biology*, **89**, 611–33.

Harpending, H. C. (1979). The population genetics of interaction. *American Naturalist* **113**, 622–30.

Harper, J. L. (1977). *Population Biology of Plants*. London: Academic Press.

Hartung, J. (1981). Transfer RNA, genome parliaments, and sex with the red queen. In *Natural Selection and Social Behavior: Recent Research and New Theory* (eds R. D. Alexander & D. W. Tinkle). New York: Chiron.

Harvey, P. H. & Mace, G. M. (1982) Comparisons between taxa and adaptive trends: problems of methodology. In *Current Problems in Sociobiology* (ed. King's College Sociobiology Group), pp. 343–61. Cambridge: Cambridge University Press.

Heinrich, B. (1979). *Bumblebee Economics*. Cambridge, Mass.: Harvard University Press.

Hickey, W. A. & Craig, G. B. (1966). Genetic distortion of sex ratio in a mosquito, *Aedes aegypti*. *Genetics* **53**, 1177–96.

Hinde, R. A. (1975). The concept of function. In *Function and Evolution of Behaviour* (eds G. Baerends, C. Beer & A. Manning), pp. 3–15. Oxford: Oxford University Press.

Hinde, R. A. (1981). Animal signals: ethological and game theory approaches are not incompatible. *Animal Behaviour* **29**, 535–42.

Hinde, R. A. & Steel, E. (1978). The influence of daylength and male vocalizations on the estrogen-dependent behavior of female canaries and budgerigars, with discussion of data from other species. In *Advances in the Study of Behavior*, Vol. 8 (eds J. S. Rosenblatt et al.), pp. 39–73. New York: Academic Press.

Hines, W. G. S. & Maynard Smith, J. (1979). Games between relatives. *Journal of Theoretical Biology* **79**, 19–30.

Hofstadter, D. R. (1979). *Gödel, Escher, Bach: An Eternal Golden Braid*. Brighton: Harvester Press.

Hölldobler, B. & Michener, C. D. (1980). Methods of identification and discrimination in social Hymenoptera. In *Evolution of Social Behavior:*

Hypotheses and Empirical Tests (ed. H. Markl), pp. 35–57. Weinheim: Verlag Chemie.

Holmes, J. C. & Bethel, W. M. (1972). Modification of intermediate host behaviour by parasites. In *Behavioural Aspects of Parasite Transmission* (eds E. U. Canning & C. A. Wright), pp. 123–49. London: Academic Press.

Howard, J. C. (1981). A tropical volute shell and the Icarus syndrome. *Nature* **290**, 441–2.

Hoyle, F. (1964). *Man in the Universe*. New York: Columbia University Press.

Hull, D. L. (1976). Are species really individuals? *Systematic Zoology* **25**, 174–91.

Hull, D. L. (1980a). The units of evolution: a metaphysical essay. In *Studies in the Concept of Evolution* (eds U. J. Jensen & R. Harré). Brighton: Harvester Press.

Hull, D. L. (1980b). Individuality and selection. *Annual Review of Ecology and Systematics* **11**, 311–32.

Huxley, J. S. (1912). *The Individual in the Animal Kingdom*. Cambridge: Cambridge University Press.

Huxley, J. S. (1932). *Problems of Relative Growth*. London: McVeagh.

Jacob, F. (1977). Evolution and tinkering. *Science* **196**, 1161–6.

Janzen, D. H. (1977). What are dandelions and aphids? *American Naturalist* **111**, 586–9.

Jensen, D. (1961). Operationism and the question 'Is this behavior learned or innate?' *Behaviour* **17**, 1–8.

Jeon, K. W. & Danielli, J. F. (1971). Micrurgical studies with large free-living amebas. *International Reviews of Cytology* **30**, 49–89.

Judson, H. F. (1979). *The Eighth Day of Creation*. London: Cape.

Kalmus, H. (1955). The discrimination by the nose of the dog of individual human odours. *British Journal of Animal Behaviour* **3**, 25–31.

Keeton, W. T. (1980). *Biological Science*, 3rd edn. New York: W. W. Norton.

Kempthorne, O. (1978). Logical, epistemological and statistical aspects of nature–nurture data interpretation. *Biometrics* **34**, 1–23.

Kerr, A. (1978). The Ti plasmid of *Agrobacterium*. *Proceedings of the 4th International Conference, Plant Pathology and Bacteriology*, Angers, 101–8.

Kettlewell, H. B. D. (1955). Recognition of appropriate backgrounds by the pale and dark phases of Lepidoptera. *Nature* **175**, 943–4.

Kettlewell, H. B. D. (1973). *The Evolution of Melanism*. Oxford: Oxford University Press.

Kirk, D. L. (1980). *Biology Today*. New York: Random House.

Kirkwood, T. B. L. & Holliday, R. (1979). The evolution of ageing and longevity. *Proceedings of the Royal Society of London* B **205**, 531–46.

Knowlton, N. & Parker, G. A. (1979). An evolutionarily stable strategy approach to indiscriminate spite. *Nature* **279**, 419–21.

Koestler, A. (1967). *The Ghost in the Machine*. London: Hutchinson.

Krebs, J. R. (1977). Simplifying sociobiology. *Nature* **267**, 869.

Krebs, J. R. (1978). Optimal foraging: decision rules for predators. In *Behavioural Ecology* (eds J. R. Krebs & N. B. Davies), pp. 23–63. Oxford: Blackwell Scientific.

Krebs, J. R. & Davies, N. B. (1978). *Behavioural Ecology*. Oxford: Blackwell Scientific.

Kuhn, T. S. (1970). *The Structure of Scientific Revolutions*, 2nd edn. Chicago: University of Chicago Press.

Kurland, J. A. (1979). Can sociality have a favorite sex chromosome? *American Naturalist* **114**, 810–17.

Kurland, J. A. (1980). Kin selection theory: a review and selective bibliography. *Ethology & Sociobiology* **1**, 255–74.

Lack, D. (1966). *Population Studies of Birds*. Oxford: Oxford University Press.

Lack, D. (1968). *Ecological Adaptations for Breeding in Birds*. London: Methuen.

Lacy, R. C. (1980). The evolution of eusociality in termites: a haplodiploid analogy? *American Naturalist* **116**, 449–51.

Lande, R. (1976). Natural selection and random genetic drift. *Evolution* **30**, 314–34.

Lawlor, L. R. & Maynard Smith, J. (1976). The coevolution and stability of competing species. *American Naturalist* **110**, 79–99.

Lehrman, D. S. (1970). Semantic and conceptual issues in the nature-nurture problem. In *Development and Evolution of Behavior* (eds L. R. Aronson et al.), pp. 17–52. San Francisco: W. H. Freeman.

Leigh, E. (1971). *Adaptation and Diversity*. San Francisco: Freeman Cooper.

Leigh, E. (1977). How does selection reconcile individual advantage with the good of the group? *Proceedings of the National Academy of Sciences, U.S.A.* **74**, 4542–6.

Levinton, J. S. & Simon, C. M. (1980). A critique of the punctuated equilibria model and implications for the detection of speciation in the fossil record. *Systematic Zoology* **29**, 130–42.

Levy, D. (1978). Computers are now chess masters. *New Scientist* **79**, 256–8.

Lewontin, R. C. (1967). Spoken remark in *Mathematical Challenges to the Neo-Darwinian Interpretation of Evolution* (eds P. S. Moorhead & M. Kaplan). *Wistar Institute Symposium Monograph* **5**, 79.

Lewontin, R. C. (1970a). The units of selection. *Annual Review of Ecology and Systematics* **1**, 1–18.

Lewontin, R. C. (1970b). On the irrelevance of genes. In *Towards a Theoretical Biology, 3: Drafts* (ed. C. H. Waddington), pp. 63–72. Edinburgh: Edinburgh University Press.

Lewontin, R. C. (1974). *The Genetic Basis of Evolutionary Change*. New York and London: Columbia University Press.

Lewontin, R. C. (1977). Caricature of Darwinism. *Nature* **266**, 283–4.

Lewontin, R. C. (1978). Adaptation. *Scientific American* **239** (3), 156–69.

Lewontin, R. C. (1979a). Fitness, survival and optimality. In *Analysis of Ecological Systems* (eds D. J. Horn, G. R. Stairs & R. D. Mitchell), pp. 3–21. Columbus: Ohio State University Press.

Lewontin, R. C. (1979b). Sociobiology as an adaptationist program. *Behavioral Science* **24**, 5–14.

Lindauer, M. (1961). *Communication among Social Bees*. Cambridge, Mass.: Harvard University Press.

Lindauer, M. (1971). The functional significance of the honeybee waggle dance. *American Naturalist* **105**, 89–96.

Linsenmair, K. E. (1972). Die Bedeutung familienspezifischer "Abzeichen" für den Familienzusammenhalt bei der sozialen Wustenassel *Hemilepistus reamuri* Audouin u. Savigny (Crustacea, Isopoda, Oniscoidea). *Zeitschrift für Tierpsychologie* **31**, 131–62.

Lloyd, J. E. (1975). Aggressive mimicry in *Photuris*: signal repertoires by femmes fatales. *Science* **187**, 452–3.

Lloyd, J. E. (1979). Mating behavior and natural selection. *Florida Entomologist* **62** (1), 17–23.

Lloyd, J. E. (1981). Firefly mate-rivals mimic predators and vice versa. *Nature* **290**, 498–500.

Lloyd, M. & Dybas, H. S. (1966). The periodical cicada problem. II. Evolution. *Evolution* **20**, 466–505.

Lorenz, K. (1937). Uber die Bildung des Instinktbegriffes. *Die Naturwissenschaften* **25**, 289–300.

Lorenz, K. (1966). *Evolution and Modification of Behavior*. London: Methuen.

Love, M. (1980). The alien strategy. *Natural History* **89** (5), 30–2.

Lovelock, J. E. (1979). *Gaia*. Oxford: Oxford University Press.

Lumsden, C. J. & Wilson, E. O. (1980). Translation of epigenetic rules of individual behavior into ethnographic patterns. *Proceedings of the National Academy of Sciences, U.S.A.* **77**, 4382–6.

Lyttle, T. W. (1977). Experimental population genetics of meiotic drive systems. I. Pseudo-Y chromosomal drive as a means of eliminating cage populations of *Drosophila melanogaster. Genetics* **86**, 413–45.

McCleery, R. H. (1978). Optimal behaviour sequences and decision making. In *Behavioural Ecology* (eds J. R. Krebs & N. B. Davies), pp. 377–410. Oxford: Blackwell Scientific.

McFarland, D. J. & Houston, A. I. (1981). *Quantitative Ethology*. London: Pitman.

McLaren, A., Chandler, P., Buehr, M., Fierz, W. & Simpson, E. (1981). Immune reactivity of progeny of tetraparental male mice. *Nature* **290**, 513–14.

Manning, A. (1971). Evolution of behavior. In *Psychobiology* (ed. J. L. McGaugh), pp. 1–52. New York: Academic Press.

Margulis, L. (1970). *Origin of Eukaryotic Cells*. New Haven: Yale University Press.

Margulis, L. (1976). Genetic and evolutionary consequences of symbiosis. *Experimental Parasitology* **39**, 277–349.

Margulis, L. (1981). *Symbiosis in Cell Evolution*. San Francisco: W. H. Freeman.

Maynard Smith, J. (1969). The status of neo-Darwinism. In *Towards a Theoretical Biology, 2: Sketches* (ed. C. H. Waddington), pp. 82–9. Edinburgh: Edinburgh University Press.

Maynard Smith, J. (1972). *On Evolution*. Edinburgh: Edinburgh University Press.

Maynard Smith, J. (1974). The theory of games and the evolution of animal conflicts. *Journal of Theoretical Biology* **47**, 209–21.

Maynard Smith, J. (1976a). Group Selection. *Quarterly Review of Biology* **51**, 277–83.

Maynard Smith, J. (1976b). What determines the rate of evolution? *American Naturalist* **110**, 331–8.

Maynard Smith, J. (1977). Parental investment: a prospective analysis. *Animal Behaviour* **25**, 1–9.

Maynard Smith, J. (1978a). *The Evolution of Sex*. Cambridge: Cambridge University Press.

Maynard Smith, J. (1978b). Optimization theory in evolution. *Annual Review of Ecology and Systematics* **9**, 31–56.

Maynard Smith, J. (1979). Game theory and the evolution of behaviour. *Proceedings of the Royal Society of London*, B **205**, 475–88.

Maynard Smith, J. (1980). Regenerating Lamarck. *Times Literary Supplement* No. 4047, 1195.

Maynard Smith, J. (1981). Macroevolution. *Nature* **289**, 13–14.

Maynard Smith, J. (1982) The evolution of social behaviour—a classification of models. In *Current Problems in Sociobiology* (ed. King's College Sociobiology Group), pp. 29–44. Cambridge: Cambridge University Press.

Maynard Smith, J. & Parker, G. A. (1976). The logic of asymmetric contests. *Animal Behaviour* **24**, 159–75.

Maynard Smith, J. & Price, G. R. (1973). The logic of animal conflict. *Nature* **246**, 15–18.

Maynard Smith, J. & Ridpath, M. G. (1972). Wife sharing in the Tasmanian native hen, *Tribonyx mortierii*: a case of kin selection? *American Naturalist* **106**, 447–52.

Mayr, E. (1963). *Animal Species and Evolution*. Cambridge, Mass.: Harvard University Press.

Medawar, P. B. (1952). *An Unsolved Problem in Biology*. London: H. K. Lewis.

Medawar, P. B. (1957). *The Uniqueness of the Individual*. London: Methuen.

Medawar, P. B. (1960). *The Future of Man*. London: Methuen.

Medawar, P. B. (1967). *The Art of the Soluble*. London: Methuen.

Medawar, P. B. (1981). Back to evolution. *New York Review of Books* **28** (2), 34–6.

Mellanby, K. (1979). Living with the Earth Mother. *New Scientist* **84**, 41.

Midgley, M. (1979). Gene-juggling. *Philosophy* **54**, 439–58.

Murray, J. & Clarke, B. (1966). The inheritance of polymorphic shell characters in *Partula* (Gastropoda). *Genetics* **54**, 1261–77.

'Nabi, I.' (1981). Ethics of genes. *Nature* **290**, 183.

Old, R. W. & Primrose, S. B. (1980). *Principles of Gene Manipulation*. Oxford: Blackwell Scientific.

Orgel, L. E. (1979). Selection *in vitro*. *Proceedings of the Royal Society of London*, B **205**, 435–42.

Orgel, L. E. & Crick, F. H. C. (1980). Selfish DNA: the ultimate parasite. *Nature* **284**, 604–7.

Orlove, M. J. (1975). A model of kin selection not invoking coefficients of relationship. *Journal of Theoretical Biology* **49**, 289–310.

Orlove, M. J. (1979). Putting the diluting effect into inclusive fitness. *Journal of Theoretical Biology* **78**, 449–50.

Oster, G. F. & Wilson, E. O. (1978). *Caste and Ecology in the Social Insects.* Princeton: Princeton University Press.

Packard, V. (1957). *The Hidden Persuaders.* London: Penguin.

Parker, G. A. (1978a). Searching for mates. In *Behavioural Ecology* (eds J. R. Krebs & N. B. Davies), pp. 214–44. Oxford: Blackwell Scientific.

Parker, G. A. (1978b). Selection on non-random fusion of gametes during the evolution of anisogamy. *Journal of Theoretical Biology* **73**, 1–28.

Parker, G. A. (1979). Sexual selection and sexual conflict. In *Sexual Selection and Reproductive Competition in Insects* (eds M. S. Blum & N. A. Blum), pp. 123–66. New York: Academic Press.

Parker, G. A. & Macnair, M. R. (1978). Models of parent-offspring conflict. I. Monogamy. *Animal Behaviour* **26**, 97–110.

Partridge, L. & Nunney, L. (1977). Three-generation family conflict. *Animal Behaviour* **25**, 785–6.

Peleg, B. & Norris, D. M. (1972). Symbiotic interrelationships between microbes and Ambrosia beetles. VII. *Journal of Invertebrate Pathology* **20**, 59–65.

Pittendrigh, C. S. (1958). Adaptation, natural selection, and behavior. In *Behavior and Evolution* (eds A. Roe & G. G. Simpson), pp. 390–416. New Haven: Yale University Press.

Pribram, K. H. (1974). How is it that sensing so much we can do so little? In *The Neurosciences, Third Study Program* (eds F. O. Schmitt & F. G. Worden), pp. 249–61. Cambridge, Mass.: MIT Press.

Pringle, J. W. S. (1951). On the parallel between learning and evolution. *Behaviour* **3**, 90–110.

Pulliam, H. R. & Dunford, C. (1980). *Programmed to Learn.* New York: Columbia University Press.

Pyke, G. H., Pulliam, H. R. & Charnov, E. L. (1977). Optimal foraging: a selective review of theory and tests. *Quarterly Review of Biology* **52**, 137–54.

Raup, D. M., Gould, S. J., Schopf, T. J. M. & Simberloff, D. S. (1973). Stochastic models of phylogeny and the evolution of diversity. *Journal of Geology* **81**, 525–42.

Reinhard, E. G. (1956). Parasitic castration of crustacea. *Experimental Parasitology* **5**, 79–107.

Richmond, M. H. (1979). 'Cells' and 'organisms' as a habitat for DNA. *Proceedings of the Royal Society of London*, B **204**, 235–50.

Richmond, M. H. & Smith, D. C. (1979). *The Cell as a Habitat*. London: Royal Society.

Ridley, M. (1980). Konrad Lorenz and Humpty Dumpty: some ethology for Donald Symons. *Behavioral and Brain Sciences* **3**, 196.

Ridley, M. (1982). Coadaptation and the inadequacy of natural selection. *British Journal for the History of Science* **15**, 45–68.

Ridley, M. & Dawkins, R. (1981). The natural selection of altruism. In *Altruism and Helping Behavior* (eds J. P. Rushton & R. M. Sorentino), pp. 19–39. Hillsdale, N.J.: Erlbaum.

Ridley, M. & Grafen, A. (1981). Are green beard genes outlaws? *Animal Behaviour* **29**, 954–5.

Rose, S. (1978). Pre-Copernican sociobiology? *New Scientist* **80**, 45–6.

Rothenbuhler, W. C. (1964). Behavior genetics of nest cleaning in honey bees. IV. Responses of F1 and backcross generations to disease-killed brood. *American Zoologist* **4**, 111–23.

Rothstein, S. I. (1980). The preening invitation or head-down display of parasitic cowbirds. II. Experimental analysis and evidence for behavioural mimicry. *Behaviour* **75**, 148–84.

Rothstein, S. I. (1981). Reciprocal altruism and kin selection are not clearly separable phenomena. *Journal of Theoretical Biology* **87**, 255–61.

Sahlins, M. (1977). *The Use and Abuse of Biology*. London: Tavistock.

Sargent, T. D. (1968). Cryptic moths: effects on background selection of painting the circumocular scales. *Science* **159**, 100–1.

Sargent, T. D. (1969a). Background selections of the pale and melanic forms of the cryptic moth *Phigalia titea* (Cramer). *Nature* **222**, 585–6.

Sargent, T. D. (1969b). Behavioural adaptations of cryptic moths. III. Resting attitudes of two bark-like species, *Melanolophia canadaria* and *Catocala ultronia*. *Animal Behaviour*, **17**, 670–2.

Schaller, G. B. (1972). *The Serengeti Lion*. Chicago: Chicago University Press.

Schell, J. + 13 others (1979). Interactions and DNA transfer between *Agrobacterium tumefaciens*, the Ti-plasmid and the plant host. *Proceedings of the Royal Society of London*, B **204**, 251–66.

Schleidt, W. M. (1973). Tonic communication: continual effects of discrete signs in animal communication systems. *Journal of Theoretical Biology* **42**, 359–86.

Schmidt, R. S. (1955). Termite (*Apicotermes*) nests—important ethological material. *Behaviour* **8**, 344–56.

Schuster, P. & Sigmund, K. (1981). Coyness, philandering and stable strategies. *Animal Behaviour* **29**, 186–92.

Schwagmeyer, P. L. (1980). The Bruce effect: an evaluation of male/female advantages. *American Naturalist* **114**, 932–8.

Seger, J. A. (1980). Models for the evolution of phenotypic responses to genotypic correlations that arise in finite populations. PhD thesis, Harvard University, Cambridge, Mass.

Shaw, G. B. (1921). *Back to Methuselah*. Reprinted 1977. Harmondsworth, Middx: Penguin.

Sherman, P. W. (1978). Why *are* people? *Human Biology* **50**, 87–95.

Sherman, P. W. (1979). Insect chromosome numbers and eusociality. *American Naturalist* **113**, 925–35.

Simon, C. (1979). Debut of the seventeen year cicada. *Natural History* **88** (5), 38–45.

Simon, H. A. (1962). The architecture of complexity. *Proceedings of the American Philosophical Society* **106**, 467–82.

Simpson, G. G. (1953). *The Major Features of Evolution*. New York: Columbia University Press.

Sivinski, J. (1980). Sexual selection and insect sperm. *Florida Entomologist* **63**, 99–111.

Slatkin, M. (1972). On treating the chromosome as the unit of selection. *Genetics* **72**, 157–68.

Slatkin, M. & Maynard Smith, J. (1979). Models of coevolution. *Quarterly Review of Biology* **54**, 233–63.

Smith, D. C. (1979). From extracellular to intracellular: the establishment of a symbiosis. *Proceedings of the Royal Society of London*, B **204**, 115–30.

Southwood, T. R. E. (1976). Bionomic strategies and population parameters. In *Theoretical Ecology* (ed. R. M. May), pp. 26–48. Oxford: Blackwell Scientific.

Spencer, H. (1864). *The Principles of Biology*, Vol. 1. London and Edinburgh: Williams and Norgate.

Staddon, J. E. R. (1981). On a possible relation between cultural transmission and genetical evolution. In *Perspectives in Ethology*, Vol. 4 (eds P. P. G. Bateson & P. H. Klopfer), pp. 135–45. New York: Plenum Press.

Stamps, J. & Metcalf, R. A. (1980). Parent–offspring conflict. In *Sociobiology: Beyond Nature/Nurture?* (eds G. W. Barlow & J. Silverberg), pp. 589–618. Boulder: Westview Press.

Stanley, S. M. (1975). A theory of evolution above the species level. *Proceedings of the National Academy of Sciences, U.S.A.* **72**, 646–50.

Stanley, S. M. (1979). *Macroevolution, Pattern and Process.* San Francisco: W. H. Freeman.

Stebbins, G. L. (1977). In defense of evolution: tautology or theory? *American Naturalist* **111**, 386–90.

Steele, E. J. (1979). *Somatic Selection and Adaptive Evolution.* Toronto: Williams and Wallace.

Stent, G. (1977). You can take the ethics out of altruism but you can't take the altruism out of ethics. *Hastings Center Report* **7** (6), 33–6.

Symons, D. (1979). *The Evolution of Human Sexuality.* New York: Oxford University Press.

Syren, R. M. & Luyckx, P. (1977). Permanent segmental interchange complex in the termite *Incisitermes schwarzi. Nature* **266**, 167–8.

Taylor, A. J. P. (1963). *The First World War.* London: Hamish Hamilton.

Temin, H. M. (1974). On the origin of RNA tumor viruses. *Annual Review of Ecology and Systematics* **8**, 155–77.

Templeton, A. R., Sing, C. F. & Brokaw, B. (1976). The unit of selection in *Drosophila mercatorium*. I. The interaction of selection and meiosis in parthenogenetic strains. *Genetics* **82**, 349–76.

Thoday, J. M. (1953). Components of fitness. *Society for Experimental Biology Symposium* **7**, 96–113.

Thomas, L. (1974). *The Lives of a Cell.* London: Futura.

Thompson, D'A. W. (1917). *On Growth and Form.* Cambridge: Cambridge University Press.

Tinbergen, N. (1954). The origin and evolution of courtship and threat display. In *Evolution as a Process* (eds J. S. Huxley, A. C. Hardy & E. B. Ford), pp. 233–50. London: Allen & Unwin.

Tinbergen, N. (1963). On aims and methods of ethology. *Zeitschrift für Tierpsychologie* **20**, 410–33.

Tinbergen, N. (1964). The evolution of signaling devices. In *Social Behavior and Organization among Vertebrates* (ed. W. Etkin), pp. 206–30. Chicago: Chicago University Press.

Tinbergen, N. (1965). Behaviour and natural selection. In *Ideas in Modern Biology* (ed. J. A. Moore), pp. 519–42. New York: Natural History Press.

Tinbergen, N., Broekhuysen, G. J., Feekes, F., Houghton, J. C. W., Kruuk, H. & Szulc, E. (1962). Egg shell removal by the black-headed gull, *Larus ridibundus*, L.; a behaviour component of camouflage. *Behaviour* **19**, 74–117.

Trevor-Roper, H. R. (1972). *The Last Days of Hitler*. London: Pan.

Trivers, R. L. (1971). The evolution of reciprocal altruism. *Quarterly Review of Biology* **46**, 35–57.

Trivers, R. L. (1972). Parental investment and sexual selection. In *Sexual Selection and the Descent of Man* (ed. B. Campbell), pp. 136–79. Chicago: Aldine.

Trivers, R. L. (1974). Parent-offspring conflict. *American Zoologist* **14**, 249–64.

Trivers, R. L. & Hare, H. (1976). Haplodiploidy and the evolution of the social insects. *Science* **191**, 249–63.

Turing, A. (1950). Computing machinery and intelligence. *Mind* **59**, 433–60.

Turnbull, C. (1961). *The Forest People*. London: Cape.

Turner, J. R. G. (1977). Butterfly mimicry: the genetical evolution of an adaptation. In *Evolutionary Biology*, Vol. 10 (eds M. K. Hecht et al.), pp. 163–206. New York: Plenum Press.

Vermeij, G. J. (1973). Adaptation, versatility and evolution. *Systematic Zoology* **22**, 466–77.

Vidal, G. (1955). *Messiah*. London: Heinemann.

Waddington, C. H. (1957). *The Strategy of the Genes*. London: Allen & Unwin.

Wade, M. J. (1978). A critical review of the models of group selection. *Quarterly Review of Biology* **53**, 101–14.

Waldman, B. & Adler, K. (1979). Toad tadpoles associate preferentially with siblings. *Nature* **282**, 611–13.

Wallace, A. R. (1866). Letter to Charles Darwin dated 2 July. In J. Marchant (1916), *Alfred Russel Wallace Letters and Reminiscences*, Vol. 1, pp. 170–4. London: Cassell.

Watson, J. D. (1976). *Molecular Biology of the Gene*. Menlo Park: Benjamin.

Weinrich, J. D. (1976). Human reproductive strategy: the importance of income unpredictability, and the evolution of non-reproduction. PhD dissertation, Harvard University, Cambridge, Mass.

Weizenbaum, J. (1976). *Computer Power and Human Reason*. San Francisco: W. H. Freeman.

Wenner, A. M. (1971). *The Bee Language Controversy: An Experience in Science*. Boulder: Educational Programs Improvement Corporation.

Werren, J. H., Skinner, S. K. & Charnov, E. L. (1981). Paternal inheritance of a daughterless sex ratio factor. *Nature* **293**, 467–8.

West-Eberhard, M. J. (1975). The evolution of social behavior by kin selection. *Quarterly Review of Biology* **50**, 1–33.

West-Eberhard, M. J. (1979). Sexual selection, social competition, and evolution. *Proceedings of the American Philosophical Society* **123**, 222–34.

White, M. J. D. (1978). *Modes of Speciation*. San Francisco: W. H. Freeman.

Whitham, T. G. & Slobodchikoff, C. N. (1981). Evolution of individuals, plant-herbivore interactions, and mosaics of genetic variability: the adaptive significance of somatic mutations in plants. *Oecologia* **49**, 287–92.

Whitney, G. (1976). Genetic substrates for the initial evolution of human sociality. I. Sex chromosome mechanisms. *American Naturalist* **110**, 867–75.

Wickler, W. (1968). *Mimicry*. London: Weidenfeld & Nicolson.

Wickler, W. (1976). Evolution-oriented ethology, kin selection, and altruistic parasites. *Zeitschrift für Tierpsychologie* **42**, 206–14.

Wickler, W. (1977). Sex-linked altruism. *Zeitschrift für Tierpsychologie* **43**, 106–7.

Williams, G. C. (1957). Pleiotropy, natural selection, and the evolution of senescence. *Evolution* **11**, 398–411.

Williams, G. C. (1966). *Adaptation and Natural Selection*. Princeton, N.J.: Princeton University Press.

Williams, G. C. (1975). *Sex and Evolution*. Princeton, N.J.: Princeton University Press.

Williams, G. C. (1979). The question of adaptive sex ratio in outcrossed vertebrates. *Proceedings of the Royal Society of London*, B **205**, 567–80.

Williams, G. C. (1980). Kin selection and the paradox of sexuality. In *Sociobiology: Beyond Nature/Nurture?* (eds G. W. Barlow & J. Silverberg), pp. 371–84. Boulder: Westview Press.

Wilson, D. S. (1980). *The Natural Selection of Populations and Communities*. Menlo Park: Benjamin/Cummings.

Wilson, E. O. (1971). *The Insect Societies*. Cambridge, Mass.: Harvard University Press.

Wilson, E. O. (1975). *Sociobiology: the New Synthesis*. Cambridge, Mass.: Harvard University Press.

Wilson, E. O. (1978). *On Human Nature*. Cambridge, Mass.: Harvard University Press.

Winograd, T. (1972). *Understanding Natural Language*. Edinburgh: Edinburgh University Press.

Witt, P. N., Reed, C. F. & Peakall, D. B. (1968). *A Spider's Web*. New York: Springer Verlag.

Wolpert, L. (1970). Positional information and pattern formation. In *Towards a Theoretical Biology, 3: Drafts* (ed. C. H. Waddington), pp. 198–230. Edinburgh: Edinburgh University Press.

Wright, S. (1932). The roles of mutation, inbreeding, crossbreeding and selection in evolution. *Proceedings of the 6th International Congress of Genetics* **1**, 356–68.

Wright, S. (1951). Fisher and Ford on the Sewall Wright effect. *American Science Monthly* **39**, 452–8.

Wright, S. (1980). Genic and organismic selection. *Evolution* **34**, 825–43.

Wu, H. M. H., Holmes, W. G., Medina, S. R. & Sackett, G. P. (1980). Kin preference in infant *Macaca nemestrina*. *Nature* **285**, 225–7.

Wynne-Edwards, V. C. (1962). *Animal Dispersion in Relation to Social Behaviour*. Edinburgh: Oliver & Boyd.

Young, J. Z. (1957). *The Life of Mammals*. Oxford: Oxford University Press.

Young, R. M. (1971). Darwin's metaphor: does nature select? *The Monist* **55**, 442–503.

Zahavi, A. (1979). Parasitism and nest predation in parasitic cuckoos. *American Naturalist* **113**, 157–9.

FURTHER READING

Allaby, M. (1982). *Animal Artisans*. London: Weidenfeld & Nicolson.

Barkow, J. H., Cosmides, L. & Tooby, J. (1992). *The Adapted Mind*. New York: Oxford University Press.

Basalla, G. (1988). *The Evolution of Technology*. Cambridge: Cambridge University Press.

Blackmore, S. (1999). *The Meme Machine*. Oxford: Oxford University Press.

Bonner, J. T. (1988). *The Evolution of Complexity*. Princeton: N.J.: Princeton University Press.

Brandon, R. N. (1990). *Adaptation and Environment*. Princeton, N.J.: Princeton University Press.

Brandon, R. N. & Burian, R. M. (1984). *Genes, Organisms, Populations: Controversies over the Units of Selection*. Cambridge, Mass.: MIT Press.

Buss, L. W. (1987). *The Evolution of Individuality*. Princeton, N.J.: Princeton University Press.

Clayton, D. & Harvey, P. (1993). Hanging nests on a phylogenetic tree. *Current Biology* **3**, 882–3.

Cronin, H. (1991). *The Ant and the Peacock*. Cambridge: Cambridge University Press.

Cziko, G. (1995). *Without Miracles*. Cambridge, Mass.: MIT Press.

Davies, N. B. (1992). *Dunnock Behaviour and Social Evolution*. Oxford: Oxford University Press.

Davis, B. D. (1986). *Storm over Biology*. Buffalo: Prometheus Books.

Dawkins, R. (1982). Universal Darwinism. In *Evolution from Molecules to Men* (ed. D. S. Bendall), pp. 403–25. Cambridge: Cambridge University Press.

Dawkins, R. (1985). Review of *Not in our Genes* (S. Rose, L. J. Kamin & R. C. Lewontin). *New Scientist* **105**, 59–60.

Dawkins, R. (1987). Universal parasitism and the extended phenotype. In *Evolution and Coadaptation in Biotic Communities* (eds S. Kawano, J. H. Connell & T. Hidaka), pp. 183–97. Tokyo: University of Tokyo Press.

Dawkins, R. (1989a). The evolution of evolvability. In *Artificial Life* (ed. C. Langton). Santa Fe, N.M.: Addison Wesley.

Dawkins, R. (1989b). *The Selfish Gene*, 2nd edn. Oxford: Oxford University Press.

Dawkins, R. (1990). Parasites, desiderata lists, and the paradox of the organism. In *The Evolutionary Biology of Parasitism* (eds A. E. Keymer & A. F. Read). Cambridge: Cambridge University Press.

Dawkins, R. (1991). Darwin triumphant: Darwinism as a universal truth. In *Man and Beast Revisited* (eds M. H. Robinson & L. Tiger), pp. 23–39. Washington, D.C.: Smithsonian Institution.

Dawkins, R. (1996). *Climbing Mount Improbable*. New York: Norton.

Dawkins, R. (1998). *Unweaving the Rainbow*. London: Penguin.

Dennett, D. (1995). *Darwin's Dangerous Idea*. New York: Simon & Schuster.

Depew, D. J. & Weber, B. H. (1996). *Darwinism Evolving*. Cambridge, Mass.: MIT Press.

de Winter, W. (1997). The beanbag genetics controversy: towards a synthesis of opposing views of natural selection. *Biology and Philosophy* **12**, 149–84.

Durham, W. H. (1991). *Coevolution: Genes, Culture and Human Diversity*. Stanford, Calif.: Stanford University Press.

Eldredge, N. (1995). *Reinventing Darwin: The Great Debate at the High Table of Evolutionary Theory*. New York: John Wiley.

Endler, J. A. (1986). *Natural Selection in the Wild*. Princeton, N.J.: Princeton University Press.

Ewald, P. (1993). *Evolution of Infectious Diseases*. Oxford: Oxford University Press.

Fletcher, D. J. C. & Michener, C. D. (1987). *Kin Recognition in Humans*. New York: Wiley.

Fox Keller, E. & Lloyd, E. A. (1992). *Keywords in Evolutionary Biology*. Cambridge, Mass.: Harvard University Press.

Futuyma, D. J. (1998). *Evolutionary Biology*, 3rd edn. Sunderland, Mass.: Sinauer.

Goodwin, B. & Dawkins, R. (1995). What is an organism? A discussion. *Perspectives in Ethology* **2**, 47–60.

Gould, S. J. & Eldredge, N. (1993). Punctuated equilibrium comes of age. *Nature* **366**, 222–227.

Grafen, A. (1984). Natural selection, kin selection and group selection. In *Behavioural Ecology: An Evolutionary Approach* (eds J. R. Krebs & N. B. Davies). Oxford: Blackwell Scientific Publications.

Grafen, A. (1985). A geometric view of relatedness. *Oxford Surveys in Evolutionary Biology* **2**, 28–89.

Grafen, A. (1990a). Biological signals as handicaps. *Journal of Theoretical Biology* **144**, 517–46.

Grafen, A. (1990b). Do animals really recognize kin? *Animal Behaviour* **39**, 42–54.

Grafen, A. (1991). Modelling in behavioural ecology. In *Behavioural Ecology: An Evolutionary Approach* (eds J. R. Krebs & N. B. Davies). Oxford: Blackwell Scientific Publications.

Guilford, T. & Dawkins, M. S. (1991). Receiver psychology and the evolution of animal signals. *Animal Behaviour* **42**, 1–14.

Haig, D. (1993). Genetic conflicts in human pregnancy. *Quarterly Review of Biology* **68**, 495–532.

Hamilton, W. D. (1996). *Narrow Roads of Gene Land: The Collected Papers of W. D. Hamilton*, i: *Evolution of Social Behaviour*. Oxford: W. H. Freeman/ Spektrum.

Hansell, M. H. (1984). *Animal Architecture and Building Behaviour*. London: Longman.

Harvey, P. H. & Pagel, M. D. (1991). *The Comparative Method in Evolutionary Biology*. Oxford: Oxford University Press.

Hecht, M. K. & Hoffman, A. (1986). Why not neo-Darwinism? A critique of paleobiological challenges. In *Oxford Surveys in Evolutionary Biology* 3, 1–47. Oxford: Oxford University Press.

Hepper, P. G. (1986). Kin recognition: functions and mechanisms. A review. *Biological Reviews* **61**, 63–93.

Hoffman, A. (1989). *Arguments on Evolution*. New York: Oxford University Press.

Hölldobler, B. & Wilson, E. O. (1990). *The Ants*. Berlin: Springer-Verlag.

Hull, D. L. (1988a). Interactors versus vehicles. In *The Role of Behaviour in Evolution* (ed. H. C. Plotkin), pp. 19–50. Cambridge, Mass.: MIT Press.

Hull, D. L. (1988b). *Science as a Process*. Chicago: University of Chicago Press.

Keymer, A. & Read, A. (1989). Behavioural ecology: the impact of parasitism. In *Parasitism: Coexistence or Conflict?* (eds C. A. Toft & A. Aeschlimann). Oxford: Oxford University Press.

Kimura, M. (1983). *The Neutral Theory of Molecular Evolution*. Cambridge: Cambridge University Press.

Kitcher, P. (1985). *Vaulting Ambition*. Cambridge, Mass.: MIT Press.

Koch, W. A. (1986). *Genes vs. Memes*. Hagen: Druck Thiebes.

Krebs, J. R. & Dawkins, R. (1984). Animal signals: mind-reading and manipulation. In *Behavioural Ecology: An Evolutionary Approach* (eds J. R. Krebs & N. B. Davies). Oxford: Blackwell Scientific Publications.

Lloyd, E. A. (1988). *The Structure and Confirmation of Evolutionary Theory*. New York: Greenwood Press.

Maynard Smith, J. (1982). *Evolution and the Theory of Games*. Cambridge: Cambridge University Press.

Maynard Smith, J. (1986a). *The Problems of Biology*. Oxford: Oxford University Press.

Maynard Smith, J. (1986b). Structuralism versus selection—is Darwinism enough? In *Science and Beyond* (eds S. Rose & L. Appignanesi). Oxford: Basil Blackwell.

Maynard Smith, J. (1988). *Did Darwin Get it Right?* London: Penguin.

Maynard Smith, J., Burian, R., Kauffman, S., Alberch, P., Campbell, J., Goodwin, B., Lande, R., Raup, D. & Wolpert, L. (1985). Developmental constraints and evolution. *Quarterly Review of Biology* **60**, 265–87.

Maynard Smith, J. & Szathmáry, E. (1995). *The Major Transitions in Evolution*. Oxford: W. H. Freeman/Spektrum.

Mayr, E. (1983). How to carry out the adaptationist program. *American Naturalist* **121**, 324–34.

Minchella, D. J. (1985). Host life-history variation in response to parasitism. *Parasitology* **90**, 205–16.

Nesse, R. & Williams, G. C. (1995). *Evolution and Healing: The New Science of Darwinian Medicine*. London: Weidenfeld & Nicolson.

Nunney, L. (1985). Group selection, altruism, and structured-deme models. *American Naturalist* **126**, 212–29.

Otte, D. & Endler, J. A. (eds) (1989). *Speciation and its Consequences*. Sunderland, Mass.: Sinauer.

Oyama, S. (1985). *The Ontogeny of Information*. Cambridge: Cambridge University Press.

Parker, G. A. & Maynard Smith, J. (1990). Optimality theory in evolutionary biology. *Nature* **348**, 27–33.

Pinker, S. (1997). *How the Mind Works*. London: Allen Lane, Penguin Press.

Queller, D. C. (1992). Quantitative genetics, inclusive fitness, and group selection. *American Naturalist* **139**, 540–58.

Ridley, M. (1986). *The Problems of Evolution*. Oxford: Oxford University Press.

Ridley, M. (1996). *Evolution*. Oxford: Blackwell Scientific.

Ruse, M. (1996). *Taking Darwin Seriously*. Oxford: Basil Blackwell.

Segerstråle, U. (1999). *Defenders of the Truth. The Battle for 'Good Science' in the Sociobiology Debate and Beyond*. Oxford: Oxford University Press.

Sherman, P. W., Jarvis, J. U. M. & Alexander, R. D. (1991). *The Biology of the Naked Mole-Rat*. Princeton, N.J.: Princeton University Press.

Sober, E. (1984). *The Nature of Selection*. Cambridge, Mass.: MIT Press.

Sober, E. & Wilson, D. S. (1998). *Unto Others*. Cambridge, Mass.: Harvard University Press.

Sterelny, K. & Kitcher, P. (1988). The return of the gene. *Journal of Philosophy* **85**, 339–61.

Sterelny, K., Smith, K. C. & Dickison, M. (1996). The extended replicator. *Biology and Philosophy* **11**, 377–403.

Stone, G. N. & Cook, J. M. (1998). The structure of cynipid oak galls: patterns in the evolution of an extended phenotype. *Proceedings of the Royal Society of London* **B 265**, 979–88.

Trivers, R. L. (1985). *Social Evolution*. Menlo Park, N.J.: Benjamin/Cummings.

Vermeij, G. J. (1987). *Evolution and Escalation: An Ecological History of Life*. Princeton, N.J.: Princeton University Press.

Vollrath, F. (1988). Untangling the spider's web. *Trends in Ecology and Evolution* **3**, 331–5.

Weiner, J. (1994). *The Beak of the Finch*. London: Cape.

Werren, J. H., Nur, U. & Wu, C.-I. (1988). Selfish genetic elements. *Trends in Ecology & Evolution* **3**, 297–302.

Williams, G. C. (1985). A defence of reductionism in evolutionary biology. In *Oxford Surveys in Evolutionary Biology* (eds R. Dawkins & M. Ridley), pp. 1–27. Oxford: Oxford University Press.

Williams, G. C. (1992). *Natural Selection: Domains, Levels and Challenges*. Oxford: Oxford University Press.

Williams, G. C. (1997). *Plan and Purpose in Nature*. London: Weidenfeld & Nicolson.

Wills, C. (1990). *The Wisdom of the Genes*. New York: Basic Books.

Zahavi, A. & Zahavi, A. (1997). *The Handicap Principle*. Oxford: Oxford University Press.

GLOSSARY

This book is primarily intended for biologists who will have no need of a glossary, but it has been suggested to me that it would be worth explaining a few technical terms to make the book more widely accessible. Many of the terms are well defined in other places (e.g. Wilson 1975; Bodmer & Cavalli-Sforza 1976). My definitions are certainly no improvement on those already available, but I have added personal asides on controversial words, or on matters of particular relevance to the thesis of this book. I have tried to avoid cluttering up the glossary with excessive numbers of explicit cross-references, but many of the words used in the definitions will be found to have their own definitions elsewhere in the glossary.

adaptation A technical term which has evolved somewhat away from its common usage as a near synonym of 'modification'. From sentences like 'cricket wings are adapted (modified from their primary function of flying) for singing' (and by implication are well designed for singing), 'an adaptation' has come to mean approximately an attribute of an organism that is 'good' for something. Good in what sense?, and good for what or for whom?, are difficult questions which are discussed at length in this book.

alleles (short for **allelomorphs**) Each gene is able to occupy only a particular region of chromosome, its locus. At any given locus there may exist, in the population, alternative forms of the gene. These alternatives are called alleles of one another. This book emphasizes that there is a sense in which alleles are competitors of each other, because over evolutionary time successful alleles achieve numerical superiority over others at the same locus, in all the chromosomes of the population.

allometry A disproportionate relationship between size of a body part and size of the whole body, the comparisons being made either across individuals or across different life stages in the same individual. For example, large ants (but small humans) tend to have relatively very large heads; the head grows at a different rate from the body as a whole. Mathematically, the size of the part is usually taken as being related to the size of the whole raised to a power, which may be fractional.

allopatric theory of speciation The widely supported view that the evolutionary divergence of populations into separate species (which no longer interbreed) takes place in geographically separate places. The alternative, *sympatric theory* gives rise to difficulties in understanding how the incipient species can separate if they are continuously in a position to interbreed with each other, and therefore to mix their gene-pools (q.v.).

altruism Biologists use the word in a restricted (some would say misleadingly so) sense, only superficially related to common usage. An entity, such as a baboon or a gene, is said to be altruistic if it has the effect (not purpose) of promoting the welfare of another entity, at the expense of its own welfare. Various shades of meaning of 'altruism' result from various interpretations of 'welfare' (see page 87). *Selfish* is used in exactly the opposite sense.

anaphase That phase of the cycle of cell division during which the paired chromosomes move apart. In meiosis (q.v.) there are two successive divisions and correspondingly two anaphases.

anisogamy A sexual system in which fusion takes place at fertilization between a large (female) and a small (male) gamete. Contrast with *isogamy* in which there is sexual fusion but no male/female separation all gametes are of roughly the same size.

antibodies Protein molecules, produced in the immune response of animals, which neutralize invading foreign bodies (antigens).

antigens Foreign bodies, usually protein molecules, which provoke the formation of antibodies.

aposematism The phenomenon whereby distasteful or dangerous organisms like wasps 'warn' enemies by bright colours or equivalent strong stimuli. These are presumed to work by making it easy for the enemies to learn to avoid them, but there are (not insuperable) theoretical difficulties over how the phenomenon might evolve in the first place.

assortative mating The tendency of individuals to choose mates that resemble (positive assortative mating or homogamy) or specifically do not resemble (negative assortative mating) themselves. Some people use the word only in the positive sense.

autosome A chromosome that is not one of the sex chromosomes.

Baldwin/Waddington Effect First proposed by Spalding in 1873. A largely hypothetical evolutionary process (also called *genetic assimilation*) whereby natural selection can create an illusion of the inheritance of acquired characteristics. Selection in favour of a genetic tendency to acquire a characteristic in response to environmental stimuli leads to the evolution of increased sensitivity to the same environmental stimuli, and eventual emancipation from the need for them. On page 67 I suggest that we might breed a race of spontaneously milk-producing male mammals by treating successive generations of males with female hormones and selecting for increased sensitivity to female hormones. The role of the hormones, or other environmental treatment, is to bring out into the open genetic variation which would otherwise lie dormant.

central dogma In molecular biology the dogma that nucleic acids act as templates for the synthesis of proteins, but never the reverse. More generally, the dogma that genes exert an influence over the form of a body, but the form of a body is never translated back into genetic code: acquired characteristics are not inherited.

chromosome One of the chains of genes found in cells. In addition to DNA itself, there is usually a complicated supporting structure of protein. Chromosomes become visible under the light microscope only at certain times in the cell cycle, but their number and linearity may be inferred by statistical reasoning from the facts of inheritance alone (*see* linkage). The chromosomes are usually present in all cells in the body, even though only a minority of them will be active in any one cell. There are usually two sex chromosomes in every diploid cell as well as a number of autosomes (44 in humans).

cistron One way of defining a gene. In molecular genetics the cistron has a precise definition in terms of a specific experimental test. More loosely it is used to refer to a length of chromosome responsible for the encoding of one chain of amino acids in a protein.

codon A triplet of units (nucleotides) in the genetic code, specifying the synthesis of a single unit (amino acid) in a protein chain.

clone In cell biology, a set of genetically identical cells, all derived from the same ancestral cell. A human body is a gigantic clone of some 10^{15} cells. The word is also used of a set of organisms all of whose cells are members of the same clone. Thus a pair of identical twins may be said to be members of the same clone.

Cope's Rule An empirical generalization that evolutionary trends towards larger body size are common.

crossing-over A complicated process whereby chromosomes, while engaged in meiosis, exchange portions of genetic material. The result is the permutation of an almost infinite variety of gametes.

D'Arcy Thompson's transformations A graphical technique demonstrating that an animal shape can be transformed into the shape of a related animal by a mathematically specifiable distortion. D'Arcy Thompson would draw one of the two shapes on ordinary graph paper, then show that it was transformed approximately into the other shape if the coordinate system were distorted in some particular way.

diploid A cell is said to be diploid if it has chromosomes in pairs, in sexual cases one from each parent. An organism is said to be diploid if its body cells are diploid. Most sexually reproducing organisms are diploid.

dominance A gene is said to be dominant to one of its alleles if it suppresses the phenotypic effect of that (recessive) allele when the two are together. For example, if brown eyes are dominant to blue, only individuals with two blue-eyed genes (homozygous recessive) would actually have blue eyes; those with one blue and one brown gene (heterozygotes) would be indistinguishable from those with two brown genes (homozygous dominant). Dominance may be incomplete, in which case heterozygotes have an intermediate phenotype. The opposite of dominant is recessive. Dominance/recessiveness is a property of a phenotypic effect, not of a gene as such: thus a gene may be dominant in one of its phenotypic effects and recessive in another (*see* pleiotropy).

epigenesis A word with a long history of controversy in embryology. As opposed to preformationism (q.v.) it is the doctrine that bodily complexity emerges by a developmental process of gene/environment interaction from a relatively simple zygote, rather than being totally mapped out in the egg. In this book it is used for the idea, which I favour, that the genetic code is more like a recipe than a blueprint. It is sometimes said that the epigenesis/preformationism distinction has been made irrelevant by modern molecular biology. I disagree, and have made much of the distinction in Chapter 9, where I claim that epigenesis, but not preformationism, implies that embryonic development is fundamentally, and in principle, irreversible (*see* central dogma).

epistasis A class of interactions between pairs of genes in their phenotypic effects. Technically the interactions are non-additive which means, roughly, that the combined effect of the two genes is not the same as the sum of their separate effects. For instance, one gene might mask the effects of the other. The word is mostly used of genes at different loci, but some authors use it to include interactions between genes at the same locus, in which case dominance/recessiveness is a special case. *See also* dominance.

eukaryotes One of the two major groups of organisms on Earth, including all animals, plants, protozoa, and fungi. Characterized by the possession of a cell nucleus, and other membrane-bounded cell organelles (analogue of 'organ' within the cell) such as mitochondria. Contrast with prokaryotes (q.v.). The prokaryote/eukaryote distinction is much more fundamental than the animal/plant distinction (not to mention the relatively negligible human/'animal' distinction!).

eusociality The most advanced of the grades of sociality recognized by éntomologists. Characterized by a complex of features, the most important of which is the presence of a caste of sterile 'workers' who assist the reproduction of their long-lived mother, the queen. It is usually considered to be confined to wasps, bees, ants, and termites, but various other kind of animals approach eusociality in interesting ways.

evolutionarily stable strategy (ESS) [Note 'evolutionarily' *not* 'evolutionary'. The latter is a common grammatical error in this context.] A strategy that does well in a population dominated by the same strategy. This definition captures the intuitive essence of the idea (see Chapter 7), but is somewhat imprecise; for a mathematical definition, see Maynard Smith (1974).

extended phenotype All effects of a gene upon the world. As always, 'effect' of a gene is understood as meaning in comparison with its alleles. The conventional phenotype is the special case in which the effects are regarded as being confined to the individual body in which the gene sits. In practice it is convenient to limit 'extended phenotype' to cases where the effects influence the survival chances of the gene, positively or negatively.

fitness A technical term with so many confusing meanings that I have devoted a whole chapter to discussing them (Chapter 10).

game theory A mathematical theory originally developed for human games, and generalized to human economics and military strategy, and to evolution in the theory of evolutionarily stable strategy (q.v.). Game theory comes into its own wherever the optimum policy is not fixed, but depends upon the policy which is statistically most likely to be adopted by opponents.

gamete One of the sex cells which fuse in sexual reproduction. Sperms and eggs are both gametes.

gemmule A discredited concept espoused by Darwin in his 'pangenesis' theory of the inheritance of acquired characteristics—probably the only serious scientific error he ever made, and an example of the 'pluralism' for which he has recently been praised. Gemmules were supposed to be small hereditary particles which carried information from all parts of the body into the germ cells.

gene A unit of heredity. May be defined in different ways for different purposes (see page 130). Molecular biologists usually employ it in the sense of cistron (q.v.). Population biologists sometimes use it in a more abstract sense. Following Williams (1966, p. 24), I sometimes use the term gene to mean 'that which segregates and recombines with appreciable frequency', and (p. 25) as 'any hereditary information for which there is a favorable or unfavorable selection bias equal to several or many times its rate of endogenous change'.

gene-pool The whole set of genes in a breeding population. The metaphor on which the term is based is a happy one for this book, for it de-emphasizes the undeniable fact that genes actually go about in discrete bodies, and emphasizes the idea of genes flowing about the world like a liquid.

genetic drift Changes in gene frequencies over generations, resulting from chance rather than selection.

genome The entire collection of genes possessed by one organism.

genotype The genetic constitution of an organism at a particular locus or set of loci. Sometimes used more loosely as the whole genetic counterpart to phenotype (q.v.).

gens (pl. **gentes**) 'Race' of female cuckoos all parasitizing one host species. There must be genetic differences between gentes, and these are presumed to be on the Y chromosome. Males have no Y chromosomes, and do not belong to gentes. The word is poorly chosen, since in the Latin it refers to a clan tracing descent through the male line.

germ-line That part of bodies which is potentially immortal in the form of reproductive copies: the genetic contents of gametes and of cells that give rise to gametes. Contrast with *soma*, the parts which are mortal and which work for the preservation of genes in the germ-line.

gradualism The doctrine that evolutionary change is gradual and does not go in jumps. In modern palaeontology it is the subject of an interesting controversy over whether the gaps in the fossil record are artefactual or real (see Chapter 6). Journalists have blown this up into a pseudo-controversy over the validity of Darwinism, which they say is a gradualist theory. It is true that all sane Darwinians are gradualists in the extreme sense that they do not believe in the *de novo* creation of very complex and therefore statistically improbable new adaptations like eyes. This is surely what Darwin understood by the aphorism 'Nature does not make leaps'. But within the spectrum of gradualism in this sense, there is room for disagreement about whether evolutionary change occurs smoothly or in small jerks punctuating long periods of *stasis*. It is this that is the subject of the modern controversy, and it does not remotely bear, one way or the other, on the validity of Darwinism.

group selection A hypothetical process of natural selection among groups of organisms. Often invoked to explain the evolution of altruism (q.v.). Sometimes confused with kin selection (q.v.). In Chapter 6 I use the replicator/vehicle distinction to distinguish group selection of altruistic traits from species selection (q.v.) resulting in macroevolutionary trends.

haplodiploid A genetic system in which males grow from unfertilized eggs and are haploid, while females grow from fertilized eggs and are diploid. Therefore males have no father and no sons. Males pass all their genes on to their daughters, while females receive only half their genes from their fathers. Haplodiploidy occurs in nearly all social and non-social Hymenoptera (ants, bees, wasps, etc.), and also a few bugs, beetles, mites, ticks, and rotifers. The complications which haplodiploidy introduces into closeness of genetic kinship have been ingeniously invoked in theories of the evolution of eusociality (q.v.) in Hymenoptera.

haploid A cell is said to be haploid if it has a single set of chromosomes. Gametes are haploid, and when they fuse in fertilization a diploid cell (q.v.) is produced. Some organisms (e.g. fungi and male bees) are haploid in all their cells, and are referred to as haploid organisms.

heterozygous The condition of having nonidentical alleles at a chromosomal locus. Is usually applied to an individual organism, in which case it refers to two alleles at a given locus. More loosely it may refer to the overall statistical within-locus heterogeneity of alleles averaged over all loci in an individual or in a population.

homeotic mutation A mutation causing one part of a body to develop in a manner appropriate to another part. For example, the homeotic mutation 'antennopedia' in *Drosophila* causes a leg to grow where an antenna normally does. This is interesting, as it shows the power of a single mutation to have elaborate and complex effects, but only when there is elaborate complexity already there to be altered.

homozygous The condition of having identical alleles at a chromosomal locus. Is usually applied to an individual organism, in which case it indicates that the individual has two identical alleles at the locus. More loosely it may refer to the overall statistical within-locus homogeneity of alleles averaged over all loci in an individual or in a population.

K-selection Selection for the qualities needed to succeed in stable, predictable environments where there is likely to be heavy competition for limited resources between individuals well-equipped to compete, at population sizes close to the maximum that the habitat can bear. A variety of qualities are thought to be favoured by *K*-selection, including large size, long life, and small numbers of intensively cared-for offspring. Contrast with *r*-selection (q.v.). The '*K*' and '*r*' are symbols in the conventional algebra of population biologists.

kin selection Selection of genes causing individuals to favour close kin, owing to the high probability that kin share those genes. Strictly speaking 'kin' includes immediate offspring, but it is unfortunately undeniable that many biologists use the phrase 'kin selection' specifically when talking about kin other than offspring. Kin selection is also sometimes confused with group selection (q.v.), from which it is logically distinct, although where species happen to go around in discrete kin groups the two may incidentally amount to the same thing—'kin group selection'.

Lamarckism Regardless of what Lamarck actually said, Lamarckism is nowadays the name given to the theory of evolution that relies on the assumption that acquired characteristics can be inherited. From the point of view of this book, the significant feature of the Lamarckian theory is the idea that new genetic variation tends to be adaptively directed, rather than 'random' (i.e. non-directed) as in the Darwinian theory. The orthodox view today is that the Lamarckian theory is completely wrong.

linkage The presence on the same chromosome of a pair (or a set) of loci. Linkage is normally recognized by the statistical tendency for alleles at linked loci to be inherited together. For example, if hair colour and eye colour are linked, a child that inherits your eye colour is likely to inherit your hair colour too, while a child that fails to inherit your eye colour is also likely to fail to inherit your hair colour. Children are relatively unlikely to inherit one but not the other, though this can come about due to crossing-over (q.v.), the probability being related to the distance apart of the loci on the chromosome. This is the basis for the technique of chromosome mapping.

linkage disequilibrium The statistical tendency for alleles to occur together, in the bodies or gametes of a population, with particular alleles at other loci. For example, if we observed a tendency for fair-haired individuals to be blue-eyed, this might indicate linkage disequilibrium. Recognized as any tendency for the frequency of combinations of alleles at different loci to depart from the frequencies that would be expected from the overall frequencies of the alleles themselves in the population.

locus The position on a chromosome occupied by a gene (or a set of alternative alleles). For instance, there might be an eye-colour locus, at which the alternative alleles code for green, brown, and red. Usually applied at the level of the cistron (q.v.), the concept of the locus can be generalized to smaller or larger lengths of chromosome.

macroevolution The study of evolutionary changes that take place over a very large time-scale. Contrast with *microevolution*, the study of evolutionary changes within populations. Microevolutionary change is change in gene frequencies in populations. Macroevolutionary change is usually recognized as change in gross morphology in a series of fossils. There is some controversy over whether macroevolutionary change is fundamentally just cumulated microevolutionary change, or whether the two are 'decoupled' and driven by fundamentally different kinds of process. The name macroevolutionist is sometimes misleadingly restricted to partisans on one side of this controversy. It should be a neutral label for anybody studying evolution on the grand time-scale.

meiosis The kind of cell division in which a cell (usually diploid) gives rise to daughter cells (usually haploid) with half as many chromosomes. Meiosis is an essential part of normal sexual reproduction. It gives rise to the gametes which subsequently fuse to restore the original chromosome number.

meiotic drive The phenomenon whereby alleles affect meiosis so that they secure for themselves a greater than 50 per cent chance of finding themselves in a successful gamete. Such genes are said to be 'driving' because they tend to spread through the population in spite of any deleterious effects they may have on organisms. *See also* segregation distorter.

meme A unit of cultural inheritance, hypothesized as analogous to the particulate gene, and as naturally selected by virtue of its 'phenotypic' consequences on its own survival and replication in the cultural environment.

Mendelian inheritance Non-blending inheritance by means of pairs of discrete hereditary factors (now identified with genes), one member of each pair coming from each parent. The main theoretical alternative is 'blending inheritance'. In Mendelian inheritance genes may blend in their effects on a body, but they themselves do not blend, and they are passed on intact to future generations.

microevolution *See* macroevolution.

mitochondria Small complex organelles within eukaryotic cells, made of membranes, the site of most of the energy-releasing biochemistry of the cells. Mitochondria have their own DNA and reproduce autonomously within cells, and, according to one theory, they originated in evolution as symbiotic prokaryotes (q.v.).

mitosis The kind of cell division in which a cell gives rise to daughter cells having a complete set of all its chromosomes. Mitosis is the ordinary cell division of bodily growth. Contrast with meiosis.

modifier gene A gene whose phenotypic effect is to modify the effect of another gene. Geneticists no longer make a distinction between two types of genes, 'major genes' and 'modifiers', but recognize that many (and perhaps most) genes modify the effects of many (and perhaps most) other genes.

monophyletic A group of organisms is said to be monophyletic if all are descended from a common ancestor which would also have been classified as a member of the group. For instance, the birds are probably a monophyletic group since the most recent common ancestor of all birds would probably have been classified as a bird. The reptiles, however, are probably *polyphyletic*, in that the most recent common ancestor of all reptiles would probably not have been classified as a reptile. Some would argue that polyphyletic groups do not deserve names, and that the Class Reptilia should not be acknowledged.

mutation An inherited change in the genetic material. In Darwinian theory mutations are said to be random. This does not mean that they are not lawfully caused, but only that there is no specific tendency for them to be directed towards improved adaptation. Improved adaptation comes about only through selection, but it needs mutation as the ultimate source of the variants among which it selects.

muton The minimum unit of mutational change. One of several alternative definitions of gene (with cistron and recon).

neo-Darwinism A term coined (actually re-coined, for the word was used in the 1880s for a very different group of evolutionists) in the middle part of this century. Its purpose was to emphasize (and in my opinion exaggerate) the distinctness of the modern synthesis of Darwinism and Mendelian genetics, achieved in the 1920s and 1930s, from Darwin's own view of evolution. I think the need for the 'neo' is fading, and Darwin's own approach to 'the economy of nature' now looks very modern.

neoteny An evolutionary slowing down of bodily development relative to the development of sexual maturity, with the result that reproduction comes to be practised by organisms which resemble the juvenile stages of ancestral forms. It is hypothesized that some major steps in evolution, for example the origin of the vertebrates, came about through neoteny.

neutral mutation A mutation that has no selective advantage or disadvantage in comparison with its allele. Theoretically, a neutral mutation may become 'fixed' (i.e. numerically predominant in the population at its locus) after a number of generations, and this would be a form of evolutionary change. There is legitimate controversy over the importance of such random fixations in evolution, but there should be no controversy over their importance in the direct production of adaptation: it is zero.

nucleotide A kind of biochemical molecule, notable as the basic building block of DNA and RNA. DNA and RNA are polynucleotides, consisting of long chains of nucleotides. The nucleotides are 'read' in triplets, each triplet being known as a codon.

ontogeny The process of individual development. In practice development is often taken to culminate in the production of the adult, but strictly it includes later stages such as senescence. The doctrine of the extended phenotype would lead us to generalize 'ontogeny' to include the 'development' of extracorporeal adaptations, for example artefacts like beaver dams.

optimon The unit of natural selection, in the sense of the unit for whose benefit adaptations may be said to exist. The thesis of this book is that the optimon is neither the individual nor the group of individuals but the gene or genetic replicator. But the dispute is in part a semantic one, whose resolution occupies portions of Chapters 5 and 6.

orthoselection Sustained selection on the members of a lineage over a long period, causing continued evolution in a given direction. Can create an appearance of 'momentum' or 'inertia' in evolutionary trends.

outlaw gene A gene which is favoured by selection at its own locus, in spite of its deleterious effects on the other genes in the organisms in which it finds itself. Meiotic drive (q.v.) provides a good example.

Paley's watch A reference to the best known of William Paley's (1743–1805) arguments for the existence of God. A watch is too complicated, and too functional, to have come about by accident: it carries its own evidence of having been purposefully designed. The argument seems to apply *a fortiori* to a living body, which is even more complicated than a watch. Darwin, as a young man, was deeply impressed by this. Although he later destroyed the God part of the argument, by showing that natural selection can play the role of watchmaker to living bodies, he did not destroy the fundamental point—still under-appreciated—that complicated design demands a very special kind of explanation. God apart, the natural selection of small inherited variations is probably the only agency capable of doing the job.

phenotype The manifested attributes of an organism, the joint product of its genes and their environment during ontogeny. A gene may be said to have phenotypic expression in, say, eye colour. In this book the concept of phenotype is *extended* to include functionally important consequences of gene differences, outside the bodies in which the genes sit.

pheromone A chemical substance secreted by an individual, and adapted to influence the nervous systems of other individuals. Pheromones are often thought of as chemical 'signals' or 'messages', and as the inter-body analogue of hormones. In this book they are more often treated as analogous to manipulative drugs.

phylogeny Ancestral history on the evolutionary time-scale.

plasmid One of a set of more or less synonymous words used for small, self-replicating fragments of genetic material, found in cells but outside chromosomes.

pleiotropy The phenomenon whereby a change at one genetic locus can bring about a variety of apparently unconnected phenotypic changes. For instance, a particular mutation might at one and the same time affect eye colour, toe length, and milk yield. Pleiotropy is probably the rule rather than the exception, and is entirely to be expected from all that we understand about the complex way in which development happens.

pluralism In modern Darwinian jargon, the belief that evolution is driven by many agencies, not just natural selection. Enthusiasts sometimes overlook the distinction between evolution (any kind of change in gene frequencies, which may well be pluralistically caused) and adaptation (which only natural selection, as far as we know, can bring about).

polygene One of a set of genes each exerting a small, cumulative effect on a quantitative trait.

polymorphism The occurrence together in the same locality of two or more discontinuous forms of a species in such proportions that the rarest of them cannot be maintained merely by recurrent mutation. Polymorphism necessarily occurs during the transient course of an evolutionary change.

Polymorphisms may also be maintained in stable balance by various special kinds of natural selection.

polyphyletic *See* monophyletic.

preformationism As opposed to epigenesis (q.v.) it is the doctrine that the form of the adult body is in some sense mapped in the zygote. One early partisan thought he could discern, with his microscope, a little man curled up in the head of a sperm. In Chapter 9 it is used for the idea that the genetic code is more like a blueprint than a recipe, implying that the processes of embryonic development are in principle reversible, in the same sense as, say, you may reconstruct its blueprint from a house.

prokaryotes One of the two major groups of organisms on Earth (contrast eukaryotes) including bacteria and blue-green algae. They have no nucleus and no membrane-bounded organelles such as mitochondria: indeed one theory has it that mitochondria and other such organelles in eukaryotic cells are, in origin, symbiotic prokaryotic cells.

propagule Any kind of reproductive particle. The word is used specifically when we wish not to commit ourselves over whether we are speaking about sexual or asexual reproduction, about gametes or spores, etc.

r-selection Selection for the qualities needed to succeed in unstable, unpredictable environments, where ability to reproduce rapidly and opportunistically is at a premium, and where there is little value in adaptations to succeed in competition. A variety of qualities are thought to be favoured by *r*-selection, including high fecundity, small size, and adaptations for long-distance dispersal. Weeds, and their animal equivalents, are examples. Contrast with *K*-selection (q.v.). It is customary to emphasize that *r*-selection and *K*-selection are the extremes of a continuum, most real cases lying somewhere between. Ecologists enjoy a curious love/hate relationship with the *r*/*K* concept, often pretending to disapprove of it while finding it indispensable.

recessiveness Opposite of dominance (q.v.).

recon The minimum unit of recombination. One of several different definitions of gene, but, like muton, it has not yet received sufficient currency to be usable without simultaneous definition.

replicator Any entity in the universe of which copies are made. Chapter 5 contains an extended discussion of replicators, and a classification of active/passive, and germ-line/dead-end replicators.

reproductive value A demographic technical term, a measure of an individual's expected number of future (female) children.

segregation distorter A gene whose phenotypic effect is to influence meiosis so that the gene has a greater than 50 per cent chance of ending up in a successful gamete. *See also* meiotic drive.

selfish *See* altruism.

sex chromosome A special chromosome concerned with the determination of sex. In mammals there are two sex chromosomes called X and Y. Males have the genotype XY, females XX. All eggs therefore bear one X chromosome, but sperms may bear either one X (in which case the sperm will give rise to a daughter) or one Y (in which case the sperm will give rise to a son). The male sex is therefore referred to as heterogametic, the female as homogametic. Birds have a very similar system, except that males are homogametic (the equivalent of XX) and females heterogametic (the equivalent of XY). Genes carried on sex chromosomes are called 'sex-linked'. This is sometimes confused (e.g. page 14) with 'sex-limited', which means having expression in one sex or the other (not necessarily carried on sex chromosomes).

somatic Literally pertaining to the body. In biology it is used for the mortal part of the body, as opposed to the germ-line.

speciation The process of evolutionary divergence whereby two species are produced from one ancestral species.

species selection The theory that some evolutionary change takes place by a form of natural selection at the level of species or lineages. If species with certain qualities are more likely to go extinct than species with other qualities, large-scale evolutionary trends in the direction of the favoured qualities may result. These favoured qualities at the species level may in theory have nothing to do with the qualities that are favoured by selection within species. Chapter 6 argues that although species selection may account for some simple major trends, it cannot account for the evolution of complex adaptation (*see* Paley's watch, and macroevolution). The theory of species selection in this sense comes from a different historical tradition from the theory of group selection (q.v.) of altruistic traits, and the two are distinguished in Chapter 6.

stasis In evolutionary theory, a period during which no evolutionary change takes place. *See also* gradualism.

strategy Like 'altruism', used by ethologists in a special sense, almost misleadingly distantly related to its common usage. It was imported from game theory into biology in the theory of evolutionarily stable strategies (q.v.), where it is essentially synonymous with 'program' in the computer sense, and means a preprogrammed rule that an animal obeys. This meaning is precise, but unfortunately strategy has become a much abused buzz-word, and is now bandied about as a trendy synonym for 'behaviour pattern'. All individuals of a population might follow the strategy 'If small flee, if large attack'; an observer would then observe two behaviour patterns, fleeing and attacking, but he would be wrong to call them two strategies: both behaviour patterns are manifestations of the same conditional strategy.

survival value The quality for which a characteristic was favoured by natural selection.

symbiosis The intimate living together (with mutual dependence) of members of different species. Some modern textbooks omit the mutual dependence proviso, and understand symbiosis to include parasitism (in parasitism, only one side, the parasite, is dependent on the other, the host, which would be better off alone). Such textbooks use *mutualism* in place of symbiosis as defined above.

symphylic substance Chemical substance used by social insect colony parasites (e.g. beetles) to influence the behaviour of their hosts.

teleonomy The science of adaptation. In effect, teleonomy is teleology made respectable by Darwin, but generations of biologists have been schooled to avoid 'teleology' as though it were an incorrect construction in Latin grammar, and many feel more comfortable with a euphemism. Not much thought has been given to what the science of teleonomy will consist of, but some of its major preoccupations will presumably be the questions of units of selection, and of costs and other constraints on perfection. This book is an essay in teleonomy.

tetraploid Having four of each chromosome type rather than the more usual two (diploid) or one (haploid). New species of plants are sometimes formed by a doubling of chromosomes to tetraploidy, but subsequently the species behaves like an ordinary diploid which happens to have twice as many chromosomes as a closely related species, and it is convenient to consider it diploid for most purposes. Chapter 11 suggests that although individual termites are diploid, the whole termite nest may be regarded as the extended phenotypic product of a tetraploid genotype.

vehicle Used in this book for any relatively discrete entity, such as an individual organism, which houses replicators (q.v.), and which can be regarded as a machine programmed to preserve and propagate the replicators that ride inside it.

Weismannism The doctrine of a rigid separation between an immortal germ-line and the succession of mortal bodies which house it. In particular the doctrine that the germ-line may influence the form of the body, but not the other way around. *See also* central dogma.

zygote The cell that is the immediate product of sexual fusion between two gametes.

AUTHOR INDEX

SUBJECT INDEX

Titles in the *Oxford Landmark Science* series

James Lovelock, *Gaia*

Roger Penrose, *The Emperor's New Mind*

Jim Baggott, *The Quantum Story*

Michio Kaku, *Hyperspace*

Daniel Nettle, *Personality*

Addy Pross, *What is Life?*

Nick Lane, *Oxygen*

Richard Dawkins, *The Selfish Gene 40th anniversary edition*

Richard Dawkins, *The Extended Phenotype*

Richard Dawkins, *The Oxford Book of Modern Science Writing*

Jerry A. Coyne, *Why Evolution is True*